Introductory
Calculus
with
Applications

J. S. Ratti

M. N. Manougian

University of
South Florida

Introductory Calculus with Applications

Houghton Mifflin Company   Boston

Atlanta   Dallas   Geneva, Illinois
Hopewell, New Jersey   Palo Alto

Printed in the United States of America

Library of Congress Catalog Card Number:
72-5647

ISBN: 0-395-13855-8

dedicated to our families

Mary Ann   *Josette*
Shammi   *Michael*
Raju

Preface

This book is the outgrowth of a course in introductory calculus for students in the social sciences and business. It is our intent to bring out clearly some of the main ideas, language, and techniques of the calculus to students who have had only a basic high school mathematics program. Also, it is our objective to relate the calculus to the social sciences, biological sciences, economic theory, and some aspects of the physical sciences.

The material in this book can be covered in a year's course, three quarters or two semesters. Portions of the text such as the review of algebra (Chapters 1 and 2), the trigonometric functions (Chapter 10), and some of the starred (*) exercises may be omitted without affecting the continuity of the book. The text may thus be used in a one-semester or two-quarter course.

The text is written for the student. An intuitive approach is used to introduce each of the basic concepts of the calculus. Although the theoretical aspect is de-emphasized, proofs of theorems, whenever used, are detailed in order that the student can follow the arguments. Some of the important and difficult theorems are stated clearly without proof. Many examples are given in nearly all sections. The text contains numerous problems ranging from direct substitution and elementary applications to more challenging problems. The problems are an integral part of the text, and the student is reminded that it is by doing mathematics that one learns mathematics.

A word about the format of the book: Each page is divided into two parts. The right side is used for the body of the text. The left margin of each page is used for figures, notes, comments, and remarks to aid the student in his learning process. The student is encouraged to use the left margin for his own notes and comments. Short biographical sketches of some of the leaders of mathematics are included. It is hoped that this will encourage the student to read further on the lives of these and other famous men and women of mathematics.

The authors wish to thank the following professors for reading portions of the manuscript and offering helpful suggestions: M. K. Ali, M. Bernkopf, A. M. Carstens, F. Dashiell, J. A. Eidswick, M. Eisenberg, G. C. Gastl, M. F. Janowitz, L. M. Johnson, G. M. Knight, C. G. Lange, S. M. Lukawecki, J. F. Ramaley, D. R. Sherbert, and J. Wagoner. Also, the authors wish to thank the many students of the University of South Florida who contributed in the preparations of the text. In particular, thanks go to Mr. I. Bello, Mrs. P. Comadoll,

Mr. W. Wanamaker, and Mr. W. Zigler for aid in preparing the answers for the exercises. The authors are very grateful to Mrs. J. Fortson and Miss W. Balliet for their outstanding work in typing the manuscript (cheerfully!). Finally, the authors wish to thank the staff of Houghton Mifflin Company for their excellent cooperation and encouragement.

Contents

Introductory Calculus with Applications

Georg F. L. P. Cantor

1

Basic

Algebraic

Concepts

An element of a set is also called a member *of the set.*

1. Sets

As the German mathematician Georg Cantor realized, the concept of *set* plays a central role in the foundations of mathematics. While we shall not plunge deeply into the formal theory of sets, a few of its basic concepts will be used to give precision to our ideas.

A mathematical system must have at its core some undefined terms, called *primitive terms*, which are not defined by other terms of the system. Indeed, if we were to attempt to define every term, we would encounter problems. Let us see what might happen. Suppose we examine a dictionary for a definition of the word "set." We find:

"Set: a collection or group of objects."

Since we have not defined the words "collection" or "group," we return to the dictionary for the meaning of these words. We find:

"Collection: a group or an assemblage,"

"Group: an assemblage or collection,"

"Assemblage: a group or collection."

It is clear that we have no way of defining a word, such as "set," without falling back on some undefined word. So for us *set* will be an undefined term; however, an intuitive feeling for the meaning of the word "set" can be obtained from the synonyms given above. We shall describe the basic property a set must have:

A set is any well-defined collection of objects. Any object in the set is called an element of the set.

The chief characteristic of a set is that it is well-defined. That is, given any particular object it must be clear whether the object is an element of the set or not. For example, the collection of all good students at your university is considered ill-defined, because it is unlikely that we all can agree on its members. On the other hand, the collection of all those students who maintained a grade point average of 3.6 or higher during 1972–73 at your university is well-defined.

Notation: We shall follow the usual custom of denoting sets by upper-case letters such as *A, B, C, X, Y, Z.* The elements of a set will be denoted by lower-case letters such as *a, b, c, x, y, z.* The elements that make up sets are listed within braces and are separated by commas. For example:

$$A = \{1, 3, 5, 7\}$$
$$B = \{\text{Mary, Sue, Janet}\}$$
$$X = \{a, e, i, o, u\}.$$

The elements of the set *A* are 1, 3, 5 and 7; the elements of the set *B* are Mary, Sue, and Janet; and the elements of the set *C* are the vowels of the English alphabet.

We indicate in symbols that *A* is a set and *x* is an element of *A* by writing

$$x \in A$$

(read "*x* is an element of *A*" or "*x* belongs to *A*").

If *A* is a set and *x* is not an element of *A*, we write

$$x \notin A$$

(read "*x* is not an element of *A*" or "*x* does not belong to *A*").

A set may have only one element, such as the set consisting of the current Governor of Florida; several elements, such as the set whose elements are the states of the U.S.A.; an infinite number of elements, such as the set of natural numbers; or no elements at all, such as the set of all girls in an all-male college.

A set that has no elements (it is included for technical reasons and there is only one such set) is called the *empty set*. It is denoted by ϕ (the Greek letter "phi"). It is also called the *null* set or the *void* set.

There are two ways to define a given set.

Here we shall assume that the student has an intuitive understanding of what we mean by "infinite number."
The natural numbers are the numbers
1, 2, 3, 4, ...

Note. {0} *is not the empty set. This set has one element, namely zero.*

A. TABULAR FORM

In this form, a set is specified by listing its elements. Some examples are

$$A = \{-5, 8, 17\}$$
$$B = \{1, 4, 7, 10, 13, \ldots, 100\}$$
$$C = \{2, 4, 6, 8, \ldots\}.$$

Here the set A consists of only three elements, so we list all three. The set B consists of a relatively large number of elements, so we employ the "three-dots" notation. This notation indicates that some elements have not been listed; to calculate these elements, use the rule illustrated by the first few elements. For the set B, it means that we calculate the sixth and succeeding elements by adding 3 to the preceding element until we reach the last element, 100. Similarly, the set C indicates all even natural numbers, since no last element is listed.

Caution. It is assumed that the reader can recognize the pattern (as conceived by the writer) from the first few numbers.

A natural number is even if it is divisible by 2.

B. SET-BUILDING FORM

In this form, a set is defined by stating a condition or a property which is satisfied by the elements of the set and by no other objects. For example, the set C defined above could be written as

$$C = \{x \mid x \text{ is an even natural number}\}.$$

Here x is a letter which merely represents an element of the set. The vertical bar $\mid$ can be translated "such that" and after the bar we write the condition on x that must be satisfied by x if it is to be included as an element of the set C. For example, $50 \in C$ and $51 \notin C$.

Definition of equality of sets

Definition 1. Two sets A and B are said to be *equal*, written

$$A = B,$$

if and only if they have the same (identical) elements. In other words, every element which belongs to the set A is also an element of the set B, and every element which belongs to the set B is also an element of the set A. If A is not equal to B, we write

$$A \neq B.$$

We illustrate these ideas in the following examples.

Example 1. Let $A = \{1, 3, 5, 7, 9\}$ and $B = \{3, 9, 1, 7, 5\}$. Are these sets equal?

Solution. Since every element which belongs to A also belongs to B and conversely, the two sets A and B are equal. That is, $A = B$, by Definition 1.

Observations: (i) A set does not change if we alter the order in which its elements are tabulated, as in Example 1. (ii) A set does not change if one or more of its elements are repeated, as in Example 2.

Example 2. Let $A = \{2, 3, 5\}$ and $B = \{2, 5, 2, 5, 3, 2\}$. Are these sets equal?

Solution. According to Definition 1, the sets A and B are equal, i.e., $A = B$.

Example 3. Let A be the set of all people in Florida with natural green hair (no wigs) who are 10 feet tall, and let B be the set of all earth-men who visited the moon during 1940. Are A and B equal?

Solution. Since both A and B are empty, we have $A = B$.

Consider the sets $A = \{1, 3, 5\}$ and $B = \{1, 2, 3, 5, 6\}$. We immediately see that every element in A is also an element of B. We say that A is a *subset* of B or that B is a *superset* of A, symbolized by writing

$$A \subset B$$

(read "A is a subset of B" or "A is contained in B") or

$$B \supset A$$

(read "B is a superset of A" or "B contains A"). In general we have

Definition of subset

Definition 2. If every element of a set A is also an element of the set B, then A is called a *subset* of B, written $A \subset B$.

When we write $A \not\subset B$, it means that A is not a subset of B. In this case, there is at least one element x such that $x \in A$ but $x \notin B$.

It is clear from the above definition that $A \subset A$ for every set A. Further, $\phi \subset A$ for every set A.

Example 4. Let $A = \{1, 3, 5\}$ and $B = \{1, 2, 3, 4, 5, 6\}$. Is

(a) $A \subset B$?
(b) $B \subset A$?

Solution. (a) Since every element of A is also an element of B, $A \subset B$.

(b) Since the element 2 of B does not belong to A, $B \not\subset A$.

It is noteworthy that $B \not\subset A$ does not imply that $A \subset B$, as can be seen by considering the sets

$$A = \{1, 3, 5\} \quad \text{and} \quad B = \{3, 5, 7\}.$$

Example 5. Let $A = \{a, b, c\}$. Write all the subsets of A.

The empty set ϕ is not written $\{\phi\}$. In fact, $\phi \neq \{\phi\}$, since the empty set ϕ has no elements, whereas the set $\{\phi\}$ has one element, namely ϕ.

Solution. The subsets of A are

$$\{a\}, \ \{b\}, \ \{c\}, \ \{a, b\}, \ \{b, c\}, \ \{a, c\}, \ \{a, b, c\}, \ \phi.$$

Exercise 1

1. Write in words the following set notations.
   (a) $P \subset Q$ (b) $y \in A$ (c) $S \not\subset T$
   (d) $x \notin Q$ (e) ϕ (f) $\{0\}$

2. Which of the following is true?
   (a) $2 \in \{1, 2, 3\}$ (b) $\{2\} \in \{1, 2, 3\}$
   (c) $\{2, 3\} = \{3, 2\}$ (d) $\{2, 3\} = \{2, 2, 3, 2, 3\}$
   (e) $\{a\} \in \{a, \{a\}, \{\{a\}\}\}$ (f) $\{\{2\}\} = \{2, \{2\}\}$

3. Give, if possible, a property which describes each of the following sets.
   (a) $\{2, 4, 6, 8\}$ (b) $\{1, 4, 9, 16, 25, 36\}$
   (c) $\{0, 1, 2, 3, 4, 5, 6, 7, 8\}$
   (d) $\{2, 3, 5, 7, 11, 13, 17, 19\}$
   (e) $\{1, 3, 5, 7, \ldots, 91\}$ (f) $\{1, 3, 5, 7, \ldots\}$
   (g) $\{25, 36, 49, 64, 81\}$ (h) $\{-1, 1\}$

4. Exhibit the elements of each of the following sets.
   (a) $\{x \mid x \text{ is an even number less than 13}\}$
   (b) $\{y \mid y \text{ is a continent}\}$
   (c) $\{z \mid z \text{ is an ocean}\}$
   (d) $\{x \mid x \text{ is a month}\}$
   (e) $\{n \mid n \text{ is a natural number between 5 and 10}\}$
   (f) $\{h \mid h \text{ is an island state of the U.S.}\}$

5. Find all possible solutions of x and y in each of the following cases.
   (a) $\{2x, y\} = \{4, 8\}$ (b) $\{2x\} = \{0\}$
   (c) $\{2x, 3\} = \{20, 3\}$ (d) $\{x, x^2\} = \{4, 2\}$

6. Are the following sets equal?
   $A = \{x \mid x \text{ is a letter of the word "follow"}\}$
   $B = \{x \mid x \text{ is a letter of the word "wolf"}\}$
   $C = \{x \mid x \text{ is a letter of the word "flow"}\}$

*7. Prove that if $a \neq b$ and $\{a, b\} = \{c, d\}$, then $c \neq d$.

*8. Show that $\{x\} = \{y, z\}$ if and only if $x = y = z$.

*9. Show that $\{\{x\}, \{x, y\}\} = \{\{x\}\}$ if and only if $x = y$.

10. Consider the following sets of figures in a plane:
    $P = \{x \mid x \text{ is a parallelogram}\}$
    $Q = \{x \mid x \text{ is a quadrilateral}\}$
    $R = \{x \mid x \text{ is a rectangle}\}$
    $A = \{x \mid x \text{ is a rhomboid}\}$
    $S = \{x \mid x \text{ is a square}\}.$

Which of the sets *P, Q, R, A, S* are subsets of one or more of the others?

11. List all the subsets of the set

$$A = \{a, b, c, d\}.$$

12. If a set *A* has *n* elements, how many different subsets of the set *A* do we have?

13. Give all the subsets of the following sets.
    (a) {1, 3, 3, 5}   (b) {0}
    (c) {ϕ}   (d) {{ϕ}}

14. Let *A* be the set consisting of John and Bob. Does Bob's ear belong to the set *A*?

2. Operations with Sets

There are various ways of building new sets from old sets. We have

Definition of union of sets

Definition 3. The **union** of a set *A* and a set *B* is the set of all elements which belong either to *A* or to *B* or to both *A* and *B*.

A. SET UNION SYMBOL

The symbol $\cup$ is used to denote the union of sets. Thus, $A \cup B$ (read "*A* union *B*") is the set of all elements that are in either *A* or *B* or both.

Geometrically, we let the elements of *A* and *B* be represented by the points inside the two circles as shown in Figure 1. Then the set $A \cup B$ is represented by the points of the shaded region. A diagram such as Figure 1 is called a *Venn diagram*.

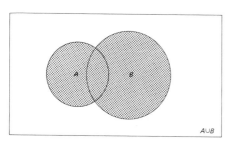

Figure 1

Named after **John Venn** (1834–1923), an English logician.

Example 1. Let $A = \{2, 4, 5, 7\}$ and $B = \{1, 3, 4, 5, 8\}$. Find $A \cup B$.

Solution. $A \cup B = \{1, 2, 3, 4, 5, 7, 8\}$.

Observe that the elements 4 and 5 appear in both sets, and yet when we form the union we list each of these elements only once (why?).

Given two sets *A* and *B*, we may be interested only in those elements which belong to both of the sets. We therefore have

Definition of intersection of sets

Definition 4. The **intersection** of a set *A* and a set *B* is the set of all elements each of which belongs to both sets *A* and *B*.

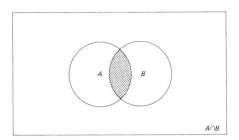

Figure 2

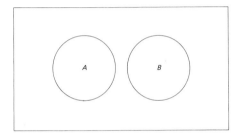

Figure 3
Disjoint sets

B. SET INTERSECTION SYMBOL

The symbol $\cap$ is used to denote the intersection of sets. Thus, $A \cap B$ (read "A intersection B") denotes the set of elements that are in both A and B.

The intersection of two sets A and B is shown in the Venn diagram in Figure 2.

If $A \cap B = \phi$, then A and B are called *disjoint* sets (see Figure 3). In this case A and B have no elements in common. The union of two or more disjoint sets will be called the disjoint union of those sets.

Example 2. For the sets A and B of Example 1, find $A \cap B$.

Solution. $A \cap B = \{4, 5\}$.

The operations of union and intersection can be performed on three or more sets by grouping them in pairs.

Example 3. Let A and B be the sets of Example 1, and let $C = \{3, 7, 9\}$. Find

 (a) $(A \cup B) \cup C$
 (b) $(A \cup B) \cap C$
 (c) $(A \cap B) \cup C$
 (d) $A \cap (B \cup C)$.

Solution. (a) $(A \cup B) \cup C = \{1, 2, 3, 4, 5, 7, 8\} \cup \{3, 7, 9\}$
$$= \{1, 2, 3, 4, 5, 7, 8, 9\}$$
 (b) $(A \cup B) \cap C = \{1, 2, 3, 4, 5, 7, 8\} \cap \{3, 7, 9\}$
$$= \{3, 7\}$$
 (c) $(A \cap B) \cup C = \{4, 5\} \cup \{3, 7, 9\}$
$$= \{3, 4, 5, 7, 9\}$$
 (d) $A \cap (B \cup C) = \{2, 4, 5, 7\} \cap \{1, 3, 4, 5, 7, 8, 9\}$
$$= \{4, 5, 7\}.$$

We are frequently interested in the set of those elements which are not in a given set A. This new set is called the *complement* of A, denoted by A'. For example, if A is the set of all numbers greater than or equal to 20, then intuitively we would take A' to be the set of all numbers less than 20. But this is not what we want. Since oranges, insects, animals, etc., are not in A, by definition all these objects must be in A'. Hence first of all it is necessary to specify the overall set to which the discussion will be confined. This set is called the *universal set of discourse* or simply the *universal* set and is denoted by U.

Definition of the universal set U

Definition 5. The *universal set U* is the set of all elements under consideration.

Therefore, any set under consideration must be a subset of U. We can now define the complement of a set.

Definition of the complement of a set

Definition 6. Let U be the universal set, and let A be a subset of U. Then the *complement* of A is the set of all elements in U that are not in A. This set is denoted by A' or $U - A$.

In Figure 4, the entire rectangle represents the universal set U and A' is represented by the shaded region.

Example 4. Let $U = \{1, 2, 3, 4, 5, 6, 7, 8, 9, 10\}$, $A = \{2, 5, 7, 10\}$, $B = \{1, 3, 5, 7, 9\}$. Find A', B', $(A \cup B)'$, $(A \cap B)'$.

Solution. By Definition 7

$$A' = \{1, 3, 4, 6, 8, 9\}, \qquad B' = \{2, 4, 6, 8, 10\}.$$

Since $A \cup B = \{1, 2, 3, 5, 7, 9, 10\}$ and $A \cap B = \{5, 7\}$, we have

$$(A \cup B)' = \{4, 6, 8\} \quad \text{and} \quad (A \cap B)' = \{1, 2, 3, 4, 6, 8, 9, 10\}.$$

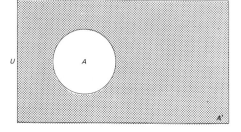

Figure 4

Exercise 2

1. Let $A = \{1, 2, 3, 4, 5, 6, 7, 8, 9\}$, $B = \{1, 2, 3, 5\}$, $C = \{4, 6, 8\}$. Find

   (a) $(A \cap B)$ (b) $A \cap C$ (c) $B \cap C$
   (d) $(A \cap B) \cap C$ (e) $(A \cup B) \cup C$
   (f) $A \cup (B \cap C)$ (g) $(A \cup B) \cap (A \cup C)$
   (h) $(A \cap B) \cup C$ (i) $A \cap (B \cup C)$.

2. Let $A = \{\{1, 2\}, 3, 4\}$, $B = \{1, 2, 3\}$, $C = \{1, 2\}$. Find

   (a) $A \cap B$ (b) $A \cap C$
   (c) $A \cup B$ (d) $A \cup C$.

3. Let

$$A = \{\{1, 2\}, \{1, 2, 3\}, 1, 2\}$$
$$B = \{\{1, 2\}, 1, 2, 3, \{2, 3\}\}.$$

Which of the following statements are true?

   (a) $\{2\} \subset A \cap B$ (b) $\{1, 2\} \in A \cap B$
   (c) $\{1, 2\} \subset A \cap B$ (d) $\{2, 3\} \subset A$
   (e) $\{1, 2, 3\} \subset A \cap B$ (f) $\{1, 2, 3\} \in A \cup B$
   (g) $\{2, \{1, 2\}\} = A \cap B$ (h) $3 \in A \cap B$

4. Give an example of a set P and a set Q such that $P \in Q$ and $P \subset Q$.

5. Given $A \cap B = \{2, 4\}$, $A \cup B = \{2, 3, 4, 5\}$, $A \cap C = \{2, 3\}$, $A \cup C = \{1, 2, 3, 4\}$. Find A, B, and C.

6. Let the universal set be $U = \{1, 2, 3, \ldots, 30\}$ and let $A = \{2, 4, 6, \ldots, 30\}$, $B = \{3, 6, 9, \ldots, 30\}$, and $C = \{1, 3, 5, \ldots, 29\}$. Describe the following sets.

(a) U' (b) A' (c) B'
(d) C' (e) $(B \cup A)'$ (f) $B' \cap A'$
(g) $B' \cup A'$ (h) $B' \cup C$ (i) $B \cup C'$

In problems 7 through 12, illustrate the given identities by drawing Venn diagrams. Here, A, B, and C are subsets of a universal set U.

7. $(A')' = A$

8. DeMorgan's Laws:
(a) $(A \cup B)' = A' \cap B'$ (b) $(A \cap B)' = A' \cup B'$

9. distributive laws:
(a) $A \cap (B \cup C) = (A \cap B) \cup (A \cap C)$
(b) $A \cup (B \cap C) = (A \cup B) \cap (A \cup C)$

10. (a) $A \cup A' = U$ (b) $A \cap A' = \phi$

11. commutative laws:
(a) $A \cup B = B \cup A$ (b) $A \cap B = B \cap A$

12. associative laws:
(a) $A \cup (B \cup C) = (A \cup B) \cup C$
(b) $A \cap (B \cap C) = (A \cap B) \cap C$

13. If U is a universal set, what is U'? What is $(U')'$? What is ϕ'?

3. The Real Numbers

There are many ways of describing the system of real numbers. We shall take here an intuitive approach. Early in our education we all become familiar with the set N of *natural numbers*. Since these numbers are used for counting, they are also called *counting numbers*. We know these numbers by their names: one, two, three, . . . , and denote them by the symbols 1, 2, 3, ... Thus

$$N = \{1, 2, 3, \ldots\}.$$

We then encountered the set I of *integers*, which contains N as a subset:

$$I = \{\ldots, -3, -2, -1, 0, 1, 2, 3, \ldots\}.$$

The next system of numbers encompassing the integers is the set of *rational* numbers.

Definition of rational numbers

Definition 7. A *rational number* is any number which can be expressed in the form p/q, where p and q are integers and $q \neq 0$.

Since every integer n can be written as $n/1$, by Definition 7 n is a rational number. Thus the set of integers is a subset of the set of rational numbers. Other examples of rational numbers are

Note. $0.25 = \dfrac{25}{100}$ and $1.378 = \dfrac{1378}{1000}$.

$$\frac{3}{5}, \quad \frac{-5}{9}, \quad \frac{3}{-16}, \quad -\frac{7}{22}, \quad 0.25, \quad 1.378.$$

One reason for extending a number system is the need to solve certain equations. Let us consider the equation

(1) $$a + x = b,$$

where a and b are natural numbers.

In equation (1) if $a = 3$ and $b = 5$, it is easy to see that x must be equal to 2, which is a natural number. However, if $a = 5$ and $b = 3$, then there is no solution in the set of natural numbers. In order that equation (1) always have a solution, we extend the set N to the set I.

Similarly, it is necessary to introduce the rational numbers if we wish to solve equations of the form

(2) $$ax = b,$$

where $a, b \in I$ and $a \neq 0$.

For example, if $a = 2$ and $b = 1$, then the equation

(3) $$2x = 1$$

has no solution in the set of integers. However, $x = \frac{1}{2}$ (a rational number) is a solution of (3).

Now suppose we wish to solve the equation

(4) $$x^2 = 2.$$

We shall show that this equation has no solution in the set of rational numbers.

Theorem 1. There is no rational number x such that $x^2 = 2$.

Proof. We shall prove this by contradiction. We shall assume that there is a rational number $x = p/q$ such that $(p/q)^2 = 2$. We also suppose that p and q do not have 2 as a common factor, because if such a factor exists, then it can be cancelled. Now $(p/q)^2 = 2$ implies

If 2 is not a factor of p, then p = 2n + 1,
where n is an integer. Thus

$$p^2 = (2n + 1)^2 = 4n^2 + 4n + 1$$
$$= 2(2n^2 + 2n) + 1,$$

which shows that 2 is also not a factor of p^2.

(5) $$p^2 = 2q^2.$$

Since q is an integer, so are q^2 and $2q^2$. Therefore p^2 is an integer divisible by 2. Now, since 2 is a factor of p^2, it must also be a factor of p. Let

(6) $$p = 2n, \quad \text{where } n \text{ is an integer.}$$

Then $p^2 = 4n^2$. Substituting this value of p^2 in (5), we have

$$4n^2 = 2q^2$$

or

$$q^2 = 2n^2.$$

Thus 2 is a factor of q^2 and as such 2 is a factor of q. Consequently, we have

(7) $$q = 2m, \quad \text{where } m \text{ is an integer.}$$

From (6) and (7) we conclude that p and q have 2 as a common factor, which is contrary to our hypothesis. Therefore, there exists no rational number whose square is 2.

We now extend the set of rational numbers to the set of *real* numbers.

A. THE REAL LINE

We present the real numbers by associating them with the points on a line. We draw a straight line (imagine the line to be extended indefinitely in both directions) and call this line the *number line* or the *real line*. We associate with each point on this line a number, and the collection of all these numbers is called the set of real numbers.

To begin, we choose any two points on this line and label them 0 and 1. By agreement we choose 1 to be to the right of 0 (see Figure 5). The direction traveled when going from 0 to 1 will be called the *positive direction*, and the opposite direction is called the *negative direction* of the line.

The fixed straight segment from 0 to 1 is called the unit segment, and its length is chosen as the unit length. By reproducing this unit of length successively on both sides of 0 and 1, we get the familiar graphical representation of the set of integers $I = \{\ldots, -3, -2, -1, 0, 1, 2, 3, \ldots\}$ on the number line (see Figure 6).

We now locate the points corresponding to the rational numbers. Let x be a positive rational number. Then $x = p/q$, where p and q are integers and $q \neq 0$. We divide the line segment from 0 to p into q equal parts (this can be done with a compass and straightedge, see Figure 7); then the endpoint of the first subdivision is the

Figure 5

Figure 6

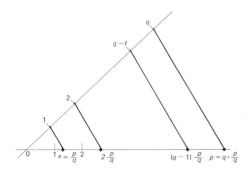

Figure 7

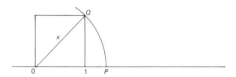

Figure 8

*The Pythagorean theorem : If a right triangle
has legs of length b and c and hypotenuse
of length a, then*

$$a^2 = b^2 + c^2.$$

number $x = p/q$. The negative rational numbers can be located similarly.

Let us now imagine that all the rational numbers have been identified, or placed on the line. It turns out that the rational numbers do not fill up the line: we exhibit a point on the line which is not a rational number. In Figure 8 let $x = 0Q =$ length of the diagonal of the square of side length 1. By using a compass we can draw an arc of the circle having radius x and center at 0. The point P where the arc intersects the number line through 0 and 1 is the point associated with the number x. It is easy to see from the Pythagorean theorem that $x^2 = 1^2 + 1^2 = 2$ and by Theorem 1, x is not rational.

Those numbers on the real line which are not rational numbers are called *irrational* numbers. Consequently, the set of real numbers is the union of two disjoint sets: the set of rational numbers and the set of irrational numbers.

At this stage it is possible to define addition and multiplication of real numbers and prove all the arithmetical properties with which the students are familiar. The set of real numbers together with the operations of addition and multiplication is called the real number system.

Exercise 3

1. Perform a geometrical construction similar to that of Figure 7 to obtain the following.
   (a) $\frac{8}{5}$ (b) $-\frac{3}{7}$ (c) $1\frac{3}{10}$
2. Prove that there is no rational number whose square is 3. Suppose we assume that there is no rational number whose square is 4 and attempt a similar proof. Where does the proof break down?
3. Prove that there is no rational number whose square is 5.
4. Construct a number line and locate the following points.
   (a) $\sqrt{5}$ (b) $2\sqrt{2}$ (c) $-\sqrt{10}$
5. Suppose x is an irrational number and a and b are rational numbers. Prove
   (a) ax is irrational if $a \neq 0$
   (b) $a + bx$ is irrational if $b \neq 0$.
6. Is it always true that the sum and product of two rational numbers is a rational number?
7. Is it always true that the sum and product of two irrational numbers is an irrational number?

4. Order

Our representation of the real numbers on the real line shows that the set of real numbers is the union of three disjoint sets:

1. The positive real numbers: those real numbers which are to the right of 0 on the number line.
2. The negative real numbers: those real numbers which are to the left of 0 on the number line.
3. Zero.

The set of positive real numbers is closed *under the operations of addition and multiplication.*

We assume that the reader knows some of the elementary properties of positive real numbers. For example, the sum of two positive numbers is a positive real number, the product of two positive real numbers is a positive real number, etc. We are ready now for the following definition.

Definition of $a > b$ and $a < b$

Definition 8. For any two real numbers a and b, we say that a is greater than b and write $a > b$ if and only if $a - b$ is positive. Further, we say a is less than b and write $a < b$ if and only if $a - b$ is negative.

The notation $a \geq b$ indicates that $a > b$ or $a = b$. Similarly, $a \leq b$ means $a < b$ or $a = b$.

To say that $a > b$ is equivalent to saying that a is to the right of b on the number line, and similarly $a < b$ if and only if a is to the left of b on the number line.

An expression involving $<$, $\leq$, $>$, or $\geq$ is called an inequality. We list here some of the important properties of inequalities, which the reader should learn thoroughly.

Property 1 is called the law of trichotomy.

Property 1. For each pair of real numbers a and b, exactly one of the following relations holds:

$$a < b, \qquad a > b, \qquad a = b.$$

Property 2 is called the law of transitivity.

Property 2. If $a > b$ and $b > c$, then $a > c$.

Property 3. For any real number c, $a > b$ if and only if $a + c > b + c$.

Property 4. For any real number $c > 0$, $a > b$ if and only if $ac > bc$.

Remember: The sense of the inequality is unchanged if we multiply both sides of the inequality by a positive number. However, if we multiply both sides by a negative number, then the sense of the inequality is reversed.

Property 5. For any real number $c < 0$, $a > b$ if and only if $ac < bc$.

We shall prove Properties 1 and 4. The student is encouraged to prove the remaining Properties.

Proof of Property 1. If a and b are real numbers, then $a - b$ is a real number. As mentioned before, the set of real numbers is the disjoint union of (i) the set of positive real numbers, (ii) the set of negative real numbers, and (iii) the set consisting of zero. Consequently, $a - b$ is either positive, negative, or zero. Hence either $a > b$, $a < b$, or $a = b$.

Proof of Property 4. Let a, b, and c be real numbers such that $a > b$ and $c > 0$. Since $a > b$, we have $a - b > 0$. Now since $a - b$ and c are positive real numbers and the product of positive real numbers is positive, we have

$$(a - b)c > 0$$

or

$$ac - bc > 0.$$

Hence,

$$ac > bc.$$

See problem 3 of Exercise 4.

Conversely, suppose $ac > bc$. Since $c > 0$, we have $1/c > 0$ (why?). Thus using the first part of this property, we have

$$(ac)\frac{1}{c} > (bc)\frac{1}{c}$$

or

$$a > b.$$

The sets described below are called *intervals*. For $a < b$, we write

1. $(a, b) = \{x \mid a < x < b\}$

2. $[a, b] = \{x \mid a \le x \le b\}$

3. $[a, b) = \{x \mid a \le x < b\}$

4. $(a, b] = \{x \mid a < x \le b\}$

5. $[a, \infty) = \{x \mid a \le x\}$

6. $(a, \infty) = \{x \mid a < x\}$

7. $(-\infty, a] = \{x \mid x \le a\}$

8. $(-\infty, a) = \{x \mid x < a\}$

9. $(-\infty, \infty) = \{x \mid x \text{ is a real number}\}$.

Figure 8

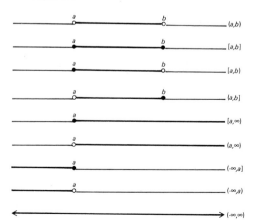

Remark 1. The numbers *a* and *b* are called the *endpoints* of the interval. A square bracket indicates that the endpoint is included in the interval, whereas a parenthesis means that the endpoint does not lie in the interval.

The intervals

$$(a, b), \quad (a, \infty), \quad (-\infty, a), \quad \text{and} \quad (-\infty, \infty)$$

are called *open intervals*.

Remark 2. We note that if *I* is an open interval, then given any *x* in *I*, there are points to the right and left of *x* in *I*. Such a point is called an *interior point*.

The intervals

$$[a, b], \quad [a, \infty), \quad (-\infty, a], \quad \text{and} \quad (-\infty, \infty)$$

are called *closed intervals*.

Remark 3. Note that $(-\infty, \infty)$ is both an open and a closed interval. In fact this is the only nonempty interval which is both open and closed.

The intervals

$$(a, b] \quad \text{and} \quad [a, b)$$

are neither open nor closed intervals. Sometimes they are called half-open or half-closed intervals.

It must be carefully noted that the symbol ∞ is not a number. It is simply part of a short form of notation for the given intervals.

A. INEQUALITIES

In many practical situations we encounter inequalities of the form

(8)
$$2x + 5 < x + 10$$

or

(9)
$$5x^2 + 8x - 2 < 0.$$

We are asked to solve (8) or (9); i.e., we are asked to find those values of *x* for which these inequalities hold. In other words, we want to rewrite the definitions of the sets $\{x \mid 2x + 5 < x + 10\}$ or $\{x \mid 5x^2 + 8x - 2 < 0\}$ in a simpler form which shows at once what the elements are. Such sets are subsets of the real numbers and generally consist of intervals or unions of intervals. They will be called solution sets.

Example 1. Solve $2x + 5 < x + 10$.

Solution. By Property 3,

$$2x + 5 - 5 < x + 10 - 5$$

or

$$2x < x + 5.$$

Again, by Property 3,

$$2x - x < x + 5 - x$$

or

$$x < 5.$$

Thus, the solution set of the given inequality is $(-\infty, 5)$. The graph of this solution set is given in Figure 9.

Figure 9

Example 2. Find all real numbers x which satisfy the compound inequality

(10) $$2 \le 5x - 3 < 7.$$

Solution. Adding 3 to each member of the inequality, we have

$$2 + 3 \le 5x - 3 + 3 < 7 + 3$$

or

(11) $$5 \le 5x < 10.$$

Now, multiplying each member of the inequality (11) by $\frac{1}{5}$, we have

(12) $$1 \le x < 2.$$

Figure 10

Since these steps are reversible, inequality (12) is equivalent to (10). Thus the solution set of (10) is the interval $[1, 2)$, as illustrated in Figure 10.

Saying that inequalities (10) and (12) are equivalent means that (10) holds if and only if (12) holds; i.e., (10) and (12) have the same solution set.

Example 3. Solve the inequality

(13) $$x^2 + x - 2 > 0.$$

Solution. Factoring this expression we have

(14) $$x^2 + x - 2 = (x + 2)(x - 1).$$

Thus, the inequality (13) will be satisfied when both factors on the right hand side of (14) have the same sign, that is, if

$$x + 2 > 0 \text{ and } x - 1 > 0 \qquad \text{or} \qquad x + 2 < 0 \text{ and } x - 1 < 0.$$

Case 1: $x + 2 > 0$ and $x - 1 > 0$.

Here we have

Note. $a \cdot b > 0$ if and only if $a > 0$ and $b > 0$, or $a < 0$ and $b < 0$.

(15) $$x > -2 \quad \text{and} \quad x > 1.$$

Both inequalities in (15) hold if $x > 1$, that is, if x is in the interval $(1, \infty)$.

Case 2: $x + 2 < 0$ and $x - 1 < 0$.

In this case we have

This involves the intersection

$$\{x \mid x > -2\} \cap \{x \mid x > 1\}.$$

(16) $$x < -2 \quad \text{and} \quad x < 1.$$

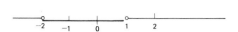

Figure 11

Both inequalities in (16) hold if $x < -2$, that is, if x is in the interval $(-\infty, -2)$.

Therefore, if we combine the solutions of case 1 and case 2, we get the two intervals $(-\infty, -2)$ and $(1, \infty)$. Since any number in either of these intervals satisfies the given inequality, the solution set of (13) is given by the union of these two intervals, as illustrated in Figure 11.

Example 4. Solve the inequality

(17) $$\frac{x}{x - 4} > 5, \qquad x \neq 4.$$

Solution. We wish to multiply both sides of (17) by $x - 4$. However, the sense of the resulting inequality depends on the sign of $x - 4$. So we must consider two cases.

Case 1: $x - 4 > 0$; that is, $x > 4$.

Multiplying both sides of (17) by $x - 4$, we have

(18) $$x > 5x - 20.$$

Adding $-5x$ to both sides of (18), we get

(19) $$-4x > -20.$$

$$\left(-\tfrac{1}{4}\right)(-4x) < \left(-\tfrac{1}{4}\right)(-20)$$
$$x < 5$$

Multiplying both sides of (19) by $-\tfrac{1}{4}$, which changes the direction of the inequality, we get

(20) $$x < 5.$$

Therefore, the solution of case 1 is the set of all real numbers x such that $x > 4$ and $x < 5$, that is, the open interval $(4, 5)$, as illustrated in Figure 12.

Figure 12

Case 2: $x - 4 < 0$; i.e., $x < 4$.

Multiplying both sides of (17) by $x - 4$, which reverses the direction of the inequality (since $x - 4$ is negative in this case), we get

(21) $$x < 5x - 20.$$

Performing operations similar to those in case 1, we get

(22) $$x > 5.$$

Thus the solution of case 2 is the set of all numbers x such that $x < 4$ and $x > 5$. But it is impossible to find a value of x which satisfies both of these requirements.

From cases 1 and 2, we conclude that the solution set of (17) is the interval $(4, 5)$.

Exercise 4

In problems 1 through 9, a, b, c, d are real numbers. Prove the inequalities by showing your work step by step.

1. If $a < c$ and $b < d$, then $a + b < c + d$.

2. If $a \neq 0$, then $a^2 > 0$.

3. $a > 0$ if and only if $\dfrac{1}{a} > 0$.

4. If $0 < a < b$, then $\dfrac{1}{a} > \dfrac{1}{b}$.

5. If $0 < a < b$ and $0 < c < d$, then $ac < bd$.

6. $a > 0$ if and only if $-a < 0$.

7. $a > b$ if and only if $-a < -b$.

8. (a) If $a > b$ and $c < 0$, then $\dfrac{a}{c} < \dfrac{b}{c}$.

   (b) If $a < b$ and $c < 0$, then $\dfrac{a}{c} > \dfrac{b}{c}$.

*9. Prove that $2ab \leq a^2 + b^2$ and that equality occurs if and only if $a = b$.

In problems 10 through 20 solve the stated inequality, and plot the solution set on the number line.

10. $2x - 3 < 5$

11. $7x - 4 < 4x + 11$

12. $3x - 2 \geq -2x + 10$

13. $1 < 7x + 4 \leq 5$

14. $3x - 7 \leq 8x + 6 \leq 11$

15. $(x - 3)(x + 5) < 0$

16. $(x - 2)(x + 3) \geq 14$

17. $7 - \dfrac{3}{x} < 12$, $x \neq 0$

18. $\dfrac{1}{2x - 7} \leq \dfrac{3}{5 - 3x}$, $x \neq \dfrac{7}{2}$, $x \neq \dfrac{5}{3}$

*19. $\dfrac{3}{x - 1} + \dfrac{1}{x - 5} \leq 1$, $x \neq 1$, $x \neq 5$

*20. $1 \leq \dfrac{2x}{x - 1} + \dfrac{3x}{x - 2} \leq 6$, $x = 1$, $x \neq 2$

5. Absolute Value

The "absolute value" of x (also called the numerical value of x), denoted by $|x|$, is an important concept often useful in the study of inequalities.

Definition of absolute value of a number. Notice that the absolute value of a number is always either a positive number or zero.

Definition 9. For each real number *x*, the *absolute value* of *x* is

$$(23) \qquad |x| = \begin{cases} x & \text{if } x \geq 0 \\ -x & \text{if } x < 0. \end{cases}$$

For example,

$$|3| = 3, \qquad |-3| = -(-3) = 3, \qquad |0| = 0,$$
$$|7 - 13| = |-6| = -(-6) = 6.$$

Now $|3| = |-3|$. Thus if we look on the number line, we observe that 3 and -3 are equidistant from 0 (see Figure 13). So geometrically, the absolute value of a number *x* is its distance from 0, without regard to direction. In general, $|a - b|$ denotes the distance between *a* and *b* without regard to direction (see Figure 14).

We have the following properties of absolute value.

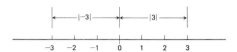

Figure 13

Figure 14

Theorem 2. For $a > 0$, $|x| < a$ if and only if $-a < x < a$.

Proof. Since the theorem is an "if and only if" statement, we have to prove it in two parts.

I. Assume $|x| < a$. We must show that $-a < x < a$.

Case 1: $x \geq 0$. In this case, $|x| = x$, so that $x = |x| < a$ implies $x < a$. Also, *a* is given to be positive. Thus $-a < 0 \leq x$. Combining, we get

$$-a < 0 \leq x = |x| < a$$

or

$$-a < x < a.$$

Case 2: $x < 0$. In this case, $|x| = -x$. Since $|x| < a$, we conclude that $-x < a$ or $x > -a$ or $-a < x$. Also, since $x < 0$, we have $0 < -x$. Therefore $-a < 0 < -x$ or $-a < -x$ or $x < a$. Combining, we get

$$-a < x < a.$$

In both cases

$$|x| < a \text{ implies } -a < x < a.$$

II. Suppose $-a < x < a$. We must show that $|x| < a$.

Case 1. Let $x \geq 0$. Then $|x| = x$, so $x < a$ implies $|x| < a$.

Case 2. Let $x < 0$. Then $|x| = -x$. Since $-a < x$, we have $-x < a$ (Exercise 4, problem 7). Consequently,

$$|x| = -x < a \qquad \text{or} \qquad |x| < a.$$

Theorem 3. **For $a > 0$, $|x| \le a$ if and only if $-a \le x \le a$.**

Theorem 4. **For $a > 0$, $|x| > a$ if and only if $x > a$ or $x < -a$.**

Theorem 5. **For $a > 0$, $|x| \ge a$ if and only if $x \ge a$ or $x \le -a$.**

The proofs of Theorems 3, 4, and 5 are similar to that of Theorem 2 and are left as exercises for the reader.

We shall now find the solution of inequalities involving absolute values.

Example 1. Solve for x:

(24) $$|3x - 5| \le 4.$$

Solution. By Theorem 3, the inequality (24) is equivalent to

(25) $$-4 \le 3x - 5 \le 4.$$

The inequality (25) can easily be seen to be equivalent to

(26) $$\tfrac{1}{3} \le x \le 3.$$

Thus the solution set for (24) is the closed interval $[\tfrac{1}{3}, 3]$ (see Figure 15).

Figure 15

Example 2. Solve for x:

(27) $$\left| \frac{2 - 3x}{x + 1} \right| \le 5, \qquad x \ne -1.$$

Solution. By Theorem 3, this inequality is equivalent to

(28) $$-5 \le \frac{2 - 3x}{x + 1} \le 5.$$

Now to solve (28), we wish to multiply each member of (28) by $x + 1$. Consequently, we consider the two cases:

Case 1: $x + 1 > 0$, which means that $x > -1$.

Multiplying both sides of (28) by $x + 1$, we get

(29) $$-5x - 5 \le 2 - 3x \le 5x + 5.$$

To solve (29), we first consider

(30) $$-5x - 5 \le 2 - 3x.$$

The solution of (30) is easily seen to be

$$(31) \qquad x \geq -\tfrac{7}{2}.$$

We now solve the second inequality in (29),

$$(32) \qquad 2 - 3x \leq 5x + 5.$$

The solution of (32) is

Figure 16

$$(33) \qquad x \geq -\tfrac{3}{8}.$$

Therefore if $x > -1$, the original inequality (27) holds if and only if $x \geq -\tfrac{7}{2}$ and $x \geq -\tfrac{3}{8}$.

Since all three inequalities $x > -1$, $x \geq -\tfrac{7}{2}$, and $x \geq -\tfrac{3}{8}$ must hold for every x in the solution set, we must have $x \geq -\tfrac{3}{8}$, or the interval $[-\tfrac{3}{8}, \infty)$.

Case 2: $x + 1 < 0$, or $x < -1$.

Multiplying each member of (28) by $x + 1$ and reversing the direction of the inequalities (since $x + 1$ is negative), we get

$$(34) \qquad -5x - 5 \geq 2 - 3x \geq 5x + 5.$$

The solution of the first inequality in (34)

$$(35) \qquad -5x - 5 \geq 2 - 3x$$

is

$$(36) \qquad x \leq -\tfrac{7}{2}.$$

The solution of the second inequality in (34)

$$(37) \qquad 2 - 3x \geq 5x + 5$$

is

$$(38) \qquad x \leq -\tfrac{3}{8}.$$

Figure 17

Since all three inequalities $x < -1$, $x \leq -\tfrac{7}{2}$, and $x \leq -\tfrac{3}{8}$ must be satisfied, the solution of (27) in this case is $x \leq -\tfrac{7}{2}$, or the closed interval $(-\infty, -\tfrac{7}{2}]$.

Combining the solutions of case 1 and case 2, we get the solution of (27): $(-\infty, -\tfrac{7}{2}] \cup [-\tfrac{3}{8}, \infty)$.

Figure 18

Example 3. Solve for x:

$$(39) \qquad |3x + 7| > 5.$$

Solution. By Theorem 4, this inequality is equivalent to

$$(40) \qquad 3x + 7 > 5 \quad \text{or} \quad 3x + 7 < -5.$$

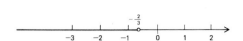

Figure 19

Figure 20

Figure 21

Note that $\sqrt{9} \neq -3$, although $(-3)^2 = 9$; $\sqrt{9}$ denotes only the positive square root. For this reason 3 is sometimes called the principal square root of 9.

The triangle inequality

Considering the first inequality in (40) we have

$$3x + 7 > 5$$

or

$$3x > -2$$

or

$$x > -\tfrac{2}{3}.$$

Thus the interval $(-\tfrac{2}{3}, \infty)$ is a solution set.

From the second inequality, we have

$$3x + 7 < -5$$

or

$$3x < -12$$

or

$$x < -4.$$

Therefore the interval $(-\infty, -4)$ is a solution set.

The given inequality (39) will be satisfied if either of the inequalities (40) are satisfied. Consequently, the solution set of (39) is $(-\infty, -4) \cup (-\tfrac{2}{3}, \infty)$.

Remark. We recall from algebra that the symbol $\sqrt{x}$ where $x \geq 0$ is the unique nonnegative real number y such that $y^2 = x$.

For example,

$$\sqrt{9} = 3, \quad \sqrt{25} = 5, \quad \sqrt{0} = 0.$$

From the above remark it is clear that

$$(41) \qquad \sqrt{x^2} = |x|.$$

We have two theorems dealing with the absolute value of a sum and of a product.

Theorem 6. **If x and y are any pair of real numbers, then**

$$(42) \qquad |xy| = |x|\,|y|,$$

and if $y \neq 0$ then

$$(43) \qquad \left|\frac{x}{y}\right| = \frac{|x|}{|y|}.$$

The proof of this theorem is trivial and is therefore omitted.

Theorem 7. **If x and y are any pair of real numbers, then**

$$(44) \qquad |x + y| \leq |x| + |y|.$$

Proof. From the definition of absolute value, it is obvious that for any real number x

(45) $$-|x| \leq x \leq |x|.$$

Similarly, for any real number y,

(46) $$-|y| \leq y \leq |y|.$$

Adding inequalities (45) and (46), we get

(47) $$-(|x| + |y|) \leq x + y \leq |x| + |y|;$$

and (47) is equivalent to

(48) $$|x + y| \leq \Big| |x| + |y| \Big|.$$ (Why?)

But

$$\Big| |x| + |y| \Big| = |x| + |y|;$$

so

$$|x + y| \leq |x| + |y|.$$

Exercise 5   In problems 1 through 6, solve for x.

1. $|3x + 2| \leq 9$

2. $|5x - 7| > 5$

3. $|7 - 3x| \geq |5x|$

4. $|3 - 4x| \leq |6x + 1|$

5. $\left| \dfrac{1}{3x - 2} \right| \leq \left| \dfrac{3}{x + 5} \right|$

6. $\left| \dfrac{1}{2x - 3} \right| \geq \left| \dfrac{5}{x - 2} \right|$

*7. Prove Theorem 3.

*8. Prove Theorem 4.

*9. Prove Theorem 5.

*10. Prove Theorem 6.

*11. Prove Theorem 7 by considering the four cases (i) $x \geq 0$ and $y \geq 0$, (ii) $x < 0$ and $y < 0$, (iii) $x \geq 0$ and $y < 0$, (iv) $x < 0$ and $y \geq 0$.

12. Prove that $|x - y| \leq |x| + |y|$.

13. Prove that $|x| - |y| \leq |x - y|$. *Hint:* Write $x = y + (x - y)$.

14. Prove that if x, y, and z are real numbers, then

(a) $|xyz| = |x|\,|y|\,|z|$

(b) $|x + y + z| \leq |x| + |y| + |z|$

(c) $|x - y| \leq |x - z| + |y - z|$.

*15. Solve for x: $|x + 2| + |x| > 3$.

*16. Solve for x: $3 \leq |2x - 1| < 8$.

René Descartes

2

Relations

and

Functions

Gottfried Wilhelm von Leibniz (1646–1716), known mainly as a philosopher and metaphysician, is one of the greatest thinkers of modern times. As a mathematician he contributed greatly to the development of the differential and the integral calculus and of symbolic logic. Leibniz was also a lawyer, theologian, diplomat, geologist, and historian; he wrote with ease in Greek, Latin, French, English, and German; he was the inventor of a calculating machine; and he founded the Academy of Sciences of Berlin.

   Leibniz, an only child, was born into a scholarly family. He was only six when his father (a professor of philosophy) died. At the age of seventeen he received his bachelor's degree in philosophy and at twenty his doctorate in law. His last years were saddened by ill health, and by controversy and neglect occasioned by bitter dispute over the discovery of the calculus. Both Newton and Leibniz discovered the calculus, by different means; Newton made his discovery a few years before Leibniz, although he published his results after Leibniz.

1. Cartesian Product

In this chapter we shall introduce an important mathematical concept—that of a *function*, a most fundamental notion for the study of the calculus. The word "function" was introduced into the language of mathematics in the late seventeenth century by Leibniz, one of the originators of the calculus. Functions have to do with pairs of numbers. A pair of numbers in which the order is specified is called an *ordered pair* of numbers. We denote an ordered pair by the symbol (a, b), not to be confused with the interval (a, b). The number a is called the *first component* and the number b is called the *second component* of the ordered pair (a, b).

   We shall introduce functions in terms of the notion of sets. The reader is familiar with certain operations involving sets. Here we shall define another operation, the *Cartesian product* of two sets A and B.

Definition 1. If **A** and **B** are two sets, then the ***Cartesian product*** of **A** and **B**, denoted by **A** × **B**, is the set of all ordered pairs **(a, b)** such that **a** ∈ **A** and **b** ∈ **B**. Thus **A** × **B** = **{(a, b)| a ∈ A, b ∈ B}**.

Example 1. Suppose $A = \{x, y, z\}$ and $B = \{o, p\}$. Find $A \times B$ and $B \times A$.

Note. If $a \neq b$, then $(a, b) \neq (b, a)$. Hence $A \times B \neq B \times A$. In general $(a, b) = (c, d)$ if and only if $a = c$ and $b = d$.

Cartesian product:

$$A \times B = \{(a, b) \mid a \in A, b \in B\}$$

René Descartes (1596–1650) was born in La Haye, France. As a child his delicate health kept him away from physical activities, forcing him into intellectual pursuits. As a young man, following disenchantment with his friends and his studies he joined the many wars plaguing Europe. He went to Holland and learned the trade of war under Prince Maurice of Orange, he enlisted under the Elector of Bavaria and fought against Bohemia, he fought in the battle of Prague, and he was at the seige of La Rochelle with the King of France.

Between wars, Descartes occupied himself with mathematics, philosophy, and almost all aspects of the sciences and theology. At thirty-two he went to Holland, corresponding extensively with the leading scientists and philosophers of Europe. In 1637 he published the masterpiece known in short as the "Method," which introduced the world to analytic geometry. In 1646 Queen Christina of Sweden brought Descartes to Stockholm as her tutor. There he died of an inflammation of the lungs in February, 1650.

Solution.

$$A \times B = \{(x, o), (x, p), (y, o), (y, p), (z, o), (z, p)\}$$

and

$$B \times A = \{(o, x), (o, y), (o, z), (p, x), (p, y), (p, z)\}.$$

Now, consider the set R consisting of all real numbers. The Cartesian product $R \times R$, which may be denoted also by R^2, is the set of all possible ordered pairs of real numbers

$$R \times R = \{(x, y) \mid x, y \in R\}.$$

Geometrically, R^2 may be interpreted as the set of all points in the Cartesian (or geometric) plane. The *Cartesian* (or *rectangular*) *coordinate system* is usually used to set up a correspondence between ordered pairs of real numbers and points in the plane. The formulation of this correspondence is attributed to the French mathematician-philosopher Descartes.

We begin by constructing a line in the plane from left to right, called the *x-axis*. Then, a vertical line is drawn, called the *y-axis*.

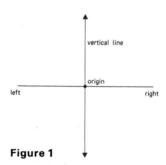

Figure 1

The two lines are referred to as the *coordinate axes* and their point of intersection is called the *origin*. A convenient unit of length is chosen and a number scale is marked off on each of the axes. Usually, the number zero is placed at the origin; on the x-axis the positive numbers are to the right (as in Chapter 1), and on the y-axis the positive numbers extend upwards. Figure 2 shows a Cartesian coordinate system including some points. The axes divide the plane into four parts called *quadrants*. The first quadrant is the set $\{(x, y) \mid x > 0$ and $y > 0\}$, the second quadrant is the set $\{(x, y) \mid x < 0$ and $y > 0\}$, the third quadrant is the set $\{(x, y) \mid x < 0$

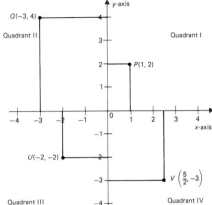

Figure 2

and $y < 0$}, and the fourth quadrant is the set {$(x, y)|\ x > 0$ and $y < 0$}. The points P, Q, U, and V are in the first, second, third, and fourth quadrants respectively. The association of these points in the plane with the corresponding ordered pairs is illustrated in Figure 2. For example, the *coordinates* of the point P are the ordered pair of numbers $(1, 2)$; the abscissa is 1 and the ordinate is 2.

To each element in R^2 there corresponds exactly one point in the plane, and conversely, to each point in the plane there corresponds exactly one element of R^2.

Example 2. Suppose $A = \{x|\ -1 \le x < 2\}$ and $B = \{1\}$. Determine $A \times B$ and $B \times A$.

Solution.

$$A \times B = \{(x, y)|\ -1 \le x < 2 \text{ and } y = 1\}.$$

A geometrical representation (or a graph, or a "picture") of $A \times B$ is given in Figure 3. Every point on the horizontal line between the points $(-1, 1)$ and $(2, 1)$ represents an element of $A \times B$. The point $(-1, 1) \in A \times B$, so we place a shaded circle at $(-1, 1)$. The point $(2, 1) \notin A \times B$, so we place an open circle there. We have

$$B \times A = \{(x, y)|\ x = 1 \text{ and } -1 \le y < 2\}.$$

A geometrical representation of $B \times A$ is given in Figure 4.

Example 3. Suppose $A = \{x|\ |x - 1| < 2\}$ and $B = \{-1, 2\}$. Determine $A \times B$ and $B \times A$.

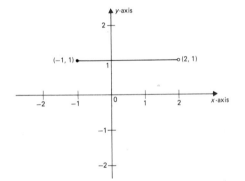

Figure 3 $A \times B$

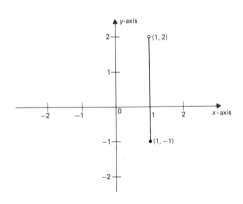

Figure 4 $B \times A$

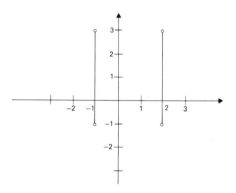

Figure 5 *A × B*

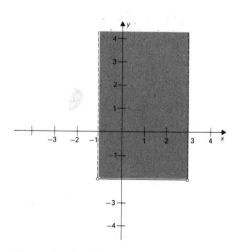

Figure 6 *B × A*

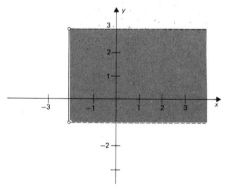

Solution. Solving for *x* in the set *A* we have

$$|x - 1| < 2 \text{ implies } -1 < x < 3.$$

Hence

$$A \times B = \{x \mid -1 < x < 3\} \times \{-1, 2\}$$
$$= \{(x, y) \mid -1 < x < 3 \text{ and } y = -1, 2\}$$

and

$$B \times A = \{(x, y) \mid x = -1, 2 \text{ and } -1 < y < 3\}.$$

For geometrical representations see Figures 5 and 6.

Example 4. If *A* = {*x*| |*x* − 1| < 2} and *B* = {*x*| *x* ≥ −2}, determine *A* × *B* and *B* × *A*.

Solution.

$$A \times B = \{(x, y) \mid -1 < x < 3 \text{ and } y \geq -2\}.$$

We picture this as the set of all points in the shaded region and on the solid line joining (−1, −2) and (3, −2) (see Figure 7). However, the set of points on the broken lines {(*x*, *y*)| *x* = −1 and *y* ≥ −2} and {(*x*, *y*)| *x* = 3 and *y* ≥ −2} is not included. Next,

$$B \times A = \{(x, y) \mid x \geq -2 \text{ and } -1 < y < 3\}$$

(see Figure 8).

Figure 8 *B × A*

The sets *A* × *B* and *B* × *A* in the example above are subsets of the set *R* × *R*. Such sets are called *relations*.

Figure 7 *A × B*

relation is more general than algebraic set.

Definition 2. Let R be the set of real numbers, then a *relation* is any subset S of $R \times R$.

We define the *domain* of the relation S to be the set of all first components of the elements of S and the *range* of the relation S to be the set of all second components of the elements of S.

The graph of a relation

The geometric representation of the set S is called the *graph* of S. Thus in Example 4 the domain of $A \times B$ is $\{x|\ -1 < x < 3\}$ and the range of $A \times B$ is $\{y|\ y \geq -2\}$. The graph is given in Figure 7.

Exercise 1

1. On a Cartesian coordinate system indicate the points with the following coordinates.
   (a) (3, 2)   (b) $(-\frac{3}{2}, 2)$ (c) (0, 5)
   (d) (5, 0)   (e) (−3, −3)   (f) (−2, −5)
   (g) (1, −3)

In problems 2–14, find the Cartesian product of the given sets. If the given sets are sets of real numbers, draw a graph of the resulting set.

2. $A = \{a, b\},\ B = \{c\}$
3. $A = \{a\},\ B = \{b, c, d\}$
4. $A = \{1, 2\},\ B = \{-1, -2\}$
5. $A = \{0, 1, 2\},\ B = \{-1, 0\}$
6. $A = \{0, 1, 2\},\ B = \{0, 1, 2\}$
7. $A = \{a, b, c\},\ B = \{b, d, e, f\}$
8. $A = \{x|\ -2 \leq x \leq 1,\ B = \{3\}$
9. $A = \{x|\ |x - 2| < 3\},\ B = \{2\}$
10. $A = \{x|\ |x + 1| < 4\},\ B = \{3, 4\}$
11. $A = \{x|\ |x + 3| < 5\}\ B = \{x|\ x > -2\}$
12. $A = \{x|\ x \geq -3\},\ B = \{x|\ |2x - 1| < 3\}$
13. $A = \{x|\ |x - 1| \leq 3\},\ B = \{x|\ |x| > 1\}$
14. $A = \{x|\ |2x - 3| > 5\},\ B = \{x|\ x \geq -1\}$
15. If $A = \{1, 2, 3, 4\},\ B = \{1, 2, 3, 4, 5\}$, how many elements are in $A \times B$?   in $B \times A$?
16. Let $A = \{1, 2, 3, \ldots, m\},\ B = \{1, 2, 3, \ldots, n\},\ m < n$.
    (a) How many elements are in $A \times B$?   in $B \times A$?
    (b) How many elements in $A \times B$ have the same first and second components?
    (c) Can you find m and n such that $A \times B$ has 4 elements?   7 elements?   9 elements?   12 elements?
17. If two vertices of a square are $A(1, 1)$ and $B(4, 1)$, determine the other two vertices. Is the solution unique?

18. If *ABCD* is a rectangle and $A = (-2, -1)$, $C = (3, 2)$, determine the coordinates of the points *B* and *D*. How many solutions are there?

2. The Distance Formula

In Chapter 1 we defined the absolute value of a number and showed that $|a - b|$ is the distance between the two points *a* and *b* on the number line. Here we shall extend the notion of distance to R^2. The distance between two points in R^2 is the length of the line segment joining them. We shall establish a formula for this distance in terms of the coordinates of the two endpoints.

Consider two points on the *x*-axis, $A(x_1, 0)$ and $B(x_2, 0)$. We have defined the distance $d(A, B)$ between *A* and *B* to be

$$d(A, B) = |x_2 - x_1|.$$

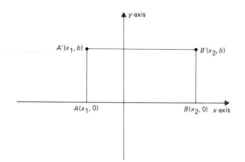

Figure 9

Consider the points $A'(x_1, b)$ and $B'(x_2, b)$, $b \neq 0$ (see Figure 9). Since the second components of *A'* and *B'* are equal, the line segment *A'B'* is parallel to the *x*-axis. The reader can easily see that the four points *A*, *B*, *B'*, and *A'* are the vertices of a rectangle, and hence the line segment *A'B'* may be assigned a length

$$d(A', B') = d(A, B)$$
$$= |x_2 - x_1|.$$

Similarly (see Figure 10), we have

$$d(C', D') = d(C, D)$$
$$= |y_2 - y_1|.$$

Recalling the Pythagorean theorem we have the following result.

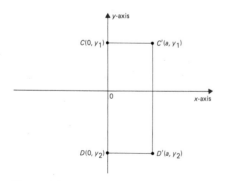

Figure 10

Theorem 1. **If $P(x_1, y_1)$ and $Q(x_2, y_2)$ are points in R^2, then**

(1) $$d(P, Q) = \sqrt{(x_2 - x_1)^2 + (y_2 - y_1)^2}.$$

Proof. Let *S* be the point (x_2, y_1). Then the points *P*, *Q*, and *S* are the vertices of a right triangle as shown in Figure 12. Since the line segment *PS* is parallel to the *x*-axis and the line segment *QS* is parallel to the *y*-axis, we have

(2) $$d(P, S) = |x_2 - x_1|$$

and

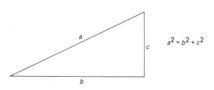

Figure 11

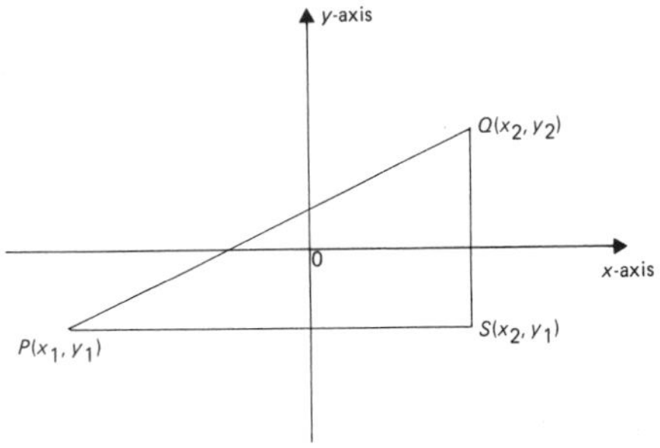

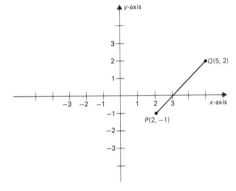

Figure 12

Figure 13

It makes no difference whether the coordinates of P or those of Q are chosen as (x_1, y_1). The reader should show that $d(P, Q) = d(Q, P)$.

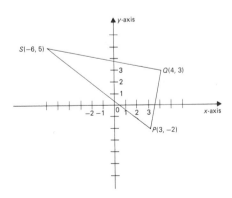

Figure 14

(3) $$d(Q, S) = |y_2 - y_1|.$$

Hence, by the Pythagorean theorem

$$[d(P, Q)]^2 = [d(P, S)]^2 + [d(Q, S)]^2$$
$$= |x_2 - x_1|^2 + |y_2 - y_1|^2$$
$$= (x_2 - x_1)^2 + (y_2 - y_1)^2$$

and

$$d(P, Q) = \sqrt{(x_2 - x_1)^2 + (y_2 - y_1)^2}.$$

This proves Theorem 1.

Example 1. Find the distance between $P(2, -1)$ and $Q(5, 2)$.

Solution. Using equation (1) we have

$$d(P, Q) = \sqrt{(x_2 - x_1)^2 + (y_2 - y_1)^2}$$
$$= \sqrt{(5 - 2)^2 + [2 - (-1)]^2}$$
$$= \sqrt{9 + 9}$$
$$= 3\sqrt{2}.$$

Example 2. Show that the triangle with vertices $P(3, -2)$, $Q(4, 3)$, and $S(-6, 5)$ is a right triangle.

Solution. Using the distance formula, we have

$$d(P, Q) = \sqrt{(4 - 3)^2 + (3 + 2)^2} = \sqrt{26}$$
$$d(Q, S) = \sqrt{(-6 - 4)^2 + (5 - 3)^2} = \sqrt{104}$$

and

$$d(P, S) = \sqrt{(-6 - 3)^2 + (5 + 2)^2} = \sqrt{130}.$$

We observe that

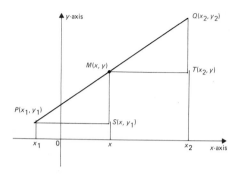

Figure 15

Congruent triangles : Two triangles are said to be congruent if and only if the corresponding sides and angles are equal.

$$(\sqrt{130})^2 = (\sqrt{104})^2 + (\sqrt{26})^2$$

or

$$[d(P, S)]^2 = [d(Q, S)]^2 + [d(P, Q)]^2.$$

Hence, by the converse of the Pythagorean theorem the triangle with vertices P, Q, and S is a right triangle with hypotenuse the line segment PS.

We shall now show how to find the coordinates of the midpoint of a line segment. Consider the line segment PQ in Figure 15. Let $M(x, y)$ be the midpoint of the line segment PQ. We wish to express x, y in terms of x_1, y_1, x_2, y_2. With the help of results in plane geometry, the reader will see that the triangles PMS and MQT are congruent. Therefore,

$$d(P, S) = d(M, T)$$

or

$$|x - x_1| = |x_2 - x|,$$

and since x is to the right of x_1 and x_2 is to the right of x, we have

$$x - x_1 = x_2 - x.$$

Solving for x we obtain

(4)
$$x = \frac{x_1 + x_2}{2}.$$

Also,

$$d(S, M) = d(T, Q)$$

or

$$y - y_1 = y_2 - y$$

(5)
$$y = \frac{y_1 + y_2}{2}.$$

Hence, the coordinates of the midpoint of the line segment PQ are given by (4) and (5).

Example 3. Find the midpoint of the line segment with endpoints $A(-4, 2)$ and $B(3, 4)$.

Solution. Let $M(x, y)$ be the midpoint of the line segment AB. Then

$$x = \frac{-4 + 3}{2} = -\frac{1}{2}$$

and

$$y = \frac{2 + 4}{2} = 3.$$

Therefore, the midpoint is $M(-\frac{1}{2}, 3)$.

Exercise 2

In problems 1–8, find the distance between the points with the given coordinates. Also, find the coordinates of the midpoint.

1. $A(1, 3)$, $B(4, 8)$
2. $A(2, 5)$, $B(-1, 4)$
3. $A(-3, -3)$, $B(-1, -6)$
4. $A(-2, 5)$, $B(-1, -2)$
5. $A(-2, 0)$, $B(-3, 1)$
6. $A(2, \frac{1}{2})$, $B(\frac{1}{2}, -3)$
7. $A(-1, 3)$, $B(-1, 7)$
8. $A(5, 2)$, $B(-3, 2)$

In the problems 9–14, show that ABC is a right triangle.

9. $A(-3, 2)$, $B(5, 5)$, $C(0, -6)$
10. $A(2, 0)$, $B(6, 3)$, $C(5, -4)$
11. $A(0, 0)$, $B(3, 2)$, $C(-4, 6)$
12. $A(4, 3)$, $B(1, 4)$, $C(-2, -5)$
13. $A(1, 2)$, $B(3, 4)$, $C(1, 6)$
14. $A(0, -3)$, $B(-1, -1)$, $C(-3, -2)$

In problems 15–18 identify the triangle ABC as isosceles, equilateral, or right.

15. $A(-6, 6)$, $B(-2, 5)$, $C(-5, 2)$
16. $A(2, 1)$, $B(5, 5)$, $C(-2, 4)$
17. $A(1, -1)$, $B(-1, 1)$, $C(-\sqrt{3}, -\sqrt{3})$
18. $A(-4, 6)$, $B(0, 5)$, $C(-3, 2)$
19. Which of the following sets of points are collinear (lie on the same line)?
    (a) $(0, 5)$, $(4, 1)$, $(12, -7)$
    (b) $(-3, 2)$, $(2, 0)$, $(1, -2)$
    (c) $(3, -5)$, $(\frac{1}{2}, 0)$, $(0, 1)$
20. The distance from the point P to the point $(2, 0)$ is 89. If the abscissa of P is -4, find the ordinate of P.
21. Show that the points $A(3, -16)$, $B(-5, -1)$, $C(10, 7)$, and $D(18, -8)$ are the vertices of a square. Find the length of the diagonals.
22. If the vertices of a triangle are $A(1, 2)$, $B(-4, 5)$, and $C(3, 7)$, find the length of each of the medians (a median is a line segment drawn from a vertex to the midpoint of the opposite side).
23. A pilot is flying from Dullsville to Middale to Pleasantville. With reference to an origin, Dullsville is located at $(2, 4)$, Middale at $(8, 12)$, and Pleasantville at $(20, 3)$, all numbers being in 100-mile units.
    (a) Locate the positions of the three cities on a coordinate system.
    (b) Compute the distance traveled by the pilot.
    (c) Compute the direct distance from Dullsville to Pleasantville.

3. Functions

An important kind of relation is the one that is called a *function*. Consider the following examples.

Example 1. In an experiment, the effect of a bactericidal agent on the number N of viable bacteria with respect to time t was measured. The following table indicates the result of the experiment.

t (minutes)	0	5	10	15	20	25	30
N = number of viable bacteria per milliliter (approx.)	10^{12}	10^{10}	10^{8}	10^{6}	10^{4}	10^{2}	1

We may express this result as a relation consisting of the set of ordered pairs of numbers

$$f = \{(0, 10^{12}), (5, 10^{10}), (10, 10^{8}),$$
$$(15, 10^{6}), (20, 10^{4}), (25, 10^{2}), (30, 1)\}.$$

Here we say that N is a *function* of time t that is described by the set f. The two sets of numbers

$$t = \{0, 5, 10, 15, 20, 25, 30\}$$

and

$$N = \{10^{12}, 10^{10}, 10^{8}, 10^{6}, 10^{4}, 10^{2}, 1\}$$

are called the *domain* and the *range* of the function f respectively.

Example 2. A mathematician working for the Fancy Shirt Company finds that the total cost c (in dollars) of producing x shirts is given by the equation

$$c = 0.001x^2 + 1.2x + 30, \quad x \geq 1.$$

In this case, the cost c is a function of the number x of shirts produced that is described by the set of ordered pairs

$$f = \{(x, c) \mid x \geq 1, c = 0.001x^2 + 1.2x + 30\}$$

where x belongs to the set of positive integers (the domain of f).

Example 3. The area A of a square is dependent on the length of a side. We have

(6) $$A = S^2$$

where S is the length of the side of a square. Thus, the area A is a function of S that is described by the set

Obviously, the company does not produce a fraction of a shirt. That is why we require that the domain of f be the set of positive integers.

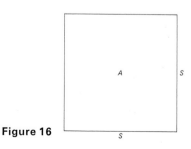

Figure 16

$$f = \{(S, A)\,|\, S > 0, A = S^2\}.$$

We observe that for the relation in (6) to be meaningful we require that the domain of f be the set $\{S|\, S > 0\}$.

In each of the examples above we see that we are dealing with

(a) two sets of numbers X and Y

and

(b) a "rule" or formula that assigns to each element of X a *single* element of Y.

We now give a precise definition of a function:

Alternate definition : A function consists of two nonempty sets X and Y and a rule f that assigns to each element of X a single element of Y.
The student would do well to compare the two definitions.

Definition 3. A **function** is a relation in which no two ordered pairs have the same first component.

The symbol ⇒ is used for the word "implies."

In other words we have

A relation f is a function if and only if

$$(x, y_1) \in f \text{ and } (x, y_2) \in f \Rightarrow y_1 = y_2.$$

Here we shall be concerned with functions that are subsets of R^2. Such functions are called *real-valued functions* and we shall refer to them simply as *functions*. The set of all first components in the ordered pairs in a function is called the *domain* of the function, and the set of all second components is called the *range* of the function. The domain of a function may be an interval, a collection of intervals, or any nonempty subset of the real numbers. If a function is given by a formula and the domain is not specified, then we shall assume that the domain consists of all real numbers for which the formula is meaningful.

Example 4. Describe the domain and range of the function

$$f = \{(x, y)\,|\, x, y \in R, y = x^2\}.$$

Solution.

$$\text{domain } f = \{x|\, x \in R\}, \quad \text{range } f = \{y|\, y \geq 0\}.$$

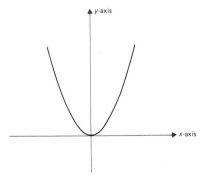

The graph of the function f is shown in Figure 17.

Notation. In general, we shall denote functions by letters such as $f, g, h, F, G, \psi, \phi$. We shall use the notation $f(x)$, read "f of x," which is the value of the function at x. This functional notation is due to the Swiss mathematician Euler.

Figure 17
Graph of $f = \{(x, y)\,|\, y = x^2\}$

Leonhard Euler (1707–1783). Few men have left a greater mark in the development of mathematics than the Swiss mathematician Leonhard Euler. He was born in Basel, Switzerland, and died in St. Petersburg (now Leningrad), Russia. Euler, one of the master analysts of the eighteenth century, has been called the most prolific mathematician in history. He was introduced to mathematics by his father, a Calvinist pastor who had studied mathematics with the first Jacob Bernoulli. At the age of seventeen, Euler received his master's degree and at nineteen he wrote his first paper for the Academy of Sciences in Paris.

From 1732–1741 Euler held a Chair of Mathematics in the Academy of Sciences in St. Petersburg. He then accepted an offer of the King of Prussia to be director of the department of mathematics at the Academy of Berlin, where he stayed twenty-five years. At the invitation of Catherine II he returned to St. Petersburg. Although the last years of his life were spent in total blindness, his productivity continued undiminished. He was the author of more than 700 papers and thirty-two books in various branches of mathematics.

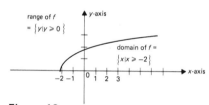

Figure 18
Graph of $f = \{(x, y) \mid y = \sqrt{2 + x}\}$

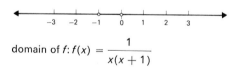

domain of $f: f(x) = \dfrac{1}{x(x + 1)}$

domain of $f: f(x) = \dfrac{1}{\sqrt{x^2 + x - 2}}$

In example 4 we may simply write

$$f(x) = x^2.$$

Here $f(x)$ plays the same role as y. Hence

$$f(-2) = (-2)^2 = 4$$
$$f(0) = 0^2 = 0$$
$$f(1) = 1^2 = 1$$
$$f(3.5) = (3.5)^2 = 12.25$$
$$f(a) = a^2$$
$$f(x + h) = (x + h)^2$$

and so on.

Example 5. Determine the domain of f if $f(x) = \sqrt{2 + x}$.

Solution. Since for a real number b, $\sqrt{b}$ is real if and only if $b \geq 0$, the domain of f is $\{x \mid 2 + x \geq 0\}$ or

$$\text{domain } f = \{x \mid x \geq -2\}.$$

Example 6. Determine the domain of the function f if

$$f(x) = \frac{1}{x(x + 1)}.$$

Solution. The question here is: For what values of x is $f(x)$ a real number? Clearly, $f(x)$ is a real number for values of x for which $x(x + 1) \neq 0$. Therefore,

$$\text{domain } f = \{x \mid x \neq 0, -1\}$$
$$= \{x \mid x < -1\} \cup \{x \mid -1 < x < 0\} \cup \{x \mid x > 0\}.$$

Example 7. Determine the domain of f if

$$f(x) = \frac{1}{\sqrt{x^2 + x - 2}}.$$

Solution. Here $f(x)$ is a real number only for values of x for which $x^2 + x - 2 > 0$. Referring to Chapter 1, Section 4, Example 3, we find that the inequality is satisfied if and only if x is in the set

$$\{x \mid x < -2\} \cup \{x \mid x > 1\}.$$

Therefore

$$\text{domain } f = \{x \mid x < -2\} \cup \{x \mid x > 1\}.$$

Example 8. Is the relation $\{(x, y) \mid x, y \in R, x^2 + y^2 = 1\}$ in R^2 a function?

Solution. Solving explicitly for y we have

$$y = \sqrt{1 - x^2} \quad \text{or} \quad y = -\sqrt{1 - x^2}.$$

There are two values of y associated with a given value of x in $|x| < 1$. Therefore, the relation is not a function.

Why do we restrict x to values such that $|x| < 1$?

Note that every vertical line intersects the graph of a function at only one point.

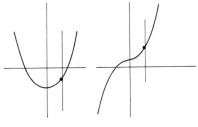

graphs of functions

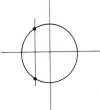

graph of a relation *not* a function

Exercise 3

In problems 1–16, determine the domain of the functions.

1. $f(x) = x + 4$

2. $f(x) = \dfrac{1}{x}$

3. $g(t) = \dfrac{1}{t^2 + 4}$

4. $g(u) = \dfrac{1}{u(u - 5)}$

5. $f(r) = \dfrac{1}{r^2 - 4}$

6. $h(s) = \sqrt{s}$

7. $h(s) = \dfrac{s}{(s - 1)(s + 2)}$

8. $F(x) = \sqrt{2 - x}$

9. $f(x) = \sqrt[3]{x}$

10. $G(z) = \sqrt{4 - z^2}$

11. $F(x) = \sqrt{x^2 - 16}$

12. $H(t) = \sqrt[3]{\dfrac{1}{1 - t}}$

13. $G(s) = \dfrac{\sqrt{s + 1}}{\sqrt{1 - s^2}}$

14. $f(t) = \sqrt{\dfrac{3t}{t + 4}}$

15. $g(t) = \sqrt{t^2 + 3t + 2}$

16. $h(x) = \dfrac{2}{\sqrt{x^2 - x - 6}}$

17. Graph the function described in Example 1.

In problems 18–22 graph each function whose value and domain are given.

18. $h(x) = 2 + x$, domain $h = \{1, 2, 3, 4, 5\}$

19. $f(x) = x^2 - 3$, domain $f = \{-1, 0, 1, 2, 3, 4\}$

20. $g(s) = \sqrt{s + 1}$, domain $g = \{-1, 0, 1, 2, 3, 4\}$

21. $f(x) = 2 + \sqrt{x + 1}$, domain $f = \{-1, 0, 1, 2, 3, 4\}$

22. $H(u) = \dfrac{u + 1}{u - 1}$, domain $H = \{3, 5, 7, 9, 11\}$

23. If $g(x) = 2x + 4$, find

(a) $g(-1)$ (b) $g(\frac{1}{2})$ (c) $g(3)$

(d) $g(a)$ (e) $g(a + 1)$ (f) $g(a - 1)$

(g) $g(x + h)$.

24. If $h(x) = 2x^2 - 2x + 3$, find

(a) $h(-2)$ (b) $h(0)$ (c) $h(\frac{1}{4})$

(d) $h(a + 1)$ (e) $h(a + 1) - h(a)$.

25. If $g(x) = 2x + 5$ defines a function g, find the number in the domain of g such that

(a) $g(x) = 2$ (b) $g(x) = -1$ (c) $g(x) = c + 4$.

26. If $f(x) = x^2 - 4$ defines a function f, find the number in the domain of f such that

(a) $f(x) = 0$ (b) $f(x) = 3$ (c) $f(x) = -5$.

27. Consider the function f defined by $f(x) = x^2$. Does $f(a) + f(b) = f(a + b)$? For what values of a and b will the equality hold?

28. Definition. A function f is said to be *even* if $f(x) = f(-x)$ for all x and $-x$ in the domain of f.

   A function f is said to be *odd* if $f(-x) = f(x)$ for all x and $-x$ in the domain of f. Which of the following functions are even and which are odd?

(a) $f(x) = 2x^2$

(b) $f(x) = x^3$

(c) $f(x) = x^6 - x^4$

(d) $f(x) = x^2 - x$

29. Can a function be both odd and even?

4. Operations on Functions

The operations of addition, multiplication, and division apply to functions as well as to numbers. If f and g are functions with domains A and B respectively, then the sum $f + g$, the difference $f - g$, the product fg, and the quotient f/g are the functions defined as follows:

(7) $$(f + g)(x) = f(x) + g(x)$$

(8) $$(f - g)(x) = f(x) - g(x)$$

(9) $$(fg)(x) = f(x)g(x)$$

and

(10) $$(f/g)(x) = f(x)/g(x)$$

respectively. The domain of $f + g$, $f - g$, and fg is $A \cap B$. The domain of f/g is $\{x|\ x \in A \cap B, g(x) \neq 0\}$. If $A \cap B = \phi$, then $f \pm g$, fg, and f/g are not defined.

Example 1. Let the functions f and g be defined by the equations

$$f(x) = 2x^2 + 3, \qquad g(x) = \sqrt{x + 1}.$$

Describe the functions $f + g$, $f - g$, fg, and f/g.

How is the domain of g obtained?

Solution. The domain of f is the set of all real numbers; the domain of g is $\{x|\ x \geq -1\}$. Therefore, the functions $f + g$, $f - g$, and fg are defined by

$$(f + g)(x) = f(x) + g(x)$$
$$= 2x^2 + 3 + \sqrt{x + 1}, \text{ domain } f + g = \{x|\ x \geq -1\}$$

$$(f - g)(x) = f(x) - g(x)$$
$$= 2x^2 + 3 - \sqrt{x + 1}, \text{ domain } f - g = \{x|\ x \geq -1\}$$

$$(fg)(x) = f(x)g(x)$$
$$= (2x^2 + 3)(\sqrt{x + 1}), \text{ domain } fg = \{x|\ x \geq -1\}.$$

Note. $g(x) = 0 \Rightarrow \sqrt{x + 1} = 0 \Rightarrow x = -1.$

Since $g(x) = 0$ when $x = -1$,

$$(f/g)(x) = f(x)/g(x)$$
$$= \frac{2x^2 + 3}{\sqrt{x + 1}}, \qquad \text{domain } f/g = \{x|\ x > -1\}.$$

Example 2. Let the functions f and g be defined by the equations

$$f(x) = \sqrt{16 - x^2}, \qquad g(x) = 1 - x^2.$$

Describe the functions f/g and g/f.

How is the domain of f obtained?

Solution. The domain of f is $\{x|\ -4 \leq x \leq 4\}$; the domain of g is the set of all real numbers. We note that $f(x) = 0$ for $x = -4, 4$ and $g(x) = 0$ for $x = 1, -1$. Thus, f/g is defined by

$$(f/g)(x) = \frac{f(x)}{g(x)}$$
$$= \frac{\sqrt{16 - x^2}}{1 - x^2}, \text{ domain } f/g = \{x|\ -4 \leq x \leq 4, x \neq 1, -1\}$$

and g/f is defined by

$$(g/f)(x) = \frac{g(x)}{f(x)}$$

$$= \frac{1 - x^2}{\sqrt{16 - x^2}}, \qquad \text{domain } g/f = \{x \mid -4 < x < 4\}.$$

Let us now examine what happens when the two components of every ordered pair in a function f are interchanged. For example, if $f = \{(-1, 1), (0, 2), (1, 2), (2, 3)\}$, then interchanging the components of these ordered pairs yields the set $\{(1, -1), (2, 0), (2, 1), (3, 2)\}$. Obviously, the resulting set is not a function, since the ordered pairs $(2, 0)$ and $(2, 1)$ have the same first component and different second components. Now suppose a function f has no two ordered pairs with the same second components. Such a function is said to be *one-to-one*, and interchanging the components of every ordered pair in f yields another function. The resulting function is called the *inverse function* of f and is denoted by f^{-1} (read "f inverse"). Clearly, the domain of f^{-1} is the range of f and the range of f^{-1} is the domain of f.

A function f is said to be one-to-one if for every $x_1, x_2 \in$ domain f, $f(x_1) = f(x_2) \Rightarrow x_1 = x_2$.

Note that the superscript -1 is not an exponent.

Example 3. Consider the function f consisting of the set $\{(1, -1), (2, -2), (3, -3), (4, -4)\}$. Find f^{-1}.

Solution. We observe that f is one-to-one. By definition therefore,

$$f^{-1} = \{(-1, 1), (-2, 2), (-3, 3), (-4, 4)\}$$

and we note that

$$\text{domain } f^{-1} = \{-1, -2, -3, -4\} = \text{range } f$$
$$\text{range } f^{-1} = \{1, 2, 3, 4\} \qquad = \text{domain } f.$$

The Celsius scale, also known as the centigrade scale, is named for the Swedish astronomer-inventor Anders Celsius (1701–1744).

Example 4. Two commonly used temperature scales are the Celsius temperature scale and the Fahrenheit temperature scale. We have two sets of numbers: representing the Celsius scale is the set of numbers $X = \{x \mid x \geq -273.15\}$; representing the Fahrenheit scale is the set of numbers $Y = \{y \mid y \geq -459.67\}$. If x is the number on the Celsius scale for a temperature, then the same temperature is represented by the number $y = \frac{9}{5}x + 32$ on the Fahrenheit scale. Thus we have a function f defined by $\{(x, y) \mid x \in X, y = \frac{9}{5}x + 32\}$. Find f^{-1}.

Solution. The function f is defined by

$$(11) \qquad y = \frac{9}{5}x + 32, \qquad x \in X.$$

To interchange the components of the ordered pairs in f as above, we interchange x and y in the equation (11) and define f^{-1} by

$$(12) \qquad x = \frac{9}{5}y + 32,$$

or solving for y in terms of x in equation (12), we obtain

(13) $$y = \tfrac{5}{9}(x - 32), \qquad x \in Y.$$

That is,

$$f^{-1} = \{(x, y)\,|\, x \in Y, y = \tfrac{5}{9}(x - 32)\}.$$

Note the interchange of the domain and range for f and f^{-1}.

Here f is a function that converts Celsius degrees into degrees Fahrenheit and f^{-1} is a function that converts degrees Fahrenheit into Celsius degrees.

Example 5. Consider the function f defined by the equation $y = x^2$. Discuss the existence of f^{-1}.

Solution. If we take the domain of f to be all real x, then f is not one-to-one and so f^{-1} does not exist. However, if we restrict the domain to $\{x\,|\, x \geq 0\}$, then the resulting function g is one-to-one, and we find g^{-1} as follows:

From the equation

$$y = x^2$$

we obtain by interchanging x and y

$$x = y^2.$$

f is not one-to-one because (x, x^2) and $(-x, x^2)$ are both in f.

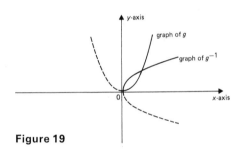

Figure 19

Solving for y in terms of x, we have

$$y = \sqrt{x} \qquad \text{or} \qquad y = -\sqrt{x}.$$

Since the domain of g is $\{x\,|\, x \geq 0\}$, this set is the range of g^{-1} defined by

$$y = \sqrt{x}.$$

Therefore, if $g = \{(x, y)\,|\, x \geq 0, y = x^2\}$, then $g^{-1} = \{(x, y)\,|\, x \geq 0, y = \sqrt{x}\}$. Now, if we take the domain of f to be $\{x\,|\, x \leq 0\}$, then the resulting function h is again one-to-one. Using the equation

$$y = x^2,$$

interchanging x and y, and solving for y, we obtain

$$y = \sqrt{x} \qquad \text{or} \qquad y = -\sqrt{x}.$$

Since the domain of h is $\{x\,|\, x \leq 0\}$, this set is the range of h^{-1}. Therefore, if $h = \{(x, y)\,|\, x \leq 0, y = x^2\}$, then $h^{-1} = \{(x, y)\,|\, x \geq 0, y = -\sqrt{x}\}$.

In later chapters the reader will encounter another useful operation on functions, called the composition of functions.

Definition 4. If f and g are two functions, the *composite function*, denoted by $f \circ g$, has values given by

(14) $$(f \circ g)(x) = f(g(x)).$$

The domain of $f \circ g$ is $\{x \mid x \in \text{domain } g,\ g(x) \in \text{domain } f\}$.

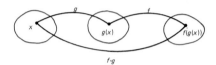

$f \cdot g$

Example 6. Let the functions f and g be defined by the equations

$$f(x) = 3x^2 + 5, \qquad g(x) = \sqrt{x - 1}.$$

Determine $f \circ g$ and $g \circ f$.

Solution. The domain of f is all real numbers; the domain of g is $\{x \mid x \geq 1\}$. Using (14) we have for $f \circ g$:

$$\begin{aligned}
(f \circ g)(x) &= f(g(x)) \\
&= f(\sqrt{x - 1}) \\
&= 3(\sqrt{x - 1})^2 + 5 \\
&= 3(x - 1) + 5 = 3x + 2
\end{aligned}$$

and

$$\begin{aligned}
\text{domain } f \circ g &= \{x \mid x \in \text{domain } g,\ g(x) \in \text{domain } f\} \\
&= \{x \mid x \geq 1,\ \sqrt{x - 1} \text{ is real}\} \\
&= \{x \mid x \geq 1\}.
\end{aligned}$$

For $g \circ f$ we have

$$\begin{aligned}
(g \circ f)(x) &= g(f(x)) \\
&= g(3x^2 + 5) \\
&= \sqrt{(3x^2 + 5) - 1} \\
&= \sqrt{3x^2 + 4}
\end{aligned}$$

and

$$\begin{aligned}
\text{domain } g \circ f &= \{x \mid x \in \text{domain } f,\ f(x) \in \text{domain } g\} \\
&= \{x \mid x \text{ is real},\ 3x^2 + 5 \geq 1\} \\
&= \{x \mid x \text{ is real}\}.
\end{aligned}$$

The reader will note that the function $f \circ g$ is different from $g \circ f$. For example, we find that

$$(f \circ g)(2) = f(g(2)) = f(\sqrt{2 - 1}) = f(1) = 8$$

and

$$(g \circ f)(2) = g(f(2)) = g(3(2)^2 + 5) = g(17) = 4.$$

Sometimes we find it convenient to express a complicated function as the composition of simpler functions.

Example 7. Express $f(x) = \sqrt{x^3 - 4}$ as a composition of two functions.

Solution. Let $g(x) = x^3 - 4$ and $h(w) = \sqrt{w}$. Then

$$(h \circ g)(x) = h(g(x))$$
$$= h(x^3 - 4)$$
$$= \sqrt{x^3 - 4}$$
$$= f(x).$$

Obviously, the functions g and h are not unique. If $G(x) = x^3$ and $H(w) = \sqrt{w - 4}$, then

$$(H \circ G)(x) = H(G(X)) = H(x^3) = \sqrt{x^3 - 4} = f(x).$$

The reader may try to write f as a composition of more than two functions.

Exercise 4

In problems 1–6, given the values of the functions f and g, describe the functions $f + g$, $f - g$, fg, and f/g.

1. $f(x) = 2x + 5,$ $g(x) = x^2$

2. $f(x) = \sqrt{x},$ $g(x) = x^2 + 1$

3. $f(x) = \sqrt{x^2 - 4},$ $g(x) = x^2 + 2x - 3$

4. $f(x) = \dfrac{1}{x + 1},$ $g(x) = \dfrac{x}{2x - 1}$

5. $f(x) = \dfrac{1}{\sqrt{x}},$ $g(x) = \dfrac{1}{x^2}$

6. $f(x) = \sqrt[3]{x},$ $g(x) = \sqrt{x^2 - 1}$

In problems 7–12, each of the equations defines a function. Describe the inverse functions.

7. $y = x - 1$ **8.** $y = 4x + 6$

9. $y + 2x = 5$ **10.** $2y + 3x = 4$

11. $y = x^3$ **12.** $y = 2x^3 - 1$

13. Let the function f be defined by the equation $y = 3x^2$. Discuss the existence of f^{-1}.

In problems 14–21, describe $f \circ g$ and $g \circ f$ for the pair of functions whose values are given.

14. $f(x) = 2 + x,$ $g(x) = 3x$

15. $f(x) = 3x^2 - 4,$ $g(x) = 2 - x$

16. $f(x) = x^2 + 4,$ $g(x) = \sqrt{x}$

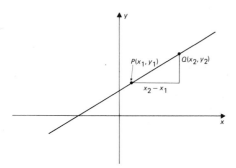

Figure 20

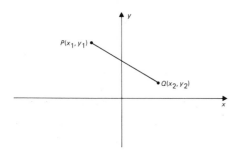

Figure 21

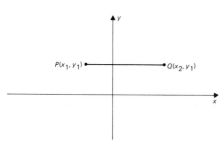

Figure 22

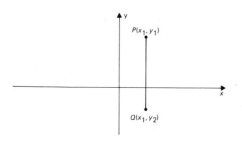

Figure 23

17. $f(x) = \sqrt[3]{x+1}$, $g(x) = 2x^3 + 1$

18. $f(x) = \dfrac{1}{x+1}$, $g(x) = x - 1$

19. $f(x) = \dfrac{x}{x^2+2}$, $g(x) = \dfrac{1}{x}$

20. $f(x) = \sqrt{2x^2+3}$, $g(x) = \dfrac{2}{x^2}$

21. $f(x) = x^{-1/2}$, $g(x) = x - 2$

22. If $f(x) = 3x + 2$, $g(x) = 2x^2$, find
   (a) $(f + g)(2)$ (b) $(f - g)(6)$ (c) $(fg)(-1)$
   (d) $(f/g)(\frac{1}{2})$ (e) $(f \circ g)(3)$ (f) $(g \circ f)(3)$.

23. Repeat problem 22 for $f(x) = \sqrt{x^2 - 16}$, $g(x) = \sqrt{x^2 + 4}$.

24. For each function f defined below, determine $f \circ f$.

   (a) $f(x) = \dfrac{1}{x}$ (b) $f(x) = 2 + 3x$

   (c) $f(x) = \sqrt{4 + x}$ (d) $f(x) = \dfrac{1}{2 + x}$

   (e) $f(x) = 2x + \dfrac{1}{x}$

In problems 25–30, express the function that is defined as a composition of two or more functions.

25. $f(x) = (x + 1)^2$

26. $f(x) = \sqrt{2x + 3}$

27. $f(x) = \dfrac{x - 1}{x + 1}$

28. $f(x) = \dfrac{1}{(x + 2)^2}$

29. $f(x) = (2x^2 - 3x + 1)^{-1/3}$

30. $f(x) = \dfrac{\sqrt{x^2 + 1}}{x^2 + 4}$

5. Linear Functions

In the next two sections we shall study a class of simple and widely applied functions called *linear* functions. We shall begin by introducing the notion of the slope of a straight line in R^2.

Let $P(x_1, y_1)$ and $Q(x_2, y_2)$ be any two points in R^2 with $x_1 \neq x_2$. We define the *slope* of the line segment PQ to be the number

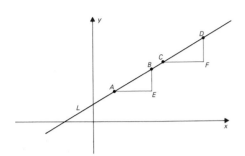

Figure 24

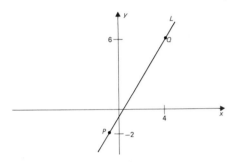

Figure 25

Note that the subscripts on the coordinates of
the points P and Q may be interchanged. Thus
if $P = (x_2, y_2)$ and $Q = (x_1, y_1)$, we again
find that

$$m = \frac{2 - 6}{-1 - 4} = \frac{-4}{-5} = \frac{4}{5}.$$

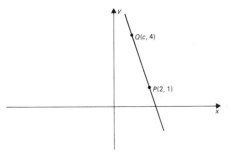

Figure 26

(15)
$$m = \frac{y_2 - y_1}{x_2 - x_1}, \qquad x_2 \neq x_1.$$

In Figure 20, since $x_2 - x_1 > 0$ and $y_2 - y_1 > 0$, the line segment PQ has slope $m > 0$. In Figure 21 however, since $x_2 - x_1 > 0$ and $y_2 - y_1 < 0$, the line segment PQ has slope $m < 0$; and for the line segment in Figure 22 which is parallel to the x-axis the slope $m = 0$. For the case where the line segment is parallel to the y-axis (see Figure 23) and hence $x_2 = x_1$, the slope is undefined.

To help you remember the sign of the slope of a line segment: If a slippery bug placed at one endpoint of a line segment slides down from right to left the slope is positive; if the bug slides down from left to right the slope is negative; if the bug fails to slide the slope is zero; and if the bug plunges vertically the slope is undefined.

We note that any two distinct points in R^2 determine a straight line. Consider the two points A and B and the line L in Figure 24. Suppose C and D are two other points in L. Now, associated with each pair of points A, B and C, D are the two right triangles ABE and CDF respectively. The reader can easily see that the two triangles are similar. Therefore, the slope of the line segment AB is equal to the slope of the line segment CD (why?). Hence the slope of the line L is equal to the slope of any of the line segments belonging to L.

Example 1. A line L contains the points $P(-1, -2)$ and $Q(4, 6)$. (See Figure 25.) Find the slope of L.

Solution. Using (15) we have

$$m = \frac{y_2 - y_1}{x_2 - x_1}$$
$$= \frac{6 - (-2)}{4 - (-1)} = \frac{8}{5}.$$

Example 2. Given the line L with slope $m = -3$, if $P(2, 1)$ and $Q(c, 4)$ are two points in L, find the number c.

Solution. Using (15) we have

$$m = \frac{y_2 - y_1}{x_2 - x_1} \qquad .$$
$$= \frac{4 - 1}{c - 2} = \frac{3}{c - 2}.$$

Setting $m = -3$ we obtain

$$-3 = \frac{3}{c - 2}$$

or

$$c - 2 = -1$$
$$c = 1.$$

Now again let $A(x_1, y_1)$ and $B(x_2, y_2)$ be two points in R^2 with $x_1 \neq x_2$. If $P(x, y)$ is any point on the line L containing A and B, then the slope of the line segment AP must equal the slope of the line segment AB. That is,

(16)
$$\frac{y - y_1}{x - x_1} = \frac{y_2 - y_1}{x_2 - x_1}.$$

Equation (16) describes the set of all points $P(x, y)$ belonging to the line L. If the slope of L is m, equation (16) may be written in the *point-slope form*

Point-slope form

(17)
$$y - y_1 = m(x - x_1).$$

Equation (17) describes a line with slope m and containing the point (x_1, y_1). If the point (x_1, y_1) is the point of intersection of L and the y-axis, then $x_1 = 0$ and equation (17) takes the form

Slope-intercept form

(18)
$$y = mx + y_1.$$

This is called the *slope-intercept form* of the equation describing L. Here m is the slope of the line and y_1 is the y-intercept.

Definition of linear function

Definition 5. A function f is said to be **linear** if it is of the form

$$f(x) = mx + b$$

where m and b are constants. The domain of a linear function is R.

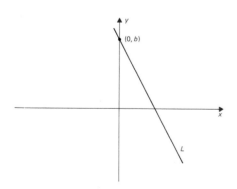

Figure 27

The graph of f is a straight line (we shall not prove this here), and hence we have the name *linear function* (see Figure 27).

Example 3. Describe the linear function f whose graph contains the points $P(-2, 1)$ and $Q(3, -4)$.

Solution. Since f is a linear function, its graph in R^2 is a straight line. Using (15) we have

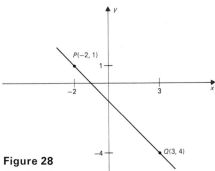

Figure 28
Graph of *f*

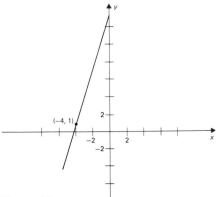

Figure 29
Graph of $f = \{(x, y) \mid y = 3x + 13\}$

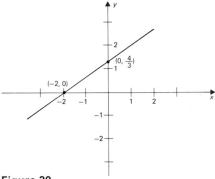

Figure 30
Graph of $f = \{(x, y) \mid 2x - 3y + 4 = 0\}$

The zeros of a function are those values of *x*
for which $f(x) = 0$. For example, the function
$f(x) = 1$ has no zeros, $f(x) = x$ has one zero,
while $f(x) = 0$ has an infinite number of zeros.

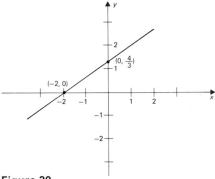

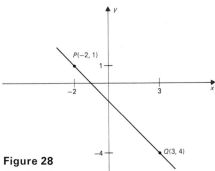

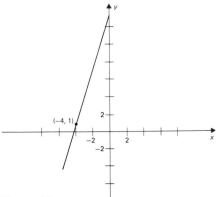

$$\frac{y - 1}{x - (-2)} = \frac{-4 - 1}{3 - (-2)}$$

or

$$\frac{y - 1}{x + 2} = \frac{-5}{5}.$$

Simplifying,

$$y = -x - 1.$$

Therefore, *f* is defined by $f(x) = -x - 1$ and its graph is sketched
in Figure 28.

Example 4. What is the linear function *f* whose domain is the set
of all real numbers such that the graph of *f* has slope 3 and contains
the point $(-4, 1)$?

Solution. Using the point-slope form of the equation of a line, we
have

$$y - y_1 = m(x - x_1).$$

Setting $m = 3$ and $(x_1, y_1) = (-4, 1)$ we obtain

$$y - 1 = 3[x - (-4)]$$

or

$$y = 3x + 13.$$

Therefore, *f* is defined by $f(x) = 3x + 13$. See Figure 29.

Example 5. Consider the linear function *f* defined by

$$\{(x, y) \mid 2x - 3y + 4 = 0\}.$$

Find the slope, *y*-intercept, and *x*-intercept of the graph of *f*.

Solution. From the equation

$$2x - 3y + 4 = 0$$

we obtain

$$y = \tfrac{2}{3}x + \tfrac{4}{3},$$

which is the slope-intercept form. Therefore,

$$m = \tfrac{2}{3} \quad \text{(slope)}$$
$$b = \tfrac{4}{3} \quad \text{(}y\text{-intercept)}.$$

To determine the *x*-intercept, or the *x*-coordinate of the point where
the line intersects the *x*-axis (also called the *zero* of the function *f*),
we ask: For what value of *x* is *y* zero? In the equation $2x - 3y + 4 = 0$, set $y = 0$ and solve for *x*:

$$2x - 3(0) + 4 = 0$$

or

$$x = -2.$$

This is the x-intercept.

In *standard form* the equation of a line is

(19) $$ax + by + c = 0$$

where a, b, and c are constants and a and b are not zero. If $b = 0$, the resulting equation does not define a function (why?). If $a = 0$ and $b \neq 0$, the resulting equation $y = -c/a$ defines a *constant function*.

A constant function is a function of the form $f(x) = k$.

One area in which linear functions have useful applications is economics. Let us consider a manufacturer producing x units of an item. Suppose the total cost y is a linear function of the number of units produced. We shall assume that the graph of the function relating y to x is given by Figure 31.

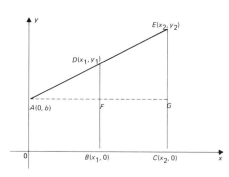

Figure 31

We interpret the graph as follows. If x_1 units are manufactured, then the total cost is y_1, which is the length of the line segment BD. Similarly, y_2 (length of the line segment CE) is the total cost of producing x_2 units of the item. If zero units of the item are produced, the cost is b. This is called the *fixed cost*.

Fixed cost may be due to such expenses as fire insurance or purchasing equipment.

The fixed cost of producing x_2 units is given by the line segment CG whose length is b, and the line segment GE represents the *variable cost*. Thus the variable cost of making x_1 units is given by the line segment FD.

Variable cost is the change in cost as the number of units made changes.

We now observe that the slope of the line segment AE is given by

$$m = \frac{y_2 - b}{x_2 - 0}$$

$$= \frac{\text{variable cost}}{\text{number of units produced}}.$$

slope = marginal cost
= change in total cost when one more
unit is produced.

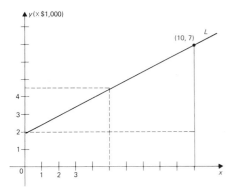

Figure 32

Economists refer to this as the *marginal cost*. When the total cost is linearly related to output, the marginal cost is constant.

Example 6. In Figure 32 the line *L* is the graph of a function that relates the total manufacturing cost *y* and the number of units *x*, of a product manufactured. Find

    (a) the fixed cost

    (b) the total cost when 10 units are produced

    (c) the variable cost when 10 units are produced

    (d) the marginal cost

    (e) the variable cost when 5 units are produced.

Solution.

    (a) Fixed cost = $2,000.

    (b) Total cost when 10 units are produced is $7,000.

    (c) Variable cost when 10 units are produced is total cost minus fixed cost = $7,000 − $2,000 = $5,000.

    (d) Marginal cost = slope of *L*

$$= \frac{7 - 2}{10 - 0}$$

$$= \frac{1}{2}.$$

    (e) First we find the total cost for any *x*, that is, we find the equation of the line *L*. Using the slope-intercept form we have

$$y = mx + b$$
$$= \tfrac{1}{2}x + 2.$$

Setting *x* = 5,

$$y = \tfrac{1}{2}(5) + 2 = 4.5.$$

That is, the total cost is 4.5 × 1,000 = $4,500. Therefore, the variable cost when 5 units are produced is

$$(\text{total cost}) - (\text{fixed cost}) = \$4,500 - \$2,000$$
$$= \$2,500.$$

Exercise 5

In problems 1–9, find the equation of the line through each of the given pair of points and sketch the graph.

    1. (3, 2), (−2, 4)   **2.** (1, 1), (2, −2)

3. $(-3, -3), (4, 9)$ **4.** $(-1, -3), (-2, 5)$
5. $(0, 0), (3, 2)$ **6.** $(5, 0), (0, 5)$
7. $(-2, -3), (5, -7)$ **8.** $(\frac{1}{2}, 2), (\frac{5}{2}, -2)$
9. $(\frac{1}{3}, \frac{1}{4}), (\frac{2}{3}, \frac{1}{3})$

In problems 10–17, find the equation of the line with the given slope and containing the given point. Sketch the graph.

10. $(\frac{1}{3}, 3), m = \frac{1}{2}$ **11.** $(-2, 5), m = -\frac{2}{3}$
12. $(-4, -3), m = 1$ **13.** $(-3, -3), m = -1$
14. $(0, 3), m = -2$ **15.** $(3, 0), m = 2$
16. $(-4, 3), m = 0$ **17.** $(1, -3), m = 0$

In problems 18–21, determine the slope, y-intercept, and x-intercept of the graph of the function defined.

18. $\{(x, y)\mid 3x + 4y - 6 = 0\}$
19. $\{(x, y)\mid x - 2y + 4 = 0\}$
20. $\{(x, y)\mid -4x + 5y + 12 = 0\}$
21. $\{(x, y)\mid 2x + y = 0\}$
22. (a) What is the equation of the line that passes through the points $(0, 3)$ and $(5, 0)$?
    (b) Show that the equation of the line that passes through the points $(0, b)$ and $(a, 0)$ can be written in the form (called the intercept form):

$$\frac{x}{a} + \frac{y}{b} = 1 \quad (a \cdot b \neq 0).$$

23. The graph of a linear function has slope $m = 2$. If $P(-1, 3)$ and $Q(c, -2)$ belong to f, find the number c.
24. Repeat problem 23 for $m = -5$ and the points $P(3, c)$ and $Q(-2, 4)$.
25. In Figure 33 the line L is the graph of a function that relates the total manufacturing cost y and the number of units x of a product manufactured.

Find
(a) the total cost when 20 units are produced
(b) the fixed cost
(c) the variable cost when 20 units are produced
(d) the marginal cost
(e) the variable cost when 7 units are produced.

26. The total cost changes from \$500 to \$800 as the number of units manufactured increases from 300 to 900. Assuming that the total cost y is a linear function of x (the number of units manufactured) determine

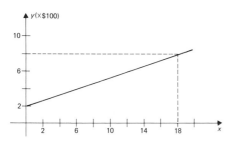

Figure 33

(a) the fixed cost

(b) the marginal cost

(c) the total cost when 500 units are manufactured

(d) the variable cost when 500 units are manufactured.

27. On an average weekday in the bustling city of Garbville, the pollution index p starts to rise at 8:00 A.M. At 10:00 A.M., $p = 30$ and at 2:00 P.M., $p = 50$. Assume that p is a linear function of time between 8:00 A.M. and 6:00 P.M.

    (a) Find an equation relating p and t.

    (b) Sketch the graph of the line.

    (c) Find p at 8:00 A.M., 12:00 noon, 6:00 P.M.

    (d) What is the "rate of increase of p with respect to time"? (That is, what is the slope of the line?)

28. Suppose that every night the pollution index at Garbville remains constant between 6:00 P.M. and 8:00 P.M. and then drops linearly to a fixed level at 8:00 A.M.

    (a) Find the equation of the line relating p and t between 8:00 P.M. and 8:00 A.M.

    (b) Calculate p at 12:00 midnight, at 2:00 A.M., and at 4:00 A.M.

    (c) What is the "rate of decrease of p with respect to t" (slope)?

6. Intersection of Lines

Many geometrical problems involve the intersection of the graphs of functions. If the equations describing linear functions are given, it is a straightforward algebraic task to find a point of intersection, if any such point exists. The reader is aware of what we mean by the intersection of lines, for example, the point P in Figure 34. Analytically, the intersection of the lines L_1 and L_2 ($L_1 \cap L_2$) is an ordered pair of numbers (x, y) such that $(x, y) \in L_1$ and $(x, y) \in L_2$. Here we shall outline the procedure for determining the point of intersection of two lines.

Let L_1 and L_2 be two lines given by

$$\{(x, y) \mid a_1 x + b_1 y + c_1 = 0\}, \qquad a_1^2 + b_1^2 \neq 0$$
$$\{(x, y) \mid a_2 x + b_2 y + c_2 = 0\}, \qquad a_2^2 + b_2^2 \neq 0.$$

Then, $L_1 \cap L_2$ is

$$\{(x, y) \mid a_1 x + b_1 y + c_1 = 0 \text{ and } a_2 x + b_2 y + c_2 = 0\}.$$

We find the point of intersection by solving the simultaneous equations

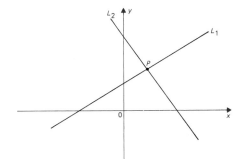

Figure 34

$a^2 + b^2 \neq 0$ is another way of saying that not both a and b are zero.

(20) $$a_1x + b_1y + c_1 = 0$$

(21) $$a_2x + b_2y + c_2 = 0.$$

That is, we wish to find a single point (x, y) which satisfies (20) and (21) simultaneously. Multiplying (20) by b_2 and (21) by b_1 we obtain

(22) $$a_1b_2x + b_1b_2y + c_1b_2 = 0$$

(23) $$a_2b_1x + b_1b_2y + c_2b_1 = 0.$$

Subtracting (23) from (22) we eliminate y to solve for x:

$$(a_1b_2 - a_2b_1)x + c_1b_2 - c_2b_1 = 0$$

or

(24) $$x = \frac{b_1c_2 - b_2c_1}{a_1b_2 - a_2b_1} \quad \text{provided } a_1b_2 \neq a_2b_1.$$

Remark. The condition $a_1b_2 \neq a_2b_1$ is necessary and sufficient for L_1 and L_2 to have a point of intersection.

Similarly, we can eliminate x and solve for y to obtain

(25) $$y = \frac{a_2c_1 - a_1c_2}{a_1b_2 - a_2b_1} \quad \text{provided } a_1b_2 \neq a_2b_1.$$

The ordered pair of numbers (x, y) defined by (24) and (25) is the intersection of L_1 and L_2. There is no need to memorize (24) and (25); the procedure outlined above can be followed for each problem.

Example 1. Determine the point of intersection of the given lines:

$$L_1 = \{(x, y) \mid x + y - 2 = 0\}, \quad L_2 = \{(x, y) \mid 2x + 3y - 1 = 0\}.$$

Solution. We find the point of intersection of L_1 and L_2 by solving simultaneously the equations

(26) $$x + y - 2 = 0$$

(27) $$2x + 3y - 1 = 0.$$

Multiplying equation (26) by -2 to obtain

(28) $$-2x - 2y + 4 = 0$$

and adding equations (27) and (28), we have

$$y + 3 = 0 \quad \text{or} \quad y = -3.$$

At this point we could follow the above procedure and eliminate y to solve for x. Or, putting $y = -3$ into either of the two equations (26) or (27) we can solve for x. Thus, from equation (26) we obtain

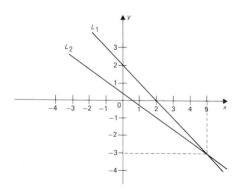

Figure 35

$$x + (-3) - 2 = 0$$

or

$$x = 5.$$

Therefore, the point of intersection is $(5, -3)$. As a check the reader should verify that $(5, -3) \in L_1$ and $(5, -3) \in L_2$.

We now illustrate how the point of intersection of two lines is used in economic theory.

Substituting $y = -3$ into equation (27) yields
$$2x + 3(-3) - 1 = 0 \Rightarrow x = 10/2 = 5.$$

Example 2. A company expects the variable cost to be \$25,000 at a sales volume (revenue) of \$40,000 with fixed cost of \$12,000. Assuming the linear model applies, find

   (a) an equation relating total cost and sales

   (b) the *break-even point*: that is, the volume at which the company experiences neither loss nor profit

   (c) the net profit if sales should turn out to be \$50,000.

Break-even point

Solution. (a) First we determine the marginal cost:

$$\text{marginal cost} = \frac{\text{variable cost}}{\text{sales volume}}$$

$$= \frac{25,000}{40,000} = \frac{5}{8}.$$

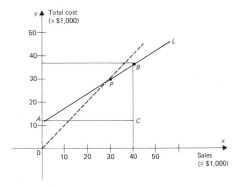

Figure 36
A break-even chart

This represents the slope of the line segment AB (see Figure 36). Using the slope-intercept form of the equation of the line L_1 we have the total cost

(29) $$y = \tfrac{5}{8}x + b,$$

where b represents the fixed cost (y-intercept). Hence

(30) $$y = \tfrac{5}{8}x + 12,000$$

is the equation relating total cost and sales.

   (b) To determine the break-even point, we are asked to find a point on L where total cost equals sales, that is, where $y = x$. In other words, we are asked to solve simultaneously the equations

(31) $$y = \tfrac{5}{8}x + 12,000$$

(32) $$y = x.$$

Subtracting (31) from (32) we obtain

$$\tfrac{3}{8}x - 12,000 = 0$$

or the break-even level of sales is

$$x = 32,000.$$

So,

$$y = 32,000.$$

This means that when the sales are at \$32,000, the total cost to the company is also \$32,000. The company's sales should be above \$32,000 to bring in a profit. If the sales are less than \$32,000, the company will have a loss.

(c) Setting $x = 50,000$ in equation (30) we obtain the total cost:

$$y = \tfrac{5}{8}(50,000) + 12,000$$
$$= 31,250 + 12,000 = 43,250$$

and the net profit (before taxes) is

$$\text{net profit} = \text{sales} - \text{cost}$$
$$= \$50,000 - \$43,250 = \$6,750.$$

Example 3. Consider the lines

$$L_1 = \{(x, y)\mid x - y + 5 = 0\}$$
$$L_2 = \{(x, y)\mid -2x + 2y - 1 = 0\}.$$

Find $L_1 \cap L_2$.

Solution. Consider the pair of equations

(33) $\qquad\qquad\qquad\qquad x - y + 5 = 0$

(34) $\qquad\qquad\qquad\qquad -2x + 2y - 1 = 0.$

We find that the method of elimination of variables employed in Example 1 above fails to yield a solution for x and y. If we multiply the first equation by 2 and add the result to the second equation, we obtain $9 = 0$, a false statement. Since the operations leading to this statement are valid, we must conclude that not both (33) and (34) can be true for the same point (x, y). This means that the two lines do not intersect. We have

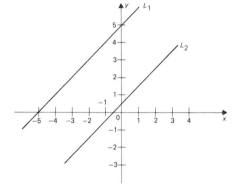

Figure 37
$L_1 \cap L_2 = \phi$

Definition 6. Two distinct lines L_1 and L_2 are said to be *parallel* if they have no point of intersection.

We state the following theorem concerning parallel lines:

Theorem 2. The two lines

$$L_1 = \{(x, y)\mid a_1x + b_1y + c_1 = 0\}, \qquad a_1^2 + b_1^2 \neq 0$$
$$L_2 = \{(x, y)\mid a_2x + b_2y + c_2 = 0\}, \qquad a_2^2 + b_2^2 \neq 0$$

are parallel (or identical) if and only if $a_1b_2 - a_2b_1 = 0.$

We could look at this from another point of view. The expression

$$a_1 b_2 - a_2 b_1 = 0$$

implies

$$\frac{a_1}{b_1} = \frac{a_2}{b_2} \qquad \text{provided } b_1 \cdot b_2 \neq 0$$

and so

$$-\frac{a_1}{b_1} = -\frac{a_2}{b_2}.$$

Since $-a_1/b_1$ is the slope of L_1 and $-a_2/b_2$ is the slope of L_2 (see Section 5), the lines L_1 and L_2 are parallel if and only if their slopes are equal.

Why is a_1 nonzero?

For the case where $b_1 = 0$, L_1 is the vertical line $\{(x, y)|\ x = -c_1/a_1\}$ and its slope is undefined. Here we find that b_2 must also be zero (why?). Hence, L_2 is also a vertical line and is parallel to L_1 (why?).

Example 4. Show that the lines

$$L_1 = \{(x, y)|\ 2x - 3y + 4 = 0\}$$
$$L_2 = \{(x, y)|\ y = \tfrac{2}{3}x + 16\}$$

are parallel.

Solution. Since $a_1 b_2 - a_2 b_1 = (2)(1) - (-\tfrac{2}{3})(-3) = 0$, by Theorem 2 the lines L_1 and L_2 are parallel.

Example 5. Determine the number c so that the line L_1 through the points $(c, 1)$ and $(2, -3)$ is parallel to the line $L_2 = \{(x, y)|\ y = x + 5\}$.

Solution. The slope of the line L_2 is

$$m_2 = 1.$$

If m_1 is the slope of L_1 and L_1 is parallel to L_2, then $m_1 = 1$. But,

$$m_1 = \frac{-3 - 1}{2 - c}.$$

Setting $m_1 = 1$ and solving for c we obtain

$$1 = \frac{-4}{2 - c}$$
$$2 - c = -4$$
$$c = 6.$$

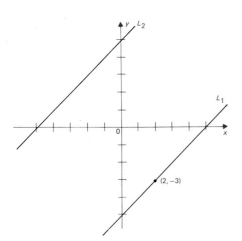

Figure 38

Next, we shall consider *perpendicular* lines. If L_1 and L_2 are two lines neither of which is parallel to an axis, the following theorem (stated without proof) gives a test for perpendicularity in terms of the slopes of the lines.

Theorem 3. The two lines L_1 and L_2 with slopes m_1 and m_2 respectively are perpendicular if and only if $m_1 m_2 = -1$.

Example 6. Show that the points $A(9, -1)$, $B(3, 1)$, and $C(4, 4)$ are the vertices of a right triangle.

Solution. The slopes of the sides of the triangle ABC are as follows:

$$\text{slope of } AB = \frac{1 - (-1)}{3 - 9} = \frac{2}{-6} = -\frac{1}{3}$$

$$\text{slope of } AC = \frac{4 - (-1)}{4 - 9} = \frac{5}{-5} = -1$$

$$\text{slope of } BC = \frac{4 - 1}{4 - 3} = 3.$$

Since

$$(\text{slope of } AB)(\text{slope of } BC) = \left(-\frac{1}{3}\right)(3)$$
$$= -1,$$

the line through A and B and the line through B and C are perpendicular. Therefore, the triangle is a right triangle with right angle at B (see Figure 39).

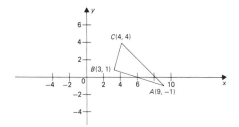

Figure 39

Exercise 6

In problems 1–10, determine the point of intersection of the given lines.

1. $\{(x, y)| \ x - 2y = 3\}$, $\{(x, y)| \ x + y + 4 = 0\}$
2. $\{(x, y)| \ 2x + y - 5 = 0\}$, $\{(x, y)| \ x - y + 16 = 0\}$
3. $\{(x, y)| \ 3x - y + 2 = 0\}$, $\{(x, y)| \ 2x - 3y + 1 = 0\}$
4. $\{(x, y)| \ 4x - 2y + 7 = 0\}$, $\{(x, y)| \ x + y + 6 = 0\}$
5. $\{(x, y)| \ -x + 5y + 2 = 0\}$, $\{(x, y)| \ 3x + y + 1 = 0\}$
6. $\{(x, y)| \ y = 5x - 6\}$, $\{(x, y)| \ y = \frac{2}{5}x + 4\}$
7. $\{(x, y)| \ y = 16x + 2\}$, $\{(x, y)| \ 3y = 4x - 2\}$
8. $\{(x, y)| \ 4x - 6y = 3\}$, $\{(x, y)| \ x + 3y = 6\}$
9. $\{(x, y)| \ x = 4y - 3\}$, $\{(x, y)| \ x + 4y = 7\}$
10. $\{(x, y)| \ 2x + 7y = 8\}$, $\{(x, y)| \ 3x - 2y = 4\}$

In problems 11–18, determine which of the given pairs of lines are parallel and which are perpendicular.

11. $\{(x, y)|\ 2x + y - 6 = 0\}$, $\{(x, y)|\ 4x + 2y + 3 = 0\}$
12. $\{(x, y)|\ y + 2 = 0\}$, $\{(x, y)|\ x + 4 = 0\}$
13. $\{(x, y)|\ 5x - y = 4\}$, $\{(x, y)|\ 2x + 10 - 7 = 0\}$
14. $\{(x, y)|\ 3x - 2y + 1 = 0\}$, $\{(x, y)|\ y = \frac{3}{2}x + 10\}$
15. $\{(x, y)|\ y + 4 = 3x\}$, $\{(x, y)|\ x + 16 = \frac{1}{3}y\}$
16. $\{(x, y)|\ x = 2y + 12\}$, $\{(x, y)|\ 4y = 2x - 3\}$
17. $\{(x, y)|\ 3x - y + 6 = 0\}$, $\{(x, y)|\ \frac{1}{3}x + y = 0\}$
18. $\{(x, y)|\ 7x + 14y - 6 = 0\}$, $\{(x, y)|\ y = -\frac{1}{2}x + 4\}$

19. A line passes through the points $(2, c)$ and $(1, -3)$. Determine the number c so that the line
    (a) has slope 3   (b) has slope zero
    (c) is parallel to the line $\{(x, y)|\ 5x + y + 6 = 0\}$
    (d) is perpendicular to the line $\{(x, y)|\ 2x + y = 0\}$.

20. A line L_1 passes through the points $(c, 1)$ and $(5, -1)$. Determine the number c so that L_1
    (a) has slope -2
    (b) is parallel to the line $L_2 = \{(x, y)|\ 2x + 3y = 1\}$
    (c) is perpendicular to L_2.
    Is it possible for L_1 to have slope zero? Why?

21. Using slopes, identify each of the figures with the given vertices.
    (a) $A(-4, 3), B(-5, 0), C(1, -2)$
    (b) $A(-1, 4), B(-2, 1), C(4, -1), D(5, 2)$
    (c) $A(4, 3), B(9, -2), C(10, 1)$
    (d) $A(-9, 5), B(-2, 13), C(6, 6), D(-1, -2)$

22. Consider the line segment with endpoints $(2, 1)$ and $(6, 5)$. Show that the point $(5, 2)$ is on the perpendicular bisector of the given line segment.

23. ABC is a triangle and D and E are the midpoints of the line segments AB and AC respectively.

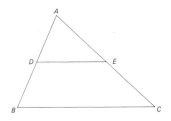

(a) Show that the line through D and E is parallel to the line through B and C.

(b) Show that the length of the line segment DE is one-half the length of the line segment BC.

24. A company expects the variable cost to be \$44,000 on sales of \$70,000 with fixed cost of \$22,000. Assuming that the linear model applies,

(a) Write the equation relating sales and total cost.

(b) Sketch the break-even chart.

(c) Find the break-even point.

(d) What will the net profit (before taxes) be on sales of \$80,000?

25. Suppose the equation

$$y = \tfrac{1}{3}x + 24{,}000$$

relates the total cost y and the sales volume x. Find

(a) the fixed cost

(b) the variable cost on sales of \$50,000

(c) the break-even point

(d) net profit (before taxes) on sales of \$40,000.

26. If the freezing point of water is measured as 0°C (Celsius) and 32°F (Fahrenheit) and the boiling point of water is measured as 100°C and 212°F,

(a) Find C as a linear function of F.

(b) Find F as a linear function of C.

(c) At what temperatures will °C = °F?

(d) Sketch a graph of each function and indicate the point of intersection.

27. Suppose f is a linear function with nonzero slope m.

(a) Show that f^{-1} is a function.

(b) Show that f^{-1} is linear.

(c) What is the slope of f^{-1}?

(d) Can the graph of f^{-1} be perpendicular to the graph of f?

(e) Can the graph of f^{-1} be parallel to the graph of f?

7. Quadratic Functions

We now consider functions defined by

(35) $$f(x) = ax^2 + bx + c$$

where a, b, and c are fixed numbers and $a \neq 0$.

Definition 7. A function *f* that can be described by (35) is called a *quadratic function*. The graph of *f* is called a *parabola*.

Consider the special case of (35),

(36) $f(x) = x^2$.

Assigning values to *x* and computing the corresponding values *f(x)*, we obtain some points belonging to *f*. For example, in tabular form we have

x	-3	-2	-1	0	1	2	3
$f(x) = x^2$	9	4	1	0	1	4	9

The graph of these points is shown in Figure 40(a). If we compute additional points belonging to *f* and connect them with a smooth curve as shown in Figure 40(b), we find that the graph of *f* forms a simple pattern. Obviously, the resulting curve is an approximation to the parabola describing the function. The graph is in the shape of the cross section of a mound opening upward (i.e., the graph is concave upward). We observe that the pairs of points

$$(1, 1) \quad \text{and} \quad (-1, 1)$$
$$(2, 4) \quad \text{and} \quad (-2, 4)$$

and in general

$$(x, x^2) \quad \text{and} \quad (-x, x^2)$$

belong to *f*. Geometrically, this means that the portion of the graph of *f* to the left of the *y*-axis is the mirror image of the graph of *f* to the right of the *y*-axis. In other words, if we fold the paper along the *y*-axis, the curve to the left of the *y*-axis will coincide with the curve to the right of the *y*-axis. Such a graph is called *symmetric*, and the *y*-axis is called the *line of symmetry*. The point of intersection of the line of symmetry and the graph of *f* is called the *vertex* of the parabola. In this case, the vertex represents the "lowest point" of the parabola. We find that

$$f(0) = 0$$

and

$$f(0) < f(x)$$

for every $x \neq 0$. We say that *f* has a *minimum* value at $x = 0$.

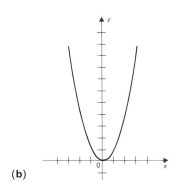

Figure 40   (a)

(b)

Note. To obtain an exact graph is an impossible task, since we are unable to locate every point belonging to f.

Line of symmetry

Vertex of a parabola

Figure 41

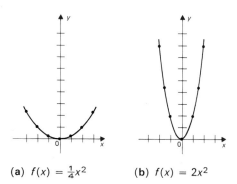

(a) $f(x) = \frac{1}{4}x^2$ **(b)** $f(x) = 2x^2$

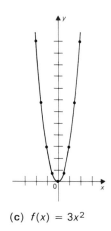

(c) $f(x) = 3x^2$

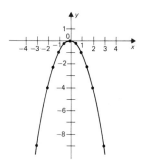

Figure 42
Graph of $f(x) = -x^2$

Example 1. Graph

(a) $f(x) = \frac{1}{4}x^2$ (b) $f(x) = 2x^2$ (c) $f(x) = 3x^2$.

Solution. First we tabulate a few points belonging to each function

	x	-3	-2	$-\frac{3}{2}$	-1	$-\frac{1}{2}$	0	$\frac{1}{2}$	1	$\frac{3}{2}$	2	3
(a)	$f(x) = \frac{1}{4}x^2$	$\frac{9}{4}$	1	$\frac{9}{16}$	$\frac{1}{4}$	$\frac{1}{16}$	0	$\frac{1}{16}$	$\frac{1}{4}$	$\frac{9}{16}$	1	$\frac{9}{4}$

	x	-2	$-\frac{3}{2}$	-1	$-\frac{1}{2}$	0	$\frac{1}{2}$	1	$\frac{3}{2}$	2
(b)	$f(x) = 2x^2$	8	$\frac{9}{2}$	2	$\frac{1}{2}$	0	$\frac{1}{2}$	2	$\frac{9}{2}$	8

	x	-2	$-\frac{3}{2}$	-1	$-\frac{1}{2}$	0	$\frac{1}{2}$	1	$\frac{3}{2}$	2
(c)	$f(x) = 3x^2$	12	$\frac{27}{4}$	3	$\frac{3}{4}$	0	$\frac{3}{4}$	3	$\frac{27}{4}$	12

In Figure 41 we illustrate the graphs of these functions. We note that as the coefficient of x^2 takes on larger values, the steepness of the curve increases. However, the vertex in each case is the point $(0, 0)$.

Next we consider (35) for $a < 0$. Suppose f is the function defined by

(37) $$f(x) = -x^2.$$

Computing some of the points belonging to the graph of f we obtain

x	-3	-2	$-\frac{3}{2}$	-1	$-\frac{1}{2}$	0	$\frac{1}{2}$	1	$\frac{3}{2}$	2	3
$f(x) = -x^2$	-9	-4	$-\frac{9}{4}$	-1	$-\frac{1}{4}$	0	$-\frac{1}{4}$	-1	$-\frac{9}{4}$	-4	-9

The graph of these points and an approximate graph of f is shown in Figure 42.

Here again, the graph of the function is in the shape of the cross section of a mound, but this time it opens downward (the curve is concave downward). The graph is symmetric with respect to the y-axis, and the vertex is at the origin. This point is called the "highest point" of the parabola. We find that

$$f(0) = 0$$

and

$$f(0) > f(x)$$

for every $x \neq 0$. We say that f attains a *maximum* value at $x = 0$. In Figure 43 we illustrate the graphs of the functions

(a) $f(x) = -\frac{1}{2}x^2$ (b) $f(x) = -\frac{3}{2}x^2$ (c) $f(x) = -3x^2$.

Notice that the steepness of the curve increases as the absolute value of the coefficients of x^2 increases.

(a) $f(x) = -\frac{1}{2}x^2$ (b) $f(x) = -\frac{3}{2}x^2$ (c) $f(x) = -3x^2$

Figure 43

Returning to (35) we now show that the graph of f has either a lowest or a highest point. We *complete the square* as follows:

Adding and subtracting $b^2/4a$ to complete the square using the first two terms

$$f(x) = ax^2 + bx + c, \quad a \neq 0$$

$$= ax^2 + bx + c + \left(\frac{b^2}{4a} - \frac{b^2}{4a}\right)$$

$$= ax^2 + bx + \frac{b^2}{4a} + c - \frac{b^2}{4a}$$

Factoring out a in the first three terms and adding the last two

$$= a\left(x^2 + \frac{b}{a}x + \frac{b^2}{4a^2}\right) + \frac{-b^2 + 4ac}{4a}$$

Note. $\left(x + \frac{b}{2a}\right)^2 = x^2 + \frac{b}{a}x + \frac{b^2}{4a^2}$

(38)

$$= a\left(x + \frac{b}{2a}\right)^2 + \frac{-b^2 + 4ac}{4a}.$$

Case 1. $a > 0$.

Now, $(x + b/2a)^2 = 0$ at $x = -b/2a$. For all other values of x, if $a > 0$ then $a(x + b/2a)^2 > 0$ (Why?). Therefore, $f(-b/2a) < f(x)$ for all $x \neq -b/2a$; hence, the graph of f has a minimum value at $x = -b/2a$. That is, the lowest point is

(39)

$$\left(-\frac{b}{2a}, \frac{-b^2 + 4ac}{4a}\right).$$

Case 2. $a < 0$.

At $x = -b/2a$, $(x + b/2a)^2 = 0$. For all other values of x, $a(x + b/2a)^2 < 0$ (Why?). Therefore, in this case $f(-b/2a) > f(x)$

for all $x \neq -b/2a$; and the function f attains a maximum value at $x = -b/2a$. In other words, the graph of f has a highest point and this point is given by (39).

We observe from (38) that if we set $f(x) = 0$ we obtain

$$a\left(x + \frac{b}{2a}\right)^2 = \frac{b^2 - 4ac}{4a}$$

$$\left(x + \frac{b}{2a}\right)^2 = \frac{b^2 - 4ac}{4a^2}$$

$$x + \frac{b}{2a} = \pm \frac{\sqrt{b^2 - 4ac}}{2a}$$

$$x = -\frac{b}{2a} \pm \frac{\sqrt{b^2 - 4ac}}{2a}.$$

Hence

Note. The student may recognize (40) as the solutions of the quadratic equation

$$ax^2 + bx + c = 0.$$

(40) $x = \dfrac{-b + \sqrt{b^2 - 4ac}}{2a}$ and $x = \dfrac{-b - \sqrt{b^2 - 4ac}}{2a}.$

These are the points where the graph of f intersects the x-axis provided $b^2 - 4ac \geq 0$. If $b^2 - 4ac < 0$, the graph of f either lies entirely above or entirely below the x-axis.

Example 2. Graph the function f where

$$f(x) = x^2 - 6x + 5.$$

Solution. First we determine the highest or lowest point of the graph. We observe that since $a = 1 > 0$, the graph has a lowest point. Here

$$a = 1, \qquad b = -6, \qquad \text{and} \qquad c = 5.$$

Therefore,

$$-\frac{b}{2a} = -\frac{(-6)}{2(1)} = 3$$

and

$$\frac{-b^2 + 4ac}{4a} = \frac{-(-6)^2 + 4(1)(5)}{4(1)} = \frac{-36 + 20}{4} = -4.$$

Hence, the lowest point is $(3, -4)$. Next we tabulate a few points belonging to f:

x	0	1	2	3	4	5
$f(x) = x^2 - 6x + 5$	5	0	-3	-4	-3	0

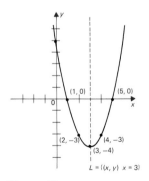

Figure 44

Figure 44 illustrates the graph of f. Note that the vertical line $L = \{(x, y) \mid x = 3\}$ is the line of symmetry. Using (40) we obtain the points of intersection of the graph of f with the x-axis: $x = 1$ and $x = 5$.

Example 3. Graph the function f where

$$f(x) = -3x^2 - 8x + 1.$$

Solution. Since $a = -3 < 0$, the graph of f has a highest point. Here

$$a = -3, \quad b = -8, \quad \text{and} \quad c = 1.$$

Therefore,

$$-\frac{b}{2a} = -\frac{(-8)}{2(-3)} = -\frac{4}{3}$$

and

$$\frac{-b^2 + 4ac}{4a} = \frac{-(-8)^2 + 4(-3)(1)}{4(-3)} = \frac{76}{12} = \frac{19}{3}.$$

The highest point on the graph is $\left(-\frac{4}{3}, \frac{19}{3}\right)$. A few points on the graph of f are

x	-2	$-\frac{3}{2}$	$-\frac{4}{3}$	-1	$-\frac{1}{2}$	0	$\frac{1}{2}$	1
$f(x) = -3x^2 - 8x + 1$	5	$\frac{25}{4}$	$\frac{19}{3}$	6	$\frac{17}{4}$	1	$-\frac{15}{4}$	-10

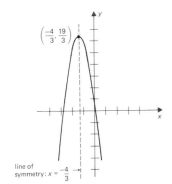

$\left(\frac{-4}{3}, \frac{19}{3}\right)$

line of symmetry: $x = \frac{-4}{3}$

Figure 45

The graph of f is shown in Figure 45. The line of symmetry is the vertical line $\{(x, y) \mid x = -\frac{4}{3}\}$.

We summarize our results for the graph of the quadratic function $f(x) = ax^2 + bx + c$:

1. If $a > 0$, the parabola is concave upward.
2. If $a < 0$, the parabola is concave downward.
3. The vertex of the parabola is the point $\left(-\dfrac{b}{2a}, \dfrac{-b^2 + 4ac}{4a}\right)$.
4. The line of symmetry is given by $x = -\dfrac{b}{2a}$.
5. The y-intercept is the point $(0, c)$.

Now we shall illustrate how certain concrete applications give rise to quadratic functions.

Figure 46

Example 4. A farmer wishes to enclose a rectangular lot of maximum area with a fence 400 feet long. Find the dimensions of the rectangle.

Solution. Suppose the length and the width of the rectangle are x and y feet respectively. Then,

$$(41) \qquad\qquad 2x + 2y = 400$$

or

$$(42) \qquad\qquad x + y = 200.$$

The area of the rectangle is

$$(43) \qquad\qquad \text{Area} = xy.$$

Solving for y in (42) and substituting into equation (43) we obtain

$$(44) \qquad\qquad A(x) = x(200 - x)$$
$$= -x^2 + 200x.$$

Equation (44) describes a quadratic function. Since the coefficient of x^2 is less than zero, the graph of the function A has a highest point. That is, A attains a maximum value and this occurs at

$$(45) \qquad\qquad x = -\frac{b}{2a} = -\frac{200}{2(-1)} = 100.$$

Setting $x = 100$ in equation (42) we find that

$$y = 200 - x = 100.$$

Therefore, the maximum area is enclosed if the rectangle is a square having side length 100 feet.

Exercise 7

In problems 1–10, sketch the graph of the given function and indicate the highest or the lowest point and the line of symmetry. Find the points of intersection (if any) of the graph of f with the x-axis.

1. $f(x) = x^2 - 4$
2. $f(x) = x^2 - 3x + 2$
3. $f(x) = x^2 - 5x + 4$
4. $f(x) = x^2 + x - 2$
5. $f(x) = -x^2 + 3x - 2$
6. $f(x) = -x^2 + 5x - 4$
7. $f(x) = -2x^2 - 3x + 2$
8. $f(x) = -\frac{1}{2}x^2 - x$
9. $f(x) = \frac{1}{4}x^2 + 3$
10. $f(x) = \frac{1}{4}x^2 + 3x$

11. Find two numbers whose sum is 10 and whose product is as large as possible.

12. A piece of wire 20 inches long is to be bent to form three sides of a rectangle. Find the dimensions of the rectangle thus formed having maximum area.

13. If the profit p in the manufacture and sale of x units of a product is given by

$$p(x) = 200x - 0.001x^2,$$

(a) Find the number x that yields maximum profit.
(b) Find the maximum profit if each item is sold at $2.50.
(c) Sketch a graph of the function p.

14. A window is to be constructed in the shape of a rectangle surmounted by a semicircle. If the perimeter of the window is 18 feet, find its dimensions for maximum area.

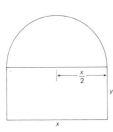

15. The Blue-Yonder Travel Club is organizing a South American trip for a group of university students. The trip will cost $300 per student if not more than 120 students make the trip; however, the cost per student will be reduced by $1.00 for each student in excess of 120.

(a) If x is the number of students in excess of 120, show that the gross income I is given by

$$I = -x^2 + 180x + 36,000.$$

(b) For what value of x is I a maximum?
(c) Find the maximum value of I.

16. The manager of an 80-unit apartment complex finds that at a rental of $110 per month all units will be occupied. However, for each $4.00 increase in rent, one unit will remain vacant. If the maintenance cost is $10 per month for an occupied unit,

(a) Show that the function relating gross profit p and number of unoccupied units x is given by $p(x) = -4x^2 + 220x + 8000$.
(b) Find the number of vacant units for which gross profit is maximum.
(c) What is the maximum gross profit?
(d) What is the rental for maximum profit?

8. Some Important Relations

In this section we shall study certain relations whose graphs have been important in many areas of astronomy, engineering, physics,

The ancient Greek mathematicians studied extensively the circle, ellipse, parabola, and hyperbola (called the conic sections). Using geometrical methods they developed a sophisticated and sound theory. Euclid (365–300 B.C.) and Apollonius of Perga (260–170 B.C.) wrote books on the subject. Euclid's work has not survived; however, parts of the books of Apollonius have survived.

Descartes introduced conic sections into the realm of algebra. He proved that every conic section can be described by a relation of the form

$$\{(x, y)\mid Ax^2 + Bxy + Cy^2 + Dx + Ey + F = 0\}$$

where A, B, C, D, E and F are constants.

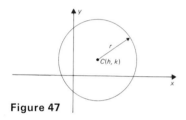

Figure 47

Using the distance formula

Squaring both sides of the equation

and space travel. These graphs, which are special geometric figures known as the circle, the parabola, and the ellipse, have been the object of study ever since the time of the ancient Greeks. Here we shall give a brief exposition.

A. THE CIRCLE

Definition 8. A *circle* is the set of all points in the plane equidistant from a fixed point.

The fixed point is called the *center* of the circle and the fixed distance is called the *radius*.

Theorem 4. The circle with center at the point $C(h, k)$ and radius r is the graph of the relation

(46) $$\{(x, y)\mid (x - h)^2 + (y - k)^2 = r^2\}.$$

Proof. The point $P(x, y)$ belongs to the circle if and only if

(47) $$d(P, C) = r,$$

that is, if

(48) $$\sqrt{(x - h)^2 + (y - k)^2} = r$$

or

(49) $$(x - h)^2 + (y - k)^2 = r^2.$$

If the point Q is not on the circle, then Q is not an element of the relation given by (46). Equation (49) is called the equation of the circle with center (h, k) and radius r.

Example 1. Find the equation of a circle with center at $(3, -4)$ and radius 6.

Solution. Here $h = 3$, $k = -4$, and $r = 6$. Substituting these into (49) we obtain

(50) $$(x - 3)^2 + [y - (-4)]^2 = 6^2$$

or

(51) $$(x - 3)^2 + (y + 4)^2 = 36.$$

Squaring and combining terms we get

(52) $$x^2 + y^2 - 6x + 8y - 11 = 0.$$

The circle is shown in Figure 48.

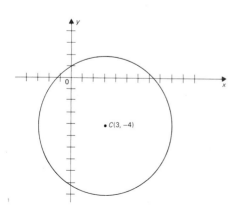

Figure 48

Note that (49) can also be written

(53) $x^2 + y^2 - 2hx - 2ky + (h^2 + k^2 - r^2) = 0$

and the equation

General form of the equation of a circle

(54) $x^2 + y^2 + Ax + By + E = 0$

where $A = -2h$, $B = -2k$, and $E = h^2 + k^2 - r^2$ is called the *general form* of the equation of a circle.

Example 2. Find the center and radius of the circle defined by

(55) $x^2 + y^2 + 4x - 6y - 12 = 0.$

Solution. The equation in (55) may be written in the form

(56) $(x^2 + 4x) + (y^2 - 6y) = 12.$

Completing the square for the terms involving x and the terms involving y, we get

(57) $(x^2 + 4x + 4) + (y^2 - 6y + 9) = 12 + 4 + 9$

or

(58) $(x + 2)^2 + (y - 3)^2 = 25.$

From equation (49) it is easily seen that we have the equation of a circle with center $(-2, 3)$ and radius $r = \sqrt{25} = 5$. The circle is shown in Figure 49.

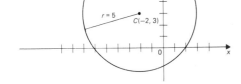

Figure 49

B. THE PARABOLA

Definition of parabola

Definition 9. A *parabola* is the set of all points in a plane equidistant from a fixed line and a fixed point.

The fixed point is not on the fixed line.

The fixed line is called the *directrix*, and the fixed point is called the *focus*.

We shall consider two special cases:
  (i) We shall assume that the focus is on the x-axis at $(c, 0)$ and the directrix is the line $\{(x, y)\mid x = -c\}$.
  (ii) We shall assume that the focus is on the y-axis at $(0, c)$ and the directrix is the line $\{(x, y)\mid y = -c\}$.
For the first case we have

Theorem 5. The parabola with focus at $(c, 0)$ and directrix the line $x = -c$ is the graph of the relation

(59) $\{(x, y)\mid y^2 = 4cx\}.$

Proof. The point $P(x, y)$ belongs to the parabola if and only if the distance from P to the focus $F(c, 0)$ is equal to the distance from P to the directrix. From P draw a line perpendicular to the directrix (see Figure 50) intersecting it at the point Q. Then the coordinates

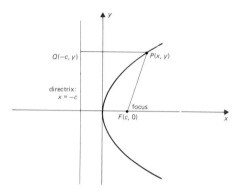

Figure 50

of Q are $(-c, y)$. Thus, we have

$$d(P, Q) = d(P, F)$$

or

Using the distance formula

$$\sqrt{(x + c)^2 + (y - y)^2} = \sqrt{(x - c)^2 + (y - 0)^2}$$

Simplifying

$$x + c = \sqrt{(x - c)^2 + y^2}.$$

Therefore

$$\{(x, y) \mid P(x, y) \text{ is on the parabola}\}$$

Squaring both sides of the equation

$$= \{(x, y) \mid (x + c)^2 = (x - c)^2 + y^2\}$$

Multiplying

$$\{(x, y) \mid x^2 + 2xc + c^2 = x^2 - 2xc + c^2 + y^2\}$$

Simplifying

$$\{(x, y) \mid y^2 = 4cx\}.$$

This proves Theorem 5.

In Figure 50, $c > 0$. If $c < 0$, we obtain the graph shown in Figure 51. In each case, the x-axis is the axis of the parabola (the line of symmetry), and the origin, which is halfway between the focus and the directrix, is the vertex of the parabola.

Example 3. Find the equation of the parabola with focus at the point $(4, 0)$ and directrix $x = -4$.

Solution. Here $c = 4$. Applying Theorem 5 we have

$$y^2 = 16x.$$

The graph is shown in Figure 52.

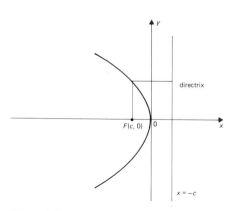

Figure 51

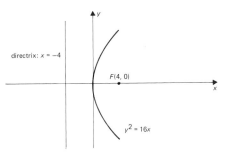

Figure 52
$y^2 = 16x$

Example 4. Find the directrix and the coordinates of the focus of the parabola whose equation is

(60) $$y^2 = -5x.$$

Solution. Equation (60) may be written as

(61) $$y^2 = 4\left(\frac{-5}{4}\right)x.$$

Comparing this with the equation (59) we find that $c = -5/4$. Therefore, the focus is the point $(-5/4, 0)$ and the directrix is $x = -(-5/4) = 5/4$. (See Figure 53.)

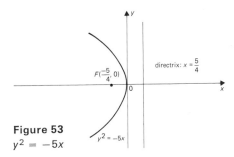

Figure 53
$y^2 = -5x$

We now return to the special case (ii): The focus is on the y-axis at $(0, c)$ and the directrix is $\{(x, y)|\, y = -c\}$. The reader can easily prove

Theorem 6. **The parabola with focus at $(0, c)$ and directrix $y = -c$ is the graph of the relation**

$$\{(x, y)|\, x^2 = 4cy\}.$$

The reader should recognize this relation as a quadratic function. We discussed this case in Section 7. In fact, we considered the more general case where the vertex of the parabola is the point (h, k) and the axis of the parabola (or the line of symmetry) is the vertical line $x = h$. (See problems 27–30 below.)

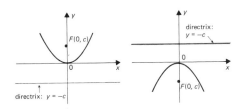

Figure 54

Example 5. Find the equation of a parabola containing the point $(2, -9)$ with vertex at the origin and whose axis is the y-axis.

Solution. Since the vertex is at the origin and the axis of the parab-

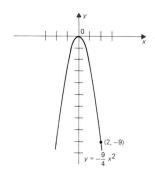

Figure 55
$y = -\frac{9}{4}x^2$

ola is the *y*-axis, the parabola is the graph of a quadratic function defined by

(62) $x^2 = 4cy.$

The point $(2, -9)$ belongs to this function. Hence, substitution into equation (62) yields

$$y = \frac{1}{4c} x^2$$

$$-9 = \frac{1}{4c} \cdot 2^2$$

or

$$-9 = \frac{1}{c}.$$

Therefore, the required function is defined by

$$y = \frac{1}{4}(-9)x^2$$

or

$$y = -\frac{9}{4}x^2.$$

Since $c = -\frac{1}{9}$, the focus is at $(-\frac{1}{9}, 0)$ and the directrix is the line $y = \frac{1}{9}$.

C. THE ELLIPSE

Definition of ellipse

Definition 10. An *ellipse* is the set of all points in the plane the sum of whose distances from two fixed points is constant.

For convenience, we shall take the two fixed points, each of which is called a *focus*, at $F_1(c, 0)$ and $F_2(-c, 0)$ and prove

Theorem 7. The ellipse with foci at $F_1(c, 0)$ and $F_2(-c, 0)$ is the graph of the relation

(63) $\left\{(x, y) \,\middle|\, \dfrac{x^2}{a^2} + \dfrac{y^2}{b^2} = 1\right\}.$

where $2a(a > 0)$ is the sum of the distances from any point on the ellipse to the two foci and $b = \sqrt{a^2 - c^2}$.

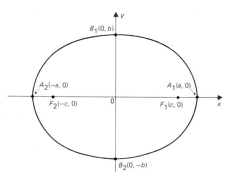

Figure 56

Proof. The point $P(x, y)$ is on the ellipse if and only if

(64) $d(P, F_1) + d(P, F_2) = 2a.$

Using the distance formula, we have

$$d(P, F_1) = \sqrt{(x - c)^2 + (y - 0)^2}$$

and

$$d(P, F_2) = \sqrt{(x + c)^2 + (y - 0)^2}.$$

Thus we have this relation describing the ellipse:

(65) $\qquad \sqrt{(x - c)^2 + y^2} + \sqrt{(x + c)^2 + y^2} = 2a.$

Performing the following algebraic manipulations, equation (65) can be written as in (63):

$$\sqrt{(x - c)^2 + y^2} = 2a - \sqrt{(x + c)^2 + y^2}$$

Squaring both sides of the equation

$$(x - c)^2 + y^2 = 4a^2 - 4a\sqrt{(x + c)^2 + y^2} + (x + c)^2 + y^2$$

Simplifying

$$a\sqrt{(x + c)^2 + y^2} = a^2 + cx$$

Squaring again

$$a^2[(x + c)^2 + y^2] = a^4 + 2a^2cx + c^2x^2$$

$$a^2x^2 + 2a^2cx + a^2c^2 + a^2y^2 = a^4 + 2a^2cx + c^2x^2$$

Simplifying

$$(a^2 - c^2)x^2 + a^2y^2 = a^2(a^2 - c^2)$$

Dividing by $a^2(a^2 - c^2)$ and letting $b^2 = a^2 - c^2$

$$\frac{x^2}{a^2} + \frac{y^2}{b^2} = 1.$$

This proves Theorem 7.

A sketch of an ellipse is shown in Figure 56. The points $A_1, A_2, B_1,$ and B_2 are called the *vertices* of the ellipse. The line segment A_1A_2 is called the *major axis;* the line segment B_1B_2 is called the *minor axis.* In this case the *center* of the ellipse is the origin. Note that the length of the major axis is $2a$ and the length of the minor axis is $2b$. (Why?)

Example 6. Find the lengths of the major and minor axes and give the coordinates of the foci and vertices of the ellipse defined by

(66) $\qquad\qquad 9x^2 + 16y^2 = 144.$

Solution. Dividing equation (66) by 144 and simplifying, we get

(67) $\qquad\qquad \dfrac{x^2}{16} + \dfrac{y^2}{9} = 1$

or the relation

(68) $\qquad\qquad \left\{ (x, y) \,\middle|\, \dfrac{x^2}{4^2} + \dfrac{y^2}{3^2} = 1 \right\}.$

Comparing this with (63) we find that $a = 4$ and $b = 3$. Hence

$$\text{length of major axis} = 2a = 8$$
$$\text{length of minor axis} = 2b = 6.$$

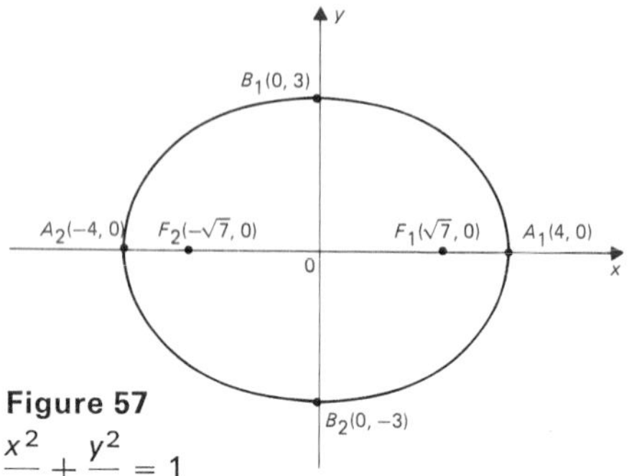

Since $b^2 = a^2 - c^2$, we find that

$$c = \sqrt{a^2 - b^2} = \sqrt{16 - 9} = \sqrt{7}.$$

Therefore, the foci are $F_1(\sqrt{7}, 0)$ and $F_2(-\sqrt{7}, 0)$, and the vertices are

$$A_1 = (a, 0) = (4, 0) \qquad B_1 = (0, b) = (0, 3)$$
$$A_2 = (-a, 0) = (-4, 0) \qquad B_2 = (0, -b) = (0, -3).$$

The graph is shown in Figure 57.

If the foci of an ellipse are at $F_1(0, c)$ and $F_2(0, -c)$, then the major axis is along the y-axis and the minor axis is along the x-axis. The vertices are $A_1(0, a)$, $A_2(0, -a)$, $B_1(b, 0)$, and $B_2(-b, 0)$. The ellipse is then given by the equation

(69)
$$\frac{x^2}{b^2} + \frac{y^2}{a^2} = 1.$$

This is illustrated in Figure 58. A simple method of determining

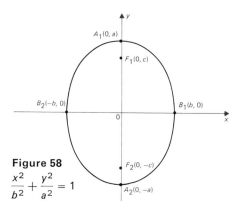

Figure 58
$$\frac{x^2}{b^2} + \frac{y^2}{a^2} = 1$$

whether the foci of the ellipse $(x^2/a^2) + (y^2/b^2) = 1$ are on the x-axis or on the y-axis is to see which relation is true: $a < b$ or $a > b$. What if $a = b$?

Example 7. Find an equation of the ellipse with foci at $F_1(0, 4)$ and $F_2(0, -4)$ and two vertices at $A_1(0, 5)$ and $A_2(0, -5)$.

Solution. We note that $a = 5$ and $c = 4$. Hence $b = \sqrt{a^2 - c^2} = \sqrt{25 - 16} = 3$. Since the foci are on the y-axis, we substitute these values into (69) to obtain

---

Figure 57
$$\frac{x^2}{16} + \frac{y^2}{9} = 1$$

Figure 59
$$\frac{x^2}{9} + \frac{y^2}{25} = 1$$

$$\frac{x^2}{3^2} + \frac{y^2}{5^2} = 1$$

or

$$\frac{x^2}{9} + \frac{y^2}{25} = 1.$$

(See Figure 59.)

The graphs of the relations we have considered in this section are called *conic sections* (or *conics*). The name derives from the fact that such curves are obtained when a plane intersects a right circular cone. In the case of a circle (Figure 60(a)) the intersecting plane is parallel to the base of the cone, in the case of a parabola (Figure 60(b)) the plane intersects the base and a side of the cone, and in the case of an ellipse the plane is not parallel to the base and intersects the cone without intersecting the base (Figure 60(c)).

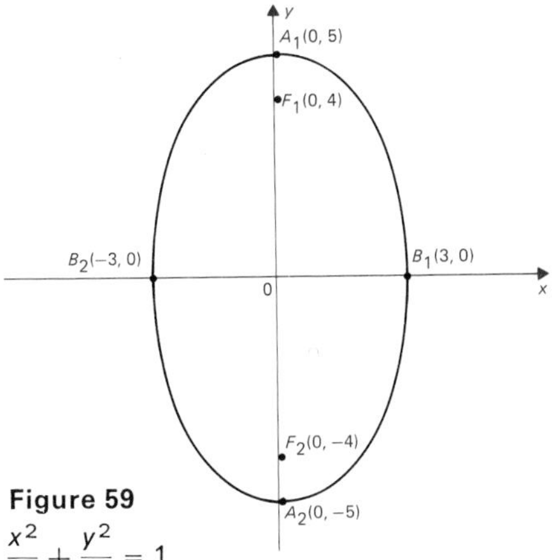

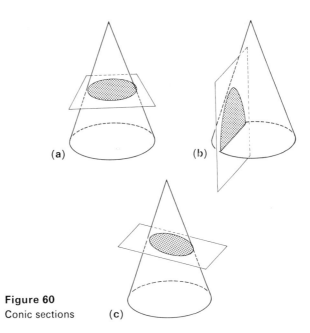

(a)

(b)

(c)

Figure 60
Conic sections

Another conic section is the *hyperbola*, which we shall not discuss here.

Exercise 8

In problems 1 through 12 identify the conic sections defined by the given equations. Determine the center and radius for the circles, the focus, vertex, and directrix for the parabolas, and the foci and vertices for the ellipses. In each case sketch a graph.

1. $x^2 + y^2 = 64$

2. $y^2 + 3x = 0$

3. $4x^2 + y^2 = 16$

4. $x^2 + 4y = 0$

5. $x^2 + y^2 - 6x = 16$

6. $2x^2 + 2y^2 = 18$

7. $2y - 5x^2 = 0$

8. $2x^2 + 5y^2 = 50$

9. $4x - y^2 = 0$

10. $x^2 + y^2 + 4x - 6y = 23$

11. $x^2 + y^2 - 6x - 8y = -25$

12. $3x^2 + 5y^2 - 30 = 0$

In problems 13 through 18 find an equation of a circle satisfying the given conditions.

13. center at $(-1, 2)$ and radius 4

14. center at $(3, -4)$ and containing the point $(-1, 5)$

15. diameter the line segment with endpoints $(3, 1)$ and $(7, 5)$

16. center at $(-1, 3)$ and touching the line $y = -1$

17. center at $(2, 2)$ and touching the line $x = 5$

*18. center at $(1, 2)$ and touching the line $x - y = 3$

In problems 19 through 25, find an equation of a parabola with vertex at the origin satisfying the given conditions.

19. focus at $(-2, 0)$

20. focus at $(0, 3)$

21. directrix $y = 4$

22. directrix $x = -\frac{2}{3}$

23. contains the point $(-3, 6)$ and its line of symmetry is the x-axis

24. contains the point $(-3, 6)$ and its line of symmetry is the y-axis

25. contains the point $(2, 4)$ and its line of symmetry is the x-axis

26. Prove Theorem 6.

In problems 27 through 30, use Definition 9 to find an equation of the parabola with the given focus and directrix. In each case, locate the vertex and the line of symmetry of the parabola.

27. focus at $F(0, 1)$, directrix $y = -5$

28. focus at $F(-1, 0)$, directrix $y = 2$

29. focus at $F(2, 3)$, directrix $y = -1$

30. focus at $(2, 3)$, directrix $y = 5$

In problems 31 through 36, find an equation of an ellipse satisfying the given conditions.

31. center at the origin, length of major axis is 8, length of minor axis is 6, and foci are on the y-axis
32. foci at (3, 0) and (−3, 0), vertices at (5, 0) and (−5, 0)
33. foci at (0, 4) and (0, −4), vertices at (0, 5) and (0, −5)
34. center at the origin, one focus at the point $(0, \sqrt{5})$, and length of major axis is 8
35. vertices at (5, 0) and (−5, 0) and contains the point (3, 3)
36. foci at (3, 0) and (−3, 0) and contains the point (4, −1)

Gottfried W. Leibniz

3

The

Limit

Concept

1. Sequences

Definition 1. A sequence is a function s whose domain is the set of natural numbers N; i.e., $s = \{(n, s(n)) \mid n \in N\}$.

If s is a sequence, the image $s(n)$ of a natural number n is usually denoted s_n. The functional value s_n is called the *nth term* of the sequence. Since the domain of a sequence s is always the set N of natural numbers, it is customary to denote the sequence s by its functional values. The classical symbol $\{s_n\}_{n=1}^{\infty}$ or simply $\{s_n\}$ is frequently used. We also sometimes write the sequence s as $s_1, s_2, s_3, \ldots s_n, \ldots$.

The terms of a sequence need not be different. The range of an infinite sequence may be a finite set. For example, the range of the sequence $\{(-1)^n\}$ is the set $\{1, -1\}$, which consists of only two elements.

The following are some examples of sequences:

(1) $1, \dfrac{1}{2}, \dfrac{1}{3}, \dfrac{1}{4}, \ldots, \dfrac{1}{n}, \ldots$. This may also be written as

$$s_n = \frac{1}{n}, n \in N.$$

(2) $\dfrac{1}{2}, \dfrac{2}{3}, \dfrac{3}{4}, \ldots, \dfrac{n}{n+1}, \ldots$; equivalently, $\left\{\dfrac{n}{n+1}\right\}$.

(3) $-1, 1, -1, 1, -1, \ldots, (-1)^n, \ldots$; equivalently, $\{(-1)^n\}$.

(4) 2, 1, 4, 3, 6, 5, 8, 7, . . . , s_n, . . . ; s_n is defined as follows:

$$s_n = \begin{cases} n + 1 & \text{when } n \text{ is odd} \\ n - 1 & \text{when } n \text{ is even.} \end{cases}$$

(5) 1, 1, 1, 2, 1, 3, 1, 4, 1, 5, 1, 6, . . . , s_n, . . . ; s_n is defined as follows:

$$s_n = \begin{cases} 1 & \text{when } n \text{ is odd} \\ \frac{1}{2}n & \text{when } n \text{ is even.} \end{cases}$$

(6) 2, 3, 5, 7, 11, 13, . . . , s_n, . . . ; s_n is the nth prime number.

In the above examples we notice that we can write out explicitly the nth term of each of the sequences (1) through (5), whereas we are not able to write out a formula for the nth term in (6). Thus, the terms of a sequence need not be given by a simple arithmetical formula.

A sequence may also be defined by specifying the first term and then stating a *recursion formula* that tells us how to find each remaining term from one or more preceding terms.

Example 1. Let $\{s_n\}$ be the sequence in which $s_1 = 3$ and $s_n = 4s_{n-1} - 2$ for $n > 1$. Find s_2, s_3, s_4.

Solution. When we set $n = 2$ in our recursion formula $s_n = 4s_{n-1} - 2$, we get

$$s_2 = 4s_1 - 2 = 4(3) - 2 = 10.$$

Now, setting $n = 3$, we get

$$s_3 = 4s_2 - 2 = 4(10) - 2 = 38.$$

Similarly,

$$s_4 = 4s_3 - 2 = 4(38) - 2 = 150.$$

It is clear how to compute the terms s_5, s_6, etc., in succession.

Another common way to specify a sequence is to list the first few terms and let the reader guess the general term. Example 2 shows that this method does not specify a sequence unambiguously.

Example 2. Find formulas for the nth term of two distinct sequences whose first four terms are

$$1, \tfrac{1}{2}, \tfrac{1}{3}, \tfrac{1}{4}, \ldots.$$

Solution. One obvious guess is that the fifth term is $\tfrac{1}{5}$, the sixth term is $\tfrac{1}{6}$, and the nth term is $1/n$. Thus one sequence is $\{s_n\}$ where $s_n = 1/n$.

Now, consider the sequence $\{t_n\}$ where

$$t_n = (n-1)(n-2)(n-3)(n-4) + \frac{1}{n}.$$

Setting $n = 1, 2, 3,$ and 4, we find that

$$t_1 = 1$$
$$t_2 = \tfrac{1}{2}$$
$$t_3 = \tfrac{1}{3}$$
$$t_4 = \tfrac{1}{4}.$$

However,

$$t_5 = (5-1)(5-2)(5-3)(5-4) + \tfrac{1}{5}$$
$$= 24 + \tfrac{1}{5}$$
$$= \tfrac{121}{5}.$$

Thus $\{s_n\}$ and $\{t_n\}$ are two distinct sequences whose first four terms are $1, \tfrac{1}{2}, \tfrac{1}{3},$ and $\tfrac{1}{4}$.

Consequently, to specify precisely a sequence $\{s_n\}$, a rule for determining the nth term s_n must be given.

Example 3. For the sequence $\{s_n\}$ where

$$s_n = 2n^2 + 4n + 1$$

find the first three terms and the tenth term.

Solution. Substituting $n = 1, 2, 3,$ and 10 in the formula, we get

$$s_1 = 2(1)^2 + 4(1) + 1 = 7$$
$$s_2 = 2(2)^2 + 4(2) + 1 = 17$$
$$s_3 = 2(3)^2 + 4(3) + 1 = 31$$
$$s_{10} = 2(10)^2 + 4(10) + 1 = 241.$$

Example 4. Suppose for a sequence $\{s_n\}$ where

$$s_n = an^2 + bn$$

the first term is 5 and the fifth term is 65. Find the first six terms of this sequence.

Solution. Setting $n = 1$ and 5 in $s_n = an^2 + bn$ we have

$$s_1 = a \cdot 1 + b \cdot 1 = a + b = 5$$
$$s_5 = a \cdot (5)^2 + b(5) = 25a + 5b = 65.$$

The solution of the equations

$$a + b = 5$$
$$25a + 5b = 65$$

is $a = 2$ and $b = 3$.

Hence the sequence is given by

$$s_n = 2n^2 + 3n.$$

To obtain the second, third, fourth, and sixth terms of this sequence, we substitute $n = 2, 3, 4,$ and 6 in $s_n = 2n^2 + 3n$. We have

$$s_1 = 5$$
$$s_2 = 14$$
$$s_3 = 27$$
$$s_4 = 44$$
$$s_5 = 65$$
$$s_6 = 90.$$

Exercise 1

In problems 1 through 6, find the second, sixth, and ninth terms for the given sequences.

1. $\{1 + (-1)^n\}$

2. $\left\{\dfrac{1}{2^n}\right\}$

3. $\left\{\dfrac{(-1)^n}{n^2}\right\}$

4. $\{2n - 1\}$

5. $\left\{\dfrac{1 \cdot 3 \cdot 5 \ldots \cdot (2n - 1)}{2 \cdot 4 \cdot 6 \ldots \cdot 2n}\right\}$

6. $\{3\}$

7. Sketch the graph of the sequences in problems 1 through 6.

In problems 8 through 10, find formulas for the nth terms of two distinct sequences whose first three terms are indicated.

8. $2, 4, 8, \ldots$

9. $1, -5, 9, \ldots$

10. $0, \frac{1}{2}, \frac{2}{3}, \ldots$

In problems 11 through 14, write the first five terms of the given sequences.

11. $s_1 = \frac{3}{5}, s_n = \frac{1}{2}s_{n-1}$ for $n > 1$

12. $s_1 = 1, s_n = 9 - 2s_{n-1}$ for $n > 1$

13. $s_1 = 2, s_n = 2s_{n-1}$ for $n > 1$

14. $s_1 = 2, s_2 = 5, s_n = 6s_{n-2} - 4s_{n-1}$

15. For the nth term of each sequence in problems 11 and 13, find an explicit formula independent of the other terms.

16. The Fibonacci sequence $F_1, F_2, F_3, \ldots$ is defined by $F_1 = 1$, $F_2 = 1$, and $F_n = F_{n-1} + F_{n-2}$ for $n > 2$. Let $s_n = F_n/F_{n+1}$. Compute the first eight terms of the sequence $\{s_n\}$.

17. In high school we learn that

$$\tfrac{1}{3} = 0.3333\ldots.$$

What sequence is involved in this equation?

18. Let $\{s_n\}$ be a sequence defined by $s_1 = 1$, $s_n =$ the number of positive integers less than n and relatively prime to n for $n > 1$. Find s_8, s_{17}, s_{20}, and s_{51}. (Two integers are said to be relatively prime if they have no common factors except 1.)

19. A bank pays 5 percent interest compounded annually. Suppose you deposit \$1000 on January 1 of this year. Let a_n be the amount in your account after the nth year (we assume no withdrawals at all). Express a_{n+1} in terms of a_n and then find a formula for a_n.

20. Suppose $\{s_n\}$ is a sequence, and define a new sequence $\{S_n\}$ by the formula $S_n = s_1 + s_2 + \ldots + s_n$.
(a) If $s_n = n$, write S_1, S_5, S_{10}. Can you give a formula for S_n?
(b) If $s_n = 3/10^n$, give a decimal representation of S_7.

2. Limit of a Sequence

We turn now to the notion of limit of a sequence. Consider the following examples.

Example 1. Sketch the graph of the following sequences and discuss the behavior of the nth term as n increases:

$$\text{(a)} \ \left\{\frac{1}{n}\right\} \qquad \text{(b)} \ \left\{1 + \frac{(-1)^n}{n}\right\}.$$

Solution. (a) Here the first few terms of the sequence are

$$1, \tfrac{1}{2}, \tfrac{1}{3}, \tfrac{1}{4}, \tfrac{1}{5}, \ldots.$$

The graph of this sequence is sketched in Figure 1. We observe that the terms of this sequence decrease as n increases. We also see intuitively that the terms of the sequence "pile up" near zero as we let n get larger and larger (although none of the terms is zero). In other words, we can make the distance between $1/n$ and 0 as small as we please simply by going far enough out into the sequence.

(b) The first few terms of this sequence are

$$0, \tfrac{3}{2}, \tfrac{2}{3}, \tfrac{5}{4}, \tfrac{4}{5}, \tfrac{7}{6}, \tfrac{6}{7}, \tfrac{9}{8}, \ldots.$$

The graph of this sequence is sketched in Figure 2. We notice that the terms of this sequence are neither decreasing steadily nor

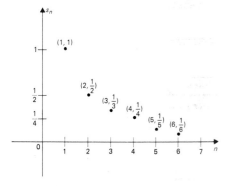

Figure 1

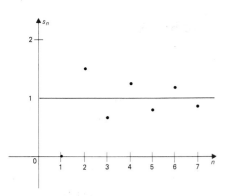

Figure 2

increasing steadily. Yet the terms of this sequence become arbitrarily close to 1. Suppose, for example, that we wish a tolerance of 1/100 as a requirement for closeness. Since the distance from s_n to 1 is

$$|s_n - 1| = \left|\frac{(-1)^n}{n}\right| = \frac{1}{n},$$

a choice of an integer n_0 such that $1/n_0 < 1/100$ or $n_0 > 100$ defines

$$s_{n_0} = 1 + \frac{(-1)^{n_0}}{n_0},$$

which differs from 1 by less than 1/100.

Furthermore, each subsequent term s_n, $n > n_0$, satisfies the specified tolerance of 1/100.

Now, suppose some positive number ϵ (epsilon), no matter how small, has been given as the tolerance of closeness. We can determine n_0 so that

$$|s_{n_0} - 1| = \left|1 + \frac{(-1)^{n_0}}{n_0} - 1\right| = \left|\frac{(-1)^{n_0}}{n_0}\right| = \frac{1}{n_0}$$

is less than ϵ. This can be accomplished by choosing the first natural number n_0 such that $1/n_0 < \epsilon$ or $n_0 > 1/\epsilon$. Now for any $n > n_0$, we have

$$\frac{1}{n} < \frac{1}{n_0}.$$

Consequently, $1/n < \epsilon$ (for $n > n_0$), so each term s_n after s_{n_0} differs from 1 by less than the tolerance ϵ. We are now ready for

---

Definition of limit of a sequence

Definition 2. The number A is said to be the *limit* of the sequence $\{s_n\}$ if, given any positive number ϵ, there exists a natural number n_0 (which depends on ϵ) such that

(7) $$|s_n - A| < \epsilon \text{ for all } n \geq n_0.$$

If the limit A exists, we say that the sequence is *convergent* and write

$$\lim_{n \to \infty} s_n = A,$$

Using this definition the reader can show that if $\lim\limits_{n \to \infty} s_n = A$ and $\lim\limits_{n \to \infty} s_n = B$, then $A = B$. That is, the limit of a sequence, if it exists, is unique.

(read "limit of s sub n equals A as n approaches infinity"). A sequence that is not convergent is said to be *divergent*.

We shall now state some useful theorems

Theorem 1. **Let c be any fixed real number.**

(i)   **If $s_n = c$ for all n, then $\displaystyle\lim_{n \to \infty} s_n = c$.**

(ii) **$\displaystyle\lim_{n \to \infty} \dfrac{c}{n} = 0$.**

The energetic student may use Definition 2 to prove this theorem.
The next two theorems are a little more difficult and we omit their
proofs. The proofs may be found in more advanced calculus texts.

Theorem 2.

(i)   **$\displaystyle\lim_{n \to \infty} \dfrac{1}{n^c} = 0$ if $c > 0$**

(ii) **$\displaystyle\lim_{n \to \infty} x^n = \begin{cases} 0 & \text{if } |x| < 1 \\ 1 & \text{if } x = 1 \\ \text{does not exist} & \text{if } x = -1 \text{ or } |x| > 1 \end{cases}$**

(iii) **$\displaystyle\lim_{n \to \infty} n^{1/n} = 1$**

$e \approx 2.71828$

(iv) **$\displaystyle\lim_{n \to \infty} \left(1 + \dfrac{a}{n}\right)^n = e^a$, where e is a constant and a is
any real number.**

Theorem 3. If $\{s_n\}$ and $\{t_n\}$ are convergent sequences, then

limit of sum = sum of limits

(i)   **$\displaystyle\lim_{n \to \infty} (s_n + t_n) = \lim_{n \to \infty} s_n + \lim_{n \to \infty} t_n$**

limit of product = product of limits

(ii) **$\displaystyle\lim_{n \to \infty} (s_n t_n) = \left(\lim_{n \to \infty} s_n\right)\left(\lim_{n \to \infty} t_n\right)$**

limit of quotient = quotient of limits

(iii) **$\displaystyle\lim_{n \to \infty} \dfrac{s_n}{t_n} = \dfrac{\displaystyle\lim_{n \to \infty} s_n}{\displaystyle\lim_{n \to \infty} t_n}$, if t_n (for all n) and $\displaystyle\lim_{n \to \infty} t_n$ differ**

from zero.

These theorems provide us with an easy method of calculating
the limits of several sequences.

Example 2. If $s_n = (2n + 1)/(3n + 5)$, show that $\displaystyle\lim_{n \to \infty} s_n = \frac{2}{3}$.

Solution. In order that we may use the above theorems, we write

Multiply numerator and denominator by $1/n$.

$$\frac{2n + 1}{3n + 5} = \frac{\dfrac{1}{n}(2n + 1)}{\dfrac{1}{n}(3n + 5)}.$$

Hence

$$\lim_{n \to \infty} \frac{2n + 1}{3n + 5} = \lim_{n \to \infty} \frac{\dfrac{1}{n}(2n + 1)}{\dfrac{1}{n}(3n + 5)}$$

$$= \lim_{n \to \infty} \frac{2 + \dfrac{1}{n}}{3 + \dfrac{5}{n}}$$

Theorem 3(iii)
$$= \frac{\lim_{n \to \infty}\left(2 + \dfrac{1}{n}\right)}{\lim_{n \to \infty}\left(3 + \dfrac{5}{n}\right)}$$

Theorem 3(i)
$$= \frac{\lim_{n \to \infty} 2 + \lim_{n \to \infty} \dfrac{1}{n}}{\lim_{n \to \infty} 3 + \lim_{n \to \infty} \dfrac{5}{n}}$$

Theorem 1
$$= \frac{2 + 0}{3 + 0}$$

$$= \frac{2}{3}.$$

Example 3. Show that $\lim_{n \to \infty} \dfrac{(n - 2)(3n + 5)}{n^2 + 1} = 3.$

Solution. $\lim_{n \to \infty} \dfrac{(n - 2)(3n + 5)}{n^2 + 1} = \lim_{n \to \infty} \dfrac{3n^2 - n - 10}{n^2 + 1}$

Divide numerator and denominator by n^2.
$$= \lim_{n \to \infty} \frac{3 - \dfrac{1}{n} - \dfrac{10}{n^2}}{1 + \dfrac{1}{n^2}}$$

Theorem 3(iii)
$$= \frac{\lim_{n \to \infty}\left(3 - \dfrac{1}{n} - \dfrac{10}{n^2}\right)}{\lim_{n \to \infty}\left(1 + \dfrac{1}{n^2}\right)}$$

Theorem 3(i), (ii)

$$= \frac{\lim\limits_{n \to \infty} 3 - \lim\limits_{n \to \infty} \frac{1}{n} - 10 \lim\limits_{n \to \infty} \frac{1}{n^2}}{\lim\limits_{n \to \infty} 1 + \lim\limits_{n \to \infty} \frac{1}{n^2}}$$

Theorems 1(i) and 2(i)

$$= \frac{3}{1}$$

$$= 3.$$

Intuitively, one can see that the sequence

$$1, 2, 3, \ldots, n, \ldots$$

does not converge to any real number. We shall use Definition 2 to show that the sequence diverges.

Example 4. Show that the sequence

$$1, 2, 3, \ldots, n, \ldots$$

does not have a limit.

Solution. Here $s_n = n$. Suppose that

$$\lim_{n \to \infty} s_n = A \qquad \text{for some real number } A.$$

Then for any $\epsilon > 0$, there exists an n_0 such that

$$|s_n - A| < \epsilon \qquad \text{for all } n \geq n_0.$$

In particular, if we take $\epsilon = 1$, we have

$$|s_n - A| < 1 \qquad \text{for all } n \geq n_0.$$

This is equivalent to

$$-1 < s_n - A < 1 \qquad \text{for all } n \geq n_0$$

or

$$-1 < n - A < 1 \qquad \text{for all } n \geq n_0$$

or

$$A - 1 < n < A + 1 \qquad \text{for all } n \geq n_0.$$

This last statement asserts that all natural numbers n greater than n_0 lie between $A - 1$ and $A + 1$. This is obviously false. Hence the sequence $\{s_n\}$ does not have a limit.

Exercise 2

In problems 1 through 12, establish the indicated limits by appealing to the theorems of this section.

1. $\lim\limits_{n \to \infty} \dfrac{n - 1}{n} = 1$

2. $\lim\limits_{n \to \infty} \dfrac{2 - n}{n} = -1$

3. $\lim\limits_{n \to \infty} \dfrac{1}{\sqrt{n}} = 0$

4. $\lim\limits_{n \to \infty} \dfrac{2n + 1}{3n - 1} = \dfrac{2}{3}$

5. $\lim\limits_{n \to \infty} \dfrac{n^2 + 1}{3n^2 - 2} = \dfrac{1}{3}$

6. $\lim\limits_{n \to \infty} \left(\dfrac{2n}{n + 1} - \dfrac{n + 1}{2n} \right) = \dfrac{3}{2}$

7. $\lim\limits_{n \to \infty} \dfrac{n}{n^2 + 1} = 0$

8. $\lim\limits_{n \to \infty} \dfrac{n^2 + n + 2}{(2n + 3)(3n + 4)} = \dfrac{1}{6}$

Note. $\lim\limits_{n \to \infty} [s_n]^{\frac{1}{2}} = [\lim\limits_{n \to \infty} s_n]^{\frac{1}{2}}$

9. $\lim\limits_{n \to \infty} \dfrac{n^2 - 3n}{\sqrt{5n^4 + 3n^2 + 1}} = \dfrac{1}{\sqrt{5}}$

10. $\lim\limits_{n \to \infty} \dfrac{2n^{-2} + 3n^{-3}}{4n^{-2} + 5n^{-3}} = \dfrac{1}{2}$

11. $\lim\limits_{n \to \infty} \dfrac{\sqrt{2n^6 + 3n^4 + 7}}{n^4 + 2} = 0$

12. $\lim\limits_{n \to \infty} (\sqrt{n^2 + n + 1} - \sqrt{n^2 + 1}) = \frac{1}{2}$

    Hint: Multiply numerator and denominator by
    $$\sqrt{n^2 + n + 1} + \sqrt{n^2 + 1}.$$

13. Suppose that the sequence
    $$\sqrt{5}, \; \sqrt{5\sqrt{5}}, \; \sqrt{5\sqrt{5\sqrt{5}}}, \; \ldots s_n, \; \ldots,$$
    where $s_n = \sqrt{5s_{n-1}}$ (s_n has n fives), converges to x. Prove that $x = 5$.

14. Suppose that the sequence
    $$\sqrt{6}, \; \sqrt{6 + \sqrt{6}}, \; \sqrt{6 + \sqrt{6 + \sqrt{6}}}, \; \ldots, s_n, \ldots$$
    (s_n has n sixes) converges to x. Prove that $x = 3$.

*15. Give examples of two divergent sequences $\{s_n\}$ and $\{t_n\}$ such that $\{s_n + t_n\}$ converges.

3. Infinite Series

Let

(8)
$$a_1, a_2, a_3, \ldots, a_n, \ldots$$

be a given infinite sequence. The sum

(9)
$$a_1 + a_2 + a_3 + \ldots + a_n + \ldots$$

has no meaning in the ordinary sense, since it is impossible to carry on the operation of addition for infinitely many numbers. However, we shall see presently that a meaning can sometimes be assigned to such an expression.

An expression of the form (9) is called an *infinite series*, and $a_1, a_2, a_3, \ldots, a_n, \ldots$ are called the terms of the series. We introduce an abbreviated notation for the operation of forming sums of the terms of sequences. The symbol $\sum$ ($\sum$ is the upper-case Greek letter sigma) is read "sigma of" or "summation of."

Notation:

(10)
$$\sum_{k=1}^{1} a_k = a_1$$

(11)
$$\sum_{k=1}^{2} a_k = a_1 + a_2$$

(12)
$$\sum_{k=1}^{n} a_k = a_1 + a_2 + a_3 + \ldots + a_n$$

(13)
$$\sum_{k=1}^{\infty} a_k = a_1 + a_2 + a_3 + \ldots + a_n + \ldots.$$

The symbol $\sum_{k=1}^{n} a_k$ is read "the summation of a_k from $k = 1$ to n."

Now, to a given infinite series

For more on the $\sum$ notation see Appendix 1.

(14)
$$\sum_{k=1}^{\infty} a_k = a_1 + a_2 + \ldots + a_n + \ldots,$$

we associate an infinite sequence $\{s_n\}$ as follows:

$$s_1 = a_1$$
$$s_2 = a_1 + a_2$$
$$\vdots$$
$$s_n = a_1 + a_2 + \ldots + a_n.$$

It is clear that s_n is the sum of the first n terms of the infinite series. The number s_n is called the *nth partial sum* of the series $\sum_{k=1}^{\infty} a_k$.

We can now give meaning to an expression of the form (14).

Definition of convergence and divergence
of series

Definition 3. An infinite series $\sum_{k=1}^{\infty} a_k$ is said to **converge** or **diverge** according as the associated sequence of partial sums $\{s_n\}$ converges or diverges. If the infinite series $\sum_{k=1}^{\infty} a_k$ converges, then the limit of the sequence $\{s_n\}$ is called the **sum or value of the series.** A divergent series has no sum.

Example 1. Find the sum of the infinite series

$$(15) \qquad \sum_{k=1}^{\infty} \frac{1}{k(k+1)} = \frac{1}{1 \cdot 2} + \frac{1}{2 \cdot 3} + \frac{1}{3 \cdot 4}$$
$$+ \ldots + \frac{1}{n(n+1)} + \ldots.$$

Solution. We first find the sequence of partial sums $\{s_n\}$.

Remember: $\dfrac{1}{n(n+1)} = \dfrac{1}{n} - \dfrac{1}{n+1}$

$$s_1 = \frac{1}{1 \cdot 2} = 1 - \frac{1}{2}$$

$$s_2 = \frac{1}{1 \cdot 2} + \frac{1}{2 \cdot 3} = \left(1 - \frac{1}{2}\right) + \left(\frac{1}{2} - \frac{1}{3}\right) = 1 - \frac{1}{3}$$

$$s_3 = \frac{1}{1 \cdot 2} + \frac{1}{2 \cdot 3} + \frac{1}{3 \cdot 4} = \left(1 - \frac{1}{2}\right) + \left(\frac{1}{2} - \frac{1}{3}\right)$$

$$+ \left(\frac{1}{3} - \frac{1}{4}\right) = 1 - \frac{1}{4} = 1 - \frac{1}{n+1}.$$

$$s_n = \left(1 - \frac{1}{2}\right) + \left(\frac{1}{2} - \frac{1}{3}\right) + \ldots + \left(\frac{1}{n} - \frac{1}{n+1}\right)$$

Now, since $s_n = 1 - 1/(n+1)$, $\lim_{n \to \infty} s_n = 1$. Therefore the series is convergent with sum 1.

Infinite geometric series

Example 2. What is the sum of the series

$$(16) \quad \sum_{k=1}^{\infty} x^{k-1} = 1 + x + x^2 + x^3 + \ldots + x^{n-1} + \ldots?$$

Solution. There are several cases to study.

Case 1. $x = 1$: In this case, we have the series

$$(17) \qquad \sum_{k=1}^{\infty} 1 = 1 + 1 + 1 + \ldots + 1 + \ldots$$

The sequence of partial sums of this series is

$$s_1 = 1$$
$$s_2 = 2$$
$$\vdots$$
$$s_n = n.$$

We saw in Section 2 that the sequence $\{s_n\}$ has no limit. It follows that the infinite series (17) diverges and hence has no sum.

Case 2. $x = -1$: In this case we have the series

$$(18) \quad \sum_{n=1}^{\infty} (-1)^{n-1} = 1 - 1 + 1 - 1 + 1 \ldots + (-1)^{n-1} + \ldots .$$

The sequence of partial sums is

$$s_1 = 1$$
$$s_2 = 0$$
$$s_3 = 1$$
$$s_4 = 0$$
$$\vdots$$
$$s_n = \begin{cases} 1 \text{ if } n \text{ is odd} \\ 0 \text{ if } n \text{ is even.} \end{cases}$$

The likely candidates for the limit of this sequence are 0 and 1. Show that neither can be the limit of this sequence. Assume that A is the limit of the sequence and obtain a contradiction.

The energetic student can see that this sequence $\{s_n\}$ does not converge. Consequently, if $x = -1$, the series (16) has no sum.

Case 3. $|x| < 1$: In this case, the sequence of partial sums is:

$$s_1 = 1$$
$$s_2 = 1 + x$$
$$\vdots$$
$$(19) \quad s_n = 1 + x + x^2 + \ldots + x^{n-1}.$$

We can find a formula for s_n as follows: Multiplying both sides of (19) by x, we get

$$(20) \quad xs_n = x + x^2 + x^3 + \ldots + x^n.$$

Subtracting (20) from (19), we have

$$(21) \quad (1 - x)s_n = 1 - x^n.$$

Since $x \neq 1$, we divide both sides of (21) by $1 - x$, and obtain

$$(22) \quad s_n = \frac{1 - x^n}{1 - x} = \frac{x^n - 1}{x - 1} .$$

Now

Note. $\dfrac{1 - x^n}{1 - x} = \dfrac{1}{1 - x} - \dfrac{x^n}{1 - x}$

$$\lim_{n \to \infty} s_n = \lim_{n \to \infty} \frac{1 - x^n}{1 - x}$$

$$= \lim_{n \to \infty} \frac{1}{1 - x} - \lim_{n \to \infty} \frac{x^n}{1 - x}$$

$$= \frac{1}{1 - x} - \frac{1}{1 - x} \lim_{n \to \infty} x^n.$$

See Theorem 2(ii)

Since $|x| < 1$, $\lim\limits_{n \to \infty} x^n = 0$.

Consequently,

$$\lim_{n \to \infty} s_n = \frac{1}{1 - x}.$$

Thus, if $|x| < 1$, the sum of the series (16) is $1/(1 - x)$.

Case 4. $|x| > 1$: As in case 3, we have

$$\lim_{n \to \infty} s_n = \frac{1}{1 - x} - \frac{1}{1 - x} \lim_{n \to \infty} x^n.$$

See Theorem 2(ii) for $\lim\limits_{n \to \infty} x^n$ *when* $|x| > 1$.

But $\lim\limits_{n \to \infty} x^n$ does not exist for $|x| > 1$. Consequently, $\lim\limits_{n \to \infty} s_n$ does not exist. Hence the series (16) diverges for $|x| > 1$.

In conclusion, the geometric series (16) converges to $1/(1 - x)$ if and only if $|x| < 1$.

Example 3. Write the nonterminating decimal

(23) $0.131313\ldots 13 \ldots$

as the ratio of two integers.

Solution. It is easy to see that (23) is a representation of the sum of the geometric series

(24) $0.13 + 0.0013 + 0.000013 + \ldots .$

Now (24) can be written as

(25) $\qquad \dfrac{13}{100} + \dfrac{13}{(100)^2} + \dfrac{13}{(100)^3} + \ldots$

$$= \frac{13}{100}\left(1 + \frac{1}{100} + \left(\frac{1}{100}\right)^2 + \ldots\right).$$

The expression inside the parentheses is of the form (16) with $x = 1/100$. Consequently, by Example 2, (25) becomes

(26) $\qquad \dfrac{13}{100}\left(\dfrac{1}{1 - \dfrac{1}{100}}\right) = \dfrac{13}{100}\left(\dfrac{100}{100 - 1}\right) = \dfrac{13}{99}.$

Thus,

$$0.131313\ldots 13 \ldots = \frac{13}{99}.$$

An alternate solution is given by letting

$$x = 0.131313\ldots 13 \ldots .$$

Then,

$$100x = 13.131313\ldots13\ldots.$$

Subtracting,

$$99x = 13$$

or

$$x = \frac{13}{99}.$$

Exercise 3

In problems 1 through 5, find the sum of the geometric series if it exists.

1. $8 + 4 + 2 + 1 + \frac{1}{2} + \ldots$
2. $3 + 2.1 + 1.47 + 1.029 + \ldots$
3. $4 - 3 + \frac{9}{4} - \frac{27}{16} + \ldots$
4. $2 - 2 + 2 - 2 + \ldots$
5. $\dfrac{1}{x + 1} + \dfrac{1}{(x + 1)^2} + \dfrac{1}{(x + 1)^3} + \ldots$

In problems 6 through 10, write the periodic decimals as the ratio of two integers.

6. $0.3333\ldots3\ldots$
7. $5.212121\ldots21\ldots$
8. $31.5373737\ldots37\ldots$
9. $0.245245245\ldots245\ldots$
10. $67.01360136\ldots0136\ldots$

*11. Show that any nonterminating decimal

$$0.c_1c_2\ldots c_kd_1d_2\ldots d_nd_1d_2\ldots d_n\ldots d_1d_2\ldots d_n\ldots$$

is a rational number.

12. Suppose a Super Ball dropped from a height of 60 feet rebounds 0.9 times the distance through which it falls. Find the total distance traveled by the ball by the time it comes to rest.

*13. Suppose $\sum_{n=1}^{\infty} a_n$ converges and $a_n > 0$ for all n. Prove that $\lim_{n \to \infty} a_n = 0$.

14. Prove that $\sum_{k=1}^{n} k = 1 + 2 + \ldots + n = \dfrac{n(n + 1)}{2}$.

Hint: Either use induction (see Appendix 1) or write $\sum_{k=1}^{n} k$ in both ascending and descending order and add the corresponding terms.

15. Find the sum of all integers between 1 and 793 which are divisible by 3. *Hint:* Use problem 14.

In problems 16 through 20, use the method of Example 1 to find the sum of the given infinite series.

*16. $\dfrac{1}{3 \cdot 5} + \dfrac{1}{5 \cdot 7} + \dfrac{1}{7 \cdot 9} + \ldots + \dfrac{1}{(2n + 1)(2n + 3)} + \ldots$

*17. $\dfrac{1}{2 \cdot 4} + \dfrac{1}{4 \cdot 6} + \dfrac{1}{6 \cdot 8} + \ldots + \dfrac{1}{2n(2n + 2)} + \ldots$

*18. $\dfrac{1}{1 \cdot 4} + \dfrac{1}{4 \cdot 7} + \dfrac{1}{7 \cdot 10} + \ldots + \dfrac{1}{(3n - 2)(3n + 1)} + \ldots$

*19. $\displaystyle\sum_{n=1}^{\infty} \dfrac{1}{(2n + 1)(2n + 5)}$

*20. $\displaystyle\sum_{n=1}^{\infty} \dfrac{1}{(2n - 1)(2n + 3)}$

4. Some Applications

Before we discuss some applications of sequences, let us define two very important sequences we encountered in a special form in Section 3.

Definition 4. Given two numbers a and d, the sequence

$$a, a + d, a + 2d, \ldots, a + (n - 1)d, \ldots$$

Arithmetic progression

is called an *arithmetic progression*. Each term in the sequence is formed from the preceding term by adding to it the *common difference d*.

Definition 5. Given two numbers a and x, the sequence

$$a, ax, ax^2, \ldots, ax^{n-1}, \ldots$$

Geometric progression

is called a *geometric progression*. Each term after the first is obtained by multiplying the preceding term by x. The number x is called the *common ratio*.

The proofs of the following two theorems are left as exercises for the students.

Theorem 4. If a and d are numbers and n is a positive integer, then

$$a + (a + d) + (a + 2d) + \ldots + [a + (n - 1)d]$$

$$= \frac{n}{2}[2a + (n - 1)d].$$

Theorem 5. If $x \neq 1$, then

$$a + ax + ax^2 + \ldots + ax^{n-1} = a\left(\frac{1 - x^n}{1 - x}\right) = a\left(\frac{x^n - 1}{x - 1}\right).$$

When the domain is restricted to a finite subset of positive integers, the sequence is called a finite sequence.

We shall now discuss several examples which show the importance of finite sequences in everyday life.

Example 1. Suppose P dollars are invested at compound interest at a rate of i per period. Show that at the end of n periods the accumulated amount P_n is given by

(27) $$P_n = P(1 + i)^n.$$

Solution. The P dollars invested at a rate of i per period will earn Pi dollars in interest at the end of the first period. Thus at the end of one period the accumulated amount P_1 is the original investment plus the earned interest. Hence

$$P_1 = P + Pi = P(1 + i).$$

If the original investment and interest are left on deposit, the interest earned at the end of the second period will be $P_1 i = P(1 + i)i$, and we have

$$\begin{aligned}
P_2 &= P_1 + P_1 i \\
&= P_1(1 + i) \\
&= P(1 + i)(1 + i) \\
&= P(1 + i)^2.
\end{aligned}$$

Similarly,

$$\begin{aligned}
P_3 &= P_2 + P_2 i \\
&= P_2(1 + i) \\
&= P(1 + i)^2(1 + i) \\
&= P(1 + i)^3.
\end{aligned}$$

In general,

$$\begin{aligned}
P_n &= P_{n-1} + P_{n-1} i \\
&= P_{n-1}(1 + i) \\
&= P(1 + i)^n.
\end{aligned}$$

Notice that the sequence $\{P_n\}$ is defined by the recursion formula

$$P_n = P_{n-1}(1 + i).$$

In business it is often necessary to find how large a fund will have been accumulated at a specified time by making certain deposits or payments at stated intervals.

Example 2. Suppose you start now to make 25 annual deposits of $100 each at the Golden Gate Bank, which pays 5% interest compounded annually. How much money will you have in your account at the end of 30 years?

Solution. Here we shall repeatedly use formula (27) developed in Example 1, with $P = 100$ and $i = 0.05$:

1st deposit with 30 years interest will amount to: $100(1 + 0.05)^{30}$
2nd deposit with 29 years interest will amount to: $100(1 + 0.05)^{29}$
3rd deposit with 28 years interest will amount to: $100(1 + 0.05)^{28}$
$$\vdots$$

25th deposit with 6 years interest will amount to: $100(1 + 0.05)^{6}$.

Thus the total amount T you will have in your account at the end of 30 years will be

$$
\begin{aligned}
T &= 100(1 + 0.05)^{30} + 100(1 + 0.05)^{29} + \ldots 100(1 + 0.05)^{6}\\
&= 100(1 + 0.05)^{6}[1 + (1 + 0.05) + \ldots + (1 + 0.05)^{23}\\
&\qquad\qquad\qquad\qquad\qquad\qquad\qquad\qquad + (1 + 0.05)^{24}]\\
&= 100(1.05)^{6}[1 + (1.05) + \ldots + (1.05)^{23} + (1.05)^{24}]\\
&= 100(1.05)^{6}\left(\frac{(1.05)^{25} - 1}{1.05 - 1}\right) \quad \text{(see formula (24))}\\
&= 6395.42 \text{ approximately (in dollars).}
\end{aligned}
$$

In formula (27) the amount P_n is usually called the *future value of an investment P.* On the other hand, the price one should pay now for an investment that will have value P_n n years from now at a given interest rate is called the *present value (P.V.).* To find the present value we reverse the compounding process used in Example 1. Thus if the value is P_n at the end of the nth time period, then from formula (27) we get

(28) $$P = \frac{P_n}{(1 + i)^{n}}.$$

Consequently the present value P of an amount P_n which is to be paid n time periods in the future is given by (28).

Example 3. $1200 now due on a video tape unit is to be paid off with interest at 15% in 24 equal monthly installments beginning 3 months from now. What must the installments be?

Note. The 25th deposit will be made at the beginning of the 24th year, or 24 years from now, and at the end of 30 years, the 25th deposit will be at interest for 6 years.

We use mathematical tables to compute these numbers.

Future value of an investment

Present value of an investment

Note. Unless otherwise mentioned the interest rate is assumed to be the annual rate.

Solution. First we find the monthly interest rate:

$$i = \frac{0.15}{12} = 0.0125.$$

Notice that the last payment will be due 23 months after the first, or 26 months from now.

Suppose each payment is A; then by (28)

P.V. of 1st payment, due 3 months from now, is $\dfrac{A}{(1 + 0.0125)^3}$

P.V. of last payment, due 26 months from now, is $\dfrac{A}{(1 + 0.0125)^{26}}$.

The sum is the total present value of all the payments and should equal the present debt of $1200. Therefore,

$$\frac{A}{(1.0125)^3} + \frac{A}{(1.0125)^4} + \cdots + \frac{A}{(1.0125)^{26}} = 1200$$

or

$$\frac{A}{(1.0125)^{26}} [1 + (1.0125) + \cdots + (1.0125)^{23}] = 1200$$

or

$$\frac{A}{(1.0125)^{26}} \left[\frac{(1.0125)^{24} - 1}{1.0125 - 1} \right] = 1200.$$

Thus

$$A = \frac{1200(0.0125)(1.0125)^{26}}{(1.0125)^{24} - 1}$$

$$= 56.79 \text{ approximately (in dollars).}$$

Example 4. A bacteriologist wishes to count the number N of bacteria in a solution s_1. He takes one-eighth of the solution s_1 and dilutes it to form a new solution s_2. He takes one-eighth of s_2 and dilutes it to form a new solution s_3, etc. If by a laboratory counting procedure he finds that the solution s_7 contains 10 bacteria, what is N, the number of bacteria in s_1?

Solution. Let a_k be the number of bacteria in the solution s_k. We are given $a_7 = 10$ and we are to find a_1.

We have

$$a_k = \tfrac{1}{8} a_{k-1}.$$

This defines a geometric progression, with common ratio $1/8$.

Thus

$$a_7 = \left(\frac{1}{8}\right)^6 a_1 = \frac{1}{(8)^6} a_1$$

or

$$\begin{aligned} a_1 &= (8)^6 a_7 \\ &= (8)^6 \times 10. \end{aligned}$$

Hence

$$N = 10 \times (8)^6.$$

Exercise 4

1. Prove Theorem 4.
2. Prove Theorem 5.
3. There are 25 terms in an arithmetic progression. The first term is 2 and the last term is 38. Find the 17th term.
4. Sum to n terms:
   (a) $9 + 99 + 999 + \ldots$
   (b) $7 + 77 + 777 + \ldots$
5. Find the sum of all positive integers less than 1000 which are not multiples of 3.
   Hint:

   $$\text{Sum} = (1 + 2 + 3 + \ldots + 999) - (3 + 6 + 9 + \ldots + 999).$$

6. Find the amount which would be obtained from an investment of $3000 compounded at an annual rate of 5% for 5 years.
7. Solve problem 6 above if the interest is compounded quarterly.
8. The *effective rate* of interest is defined to be the rate which when compounded annually gives the same amount of interest as a nominal rate compounded several times each year. Find the effective rates equivalent to the following nominal rates:
   (a) 5% compounded semiannually
   (b) 12% compounded monthly
   (c) 15% compounded monthly
9. Find the present value of a loan which will amount to $5000 in 10 years if the money is worth 5%.
10. How much should Mr. X invest for his son at 6% compounded semiannually in order to have $4000 at the end of 10 years?
11. A lot for a house is sold for $500 cash and $2000 a year for the next five years. Find the cash value of the lot if money is worth 6% compounded quarterly.
12. A balance of $4500 now due on a house is to be paid off in 20 monthly installments beginning two months hence, with interest at 8%. Find the installment payment.
13. Find the installment payment if a balance of $3000 on an

automobile is to be paid off in 18 monthly installments beginning one month hence, with interest at 6%.

14. Polluteville had a population of 50,000 in 1960, and the population has decreased by 5% each year since that time. What was
the population of the city in 1971? Draw a graph which shows
the population of each succeeding year beginning with 1960.

15. In Example 4, suppose the bacteriologist takes one-fifth of the
solution s_{i-1} and dilutes it to form the solution s_i. Find N if
$a_{10} = 4$.

Leonhard Euler

4

More

on

Limits

Both Newton and Leibniz considered the idea of limit, but their notion of the concept was vague and lacking in mathematical precision. It was the great mathematician Weierstrass who finally gave the notion of limit a rigorous mathematical foundation.

1. The Limit of a Function

In the previous chapter we considered the limit concept for functions whose domain is the set of natural numbers. Here we shall extend the notion of limit to functions whose domain consists of real numbers. We shall begin with an intuitive discussion of limits and then give a formal definition. (Historically, this is the way the limit concept was developed.) The concept of the limit of a function will then be used to introduce the reader to the notion of "continuity" of functions.

Consider the following examples:

Example 1. Let the function f be defined by

$$f(x) = x + 2.$$

The reader can easily determine the graph of f, which is given in Figure 1. Let us see what happens to $f(x)$ as x gets closer and closer to 2 as shown in the following table.

x	1	1.5	1.75	1.80	1.90	1.97	1.99	1.9999
$f(x) = x + 2$	3	3.5	3.75	3.80	3.90	3.97	3.99	3.9999

Obviously, as x approaches 2, $f(x)$ approaches 4. Of course, x could approach 2 through values greater than 2. Thus,

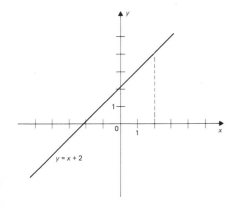

Figure 1

$f(x) = x + 2$

99

x	3	2.50	2.40	2.30	2.20	2.10	2.05	2.001	2.0001
$f(x) = x + 2$	5	4.50	4.40	4.30	4.20	4.10	4.05	4.001	4.0001

Again, $f(x)$ approaches 4 as x approaches 2 through values greater than two. In the first case we say that $f(x)$ approaches 4 as x approaches 2 from the left and write

Left-hand limit

$$f(x) \rightarrow 4 \text{ as } x \rightarrow 2^-$$

and in the second case we say that $f(x)$ approaches 4 as x approaches 2 from the right and we write

Right-hand limit

$$f(x) \rightarrow 4 \text{ as } x \rightarrow 2^+.$$

Example 2. Consider the function f defined by Figure 2.

$$f(x) = x^2, x \neq -1.$$

To illustrate a point we have removed the number -1 from the domain of f; $f(-1)$ is not defined, but we have

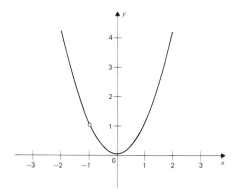

x	-2	-1.50	-1.40	-1.30	-1.20	-1.10	-1.01	-1.001
$f(x) = x^2$	4	2.25	1.96	1.69	1.44	1.21	1.0201	1.002001

That is,

$$f(x) \rightarrow 1 \text{ as } x \rightarrow -1^-.$$

Also, we have the following:

Figure 2
Graph of $f(x) = x^2, x \neq -1$

x	0	-0.50	-0.60	-0.70	-0.80	-0.90	-0.99	-0.999
$f(x) = x^2$	0.00	0.25	0.36	0.49	0.64	0.81	0.9801	0.998001

or

$$f(x) \rightarrow 1 \text{ as } x \rightarrow -1^+.$$

Example 3. Let f be the function defined by

$$f(x) = \begin{cases} x - 1 & x > 3 \\ x^2 & x \leq 3. \end{cases}$$

A graph of f is shown in Figure 3.

If the reader moves a pencil on the curve just to the left of the line $x = 3$, he will find that as the pencil gets closer and closer to the line $x = 3$, $f(x)$ approaches 9. Similarly, if the pencil is moved along the curve just to the right of the line $x = 3$, one finds that as

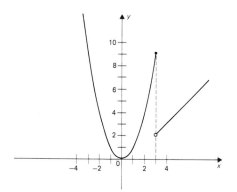

Figure 3

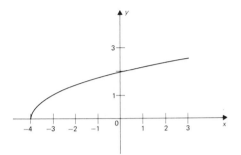

Figure 4

Graph of $f(x) = \sqrt{x + 4}$

This is another notation for

$$f(x) \rightarrow L \text{ as } x \rightarrow c^-.$$

Also written $f(x) \rightarrow L$ *as* $x \rightarrow c^-$

the pencil approaches the line $x = 3$, $f(x)$ approaches 2. Thus

$$f(x) \rightarrow 9 \text{ as } x \rightarrow 3^-,$$

whereas

$$f(x) \rightarrow 2 \text{ as } x \rightarrow 3^+.$$

Example 4. Consider the function f defined by

$$f(x) = \sqrt{x + 4}.$$

Here the domain of f is $\{x \mid x \geq -4\}$. The graph of f is shown in Figure 4. We find that

$$f(x) \rightarrow 0 \text{ as } x \rightarrow -4^+.$$

However, when $x < -4$, $f(x)$ is not defined.

Notation. We have used the expression "$f(x)$ approaches the number L as x approaches the number c from the left." A common notation is

$$\lim_{x \to c^-} f(x) = L,$$

read "the limit of $f(x)$ as x approaches c from the left is L." In the case where x approaches c from the right we write

$$\lim_{x \to c^+} f(x) = L.$$

Thus in Examples 1 through 4 above we have

Example 1. $\lim_{x \to 2^-} f(x) = 4,$ $\lim_{x \to 2^+} f(x) = 4$

Example 2. $\lim_{x \to -1^-} f(x) = 1,$ $\lim_{x \to -1^+} f(x) = 1$

Example 3. $\lim_{x \to 3^-} f(x) = 9,$ $\lim_{x \to 3^+} f(x) = 2$

Example 4. $\lim_{x \to -4^-} f(x)$ does not exist, $\lim_{x \to -4^+} f(x) = 0.$

We now give the concepts discussed above a precise mathematical formulation.

Definition 1. Let f be a function defined on an interval containing c (except possibly at c) and let L be a real number. We say that the limit of $f(x)$ as x approaches c from the left is L and write

$$\lim_{x \to c^-} f(x) = L$$

if and only if for *every* sequence $\{x_n\}$, $x_n < c$ with

$$\lim_{n \to \infty} x_n = c,$$

Note. The sequences $\{x_n\}$ chosen must be in the domain of f for this definition to be meaningful. Also, if no such sequence exists, then $\lim_{x \to c^-} f(x)$ does not exist (as in Example 4 above).

it follows that

$$\lim_{n \to \infty} f(x_n) = L.$$

We say that the limit of $f(x)$ as x approaches c from the right is L and write

$$\lim_{x \to c^+} f(x) = L$$

if and only if for *every* sequence $\{x_n\}$ with $x_n > c$ and

$$\lim_{n \to \infty} x_n = c,$$

it follows that

$$\lim_{n \to \infty} f(x_n) = L.$$

We now apply the definition to Example 3 above.

Let $\{x_n\}$ be a sequence of numbers in the domain of f such that $x_n > 3$ and

$$\lim_{n \to \infty} x_n = 3.$$

One such sequence is given by $x_n = 3 + \dfrac{1}{n}$ where $n \in N$.

Since for $x > 3$, $f(x) = x - 1$,

$$f(x_n) = x_n - 1$$

and

Application of Theorems 1 and 3 of Chapter 3

$$\lim_{n \to \infty} f(x_n) = \lim_{n \to \infty} (x_n - 1)$$
$$= \lim_{n \to \infty} x_n - 1$$
$$= 3 - 1 = 2.$$

Hence,

$$\lim_{x \to 3^+} f(x) = 2.$$

Now, for $x \le 3$, $f(x) = x^2$. If $\{x_n\}$ is a sequence in the domain of f such that $x_n < 3$ and

One such sequence is given by $x_n = 3 - \dfrac{1}{2^n}$ where $n \in N$.

$$\lim_{n \to \infty} x_n = 3,$$

then

$$f(x_n) = x_n^2$$

and

$$\lim_{n \to \infty} f(x_n) = \lim_{n \to \infty} x_n^2$$

$$= \lim_{n \to \infty} x_n \cdot \lim_{n \to \infty} x_n$$

$$= (3)(3) = 9.$$

Application of Theorem 3, Chapter 3

Hence, by definition

$$\lim_{x \to 3^-} f(x) = 9.$$

We say that the limit of $f(x)$ as x approaches c is L and write

$$\lim_{x \to c} f(x) = L$$

Note that in this case we do not restrict x to approach from either left or right.

if and only if

$$\lim_{x \to c^-} f(x) = \lim_{x \to c^+} f(x) = L.$$

Thus in the examples above we have

Example 1. $\lim_{x \to 2} f(x) = \lim_{x \to 2} (x + 2) = 4$

Example 2. $\lim_{x \to -1} f(x) = \lim_{x \to -1} x^2 = 1$

Example 3. $\lim_{x \to 3} f(x)$ does not exist

Example 4. $\lim_{x \to 4} f(x)$ does not exist.

The uniqueness of the number L, if it exists, follows from the uniqueness of the limit of of a sequence of numbers (see the note on p. 82).

Definition 2. Let f be a function defined on an interval containing c (except possibly at c) and let L be a real number. We say that the limit of $f(x)$ as x approaches c is L and write

$$\lim_{x \to c} f(x) = L$$

if and only if for *every* sequence $\{x_n\}$, $x_n \neq c$ with $\lim_{n \to \infty} x_n = c$,

$$\lim_{n \to \infty} f(x_n) = L.$$

Example 5. Let $f(x) = x/(x + 1)$, $x \neq 1$. Show directly from Definition 2 that

$$\lim_{x \to 2} \frac{x}{x + 1} = \frac{2}{3}.$$

One choice is $\{x_n\} = \left\{2 + \dfrac{(-1)^n}{n}\right\}$, $n \in N$.

Solution. Let $\{x_n\}$ be any sequence, $x_n \neq -1$, and $\lim_{n \to \infty} x_n = 2$.

Then,

$$f(x_n) = \frac{x_n}{x_n + 1}$$

and

$$\lim_{n \to \infty} f(x_n) = \lim_{n \to \infty} \frac{x_n}{x_n + 1}$$

$$= \frac{\lim\limits_{n \to \infty} x_n}{\lim\limits_{n \to \infty} (x_n + 1)}$$

$$= \frac{2}{2 + 1} = \frac{2}{3}.$$

Application of Theorems 1 and 3, Chapter 3

Hence, by Definition 2

$$\lim_{x \to 2} f(x) = \lim_{x \to 2} \frac{x}{x + 1} = \frac{2}{3}.$$

Example 6. Let $f(x) = (x^2 - 1)/(x - 1)$, $x \neq 1$. Using Definition 2 show that

$$\lim_{x \to 1} \frac{x^2 - 1}{x - 1} = 2.$$

Solution. A graph of f is shown in Figure 5. Let $\{x_n\}$ be any sequence such that $x_n \neq 1$ and

$$\lim_{n \to \infty} x_n = 1.$$

Then,

$$f(x_n) = \frac{x_n^2 - 1}{x_n - 1}$$

$$= \frac{(x_n - 1)(x_n + 1)}{x_n - 1}$$

$$= x_n + 1$$

and

$$\lim_{n \to \infty} f(x_n) = \lim_{n \to \infty} (x_n + 1)$$

$$= \lim_{n \to \infty} x_n + 1$$

$$= 1 + 1 = 2.$$

Hence, by Definition 2, we have

$$\lim_{x \to 1} f(x) = \lim_{x \to 1} \frac{x^2 - 1}{x - 1} = 2.$$

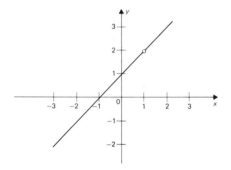

Figure 5

Graph of $f(x) = \dfrac{x^2 - 1}{x - 1}$, $x \neq 1$

Division is permissible since $x_n \neq 1$.

Application of Theorems 1 and 3, Chapter 3

Exercise 1

In problems 1–10, sketch the graphs of the functions defined and intuitively determine the indicated limits.

1. $f(x) = 3x - 5$: (a) $\lim_{x \to -1^+} f(x)$ (b) $\lim_{x \to 2^-} f(x)$

2. $f(x) = 3x^2$: (a) $\lim_{x \to 0^+} f(x)$ (b) $\lim_{x \to 0^-} f(x)$

3. $f(x) = 2x^2 + 4$: (a) $\lim_{x \to 2^+} f(x)$ (b) $\lim_{x \to 3^-} f(x)$

4. $f(x) = \begin{cases} 2x + 3, & x \le 2 \\ 3x - 1, & x > 2 \end{cases}$ (a) $\lim_{x \to 2^+} f(x)$ (b) $\lim_{x \to 2^-} f(x)$

5. $f(x) = \dfrac{x^2 - 1}{x + 1}, x \ne -1$: (a) $\lim_{x \to -1^+} f(x)$ (b) $\lim_{x \to -1^-} f(x)$

6. $f(x) = \begin{cases} \dfrac{x^2 - 4}{x - 2}, & x \ne 2 \\ 5, & x = 2 \end{cases}$ (a) $\lim_{x \to 2^+} f(x)$ (b) $\lim_{x \to 2^-} f(x)$

7. $f(x) = \dfrac{x^2 - 5x + 4}{x - 1}, x \ne 1$: (a) $\lim_{x \to 1^+} f(x)$

   (b) $\lim_{x \to 1^-} f(x)$

8. $f(x) = \sqrt{x + 3}$: (a) $\lim_{x \to -3^+} f(x)$ (b) $\lim_{x \to -3^-} f(x)$

9. $f(x) = 3 + \sqrt{x - 1}$: (a) $\lim_{x \to 1^+} f(x)$ (b) $\lim_{x \to 1^-} f(x)$

10. $f(x) = \sqrt[3]{x - 1}$: (a) $\lim_{x \to 1^+} f(x)$ (b) $\lim_{x \to 1^-} f(x)$

In problems 11–22, use Definition 2 to evaluate the indicated limits.

11. $\lim_{x \to 2} (3x - 5)$

12. $\lim_{x \to -1} x^3$

13. $\lim_{x \to 2} (2x^3 + 4)$

14. $\lim_{x \to -5} \dfrac{x^2}{2x - 3}$

15. $\lim_{x \to 3} \dfrac{x^3 + 4}{2x^2 - 1}$

16. $\lim_{x \to 2} \dfrac{x + 4}{2x^2 - 5x + 7}$

17. $\lim_{x \to 2} \dfrac{x^3 - 8}{x - 2}$

18. $\lim_{x \to -2} \dfrac{x^3 + 8}{x + 2}$

19. $\lim_{x \to 1} \sqrt{\dfrac{3x + 5}{2x - 1}}$

20. $\lim_{x \to 2} \sqrt{\dfrac{x^2 - 4}{x^2 - 3x + 2}}$

21. $\lim_{x \to 1} \sqrt{\dfrac{x^4 - 1}{x^3 - 1}}$

22. $\lim_{x \to -8} \sqrt[3]{x}$

In problems 23–26, form the new function F where

$$F(h) = \frac{f(x + h) - f(x)}{h}, h \neq 0$$

and evaluate $\lim\limits_{h \to 0} F(h)$.

23. $f(x) = 5x - 1$ **24.** $f(x) = 2x^2$

25. $f(x) = x^2 - 3x + 5$ **26.** $f(x) = \sqrt{x}$

**27. Definition 2 is called the *sequential* approach to the limit of a function. A second definition, the ϵ-δ definition, is the following:

Let f be a function and I be an open interval in the domain of f. If c and L are real numbers and c is in I, we say that the limit of $f(x)$ as x approaches c is L and write

$$\lim_{x \to c} f(x) = L$$

if and only if for each positive number ϵ there is a positive number δ such that

$$|f(x) - L| < \epsilon \text{ whenever } 0 < |x - c| < \delta.$$

Prove that the two definitions are equivalent.

2. Some Theorems on Limits

The following theorems will help the student find the limits of complicated functions without having to apply the definition. Using the results in Section 2 of Chapter 3, we shall see that the proofs of these theorems are very simple.

Theorem 1. **If a and b are real numbers and f is a function defined by**

$$f(x) = ax + b,$$

then for any number c

$$\lim_{x \to c} f(x) = ac + b.$$

Proof. Let $\{x_n\}$ be any sequence of real numbers in the domain of f such that

$$\lim_{n \to \infty} x_n = c, x_n \neq c.$$

Then
$$f(x_n) = ax_n + b$$

and
$$\lim_{n \to \infty} f(x_n) = \lim_{n \to \infty} (ax_n + b)$$

Application of Theorem 3, Chapter 3
$$= a \lim_{n \to \infty} x_n + \lim_{n \to \infty} b$$

Application of Theorem 1, Chapter 3
$$= ac + b.$$

Hence, by Definition 2 we have

Note. If a = 0 we obtain $\lim_{x \to c} b = b$.
$$\lim_{x \to c} f(x) = ac + b.$$

This proves Theorem 1.

   We shall prove part (ii) of the next theorem and leave parts (i), (iii), and (iv) as exercises for the student.

Theorem 2. **If f and g are two functions such that**

Observe that for both limits x → c.
$$\lim_{x \to c} f(x) = L_1 \text{ and } \lim_{x \to c} g(x) = L_2$$

where L_1, L_2 and c are numbers, then

(i) $\lim_{x \to c} (f + g)(x) = \lim_{x \to c} f(x) + \lim_{x \to c} g(x) = L_1 + L_2$

(ii) $\lim_{x \to c} (f \cdot g)(x) = \lim_{x \to c} f(x) \cdot \lim_{x \to c} g(x) = L_1 \cdot L_2$

(iii) $\lim_{x \to c} (f/g)(x) = \dfrac{\lim_{x \to c} f(x)}{\lim_{x \to c} g(x)} = \dfrac{L_1}{L_2}$ **provided $L_2 \neq 0$**

and

(iv) $\lim_{x \to c} \sqrt[p]{f(x)} = \sqrt[p]{L_1}$ **provided $f(x) \geq 0$ when p is an even integer.**

Proof of (ii). Let $\{x_n\}$ be any sequence in the domains of f and g such that
$$\lim_{n \to \infty} x_n = c, \qquad x_n \neq c.$$

Then
$$(fg)(x_n) = f(x_n)g(x_n)$$

and

Note that $\{f(x_n)\}$ and $\{g(x_n)\}$ are sequences of numbers. Application of Theorem 3, Chapter 3, yields this result.
$$\lim_{n \to \infty} (fg)(x_n) = \lim_{n \to \infty} [f(x_n)g(x_n)]$$
$$= \lim_{n \to \infty} f(x_n) \lim_{n \to \infty} g(x_n)$$
$$= L_1 L_2.$$

Hence by Definition 2

$$\lim_{x \to c} (fg)(x) = \lim_{x \to c} f(x) \lim_{x \to c} g(x) = L_1 L_2.$$

We shall apply Theorems 1 and 2 in the following examples.

Example 1. Show that

$$\lim_{x \to c} x^2 = c^2.$$

Solution. Applying Theorem 1 where $a = 1$ and $b = 0$ we have

$$\lim_{x \to c} x = c$$

and by Theorem 2(ii)

$$\lim_{x \to c} x^2 = \lim_{x \to c} x \lim_{x \to c} x = c^2.$$

Example 2. Show that

$$\lim_{x \to 2} \frac{x^2}{2x - 1} = \frac{4}{3}, \qquad x \neq \frac{1}{2}.$$

Solution. From Example 1 we have

$$\lim_{x \to 2} x^2 = 4$$

and from Theorem 1 with $a = 2$ and $b = -1$ we obtain

$$\lim_{x \to 2} (2x - 1) = 2(2) - 1 = 3.$$

Application of Theorem 2(iii) yields

$$\lim_{x \to 2} \frac{x^2}{2x - 1} = \frac{\lim_{x \to 2} x^2}{\lim_{x \to 2} (2x - 1)}$$

$$= \frac{4}{3}.$$

Example 3. Let f be the function defined by

$$f(x) = \frac{x^4 + 3}{x^2 + 1}.$$

Evaluate $\lim_{x \to -2} f(x)$.

Solution. Using Example 1 and Theorem 2(ii) we have

$$\lim_{x \to -2} x^4 = \lim_{x \to -2} (x^2 \cdot x^2) = \lim_{x \to -2} x^2 \lim_{x \to -2} x^2 = (-2)^2(-2)^2 = 16$$

and by Theorem 2(i) and Theorem 1

$$\lim_{x \to -2} (x^4 + 3) = \lim_{x \to -2} x^4 + \lim_{x \to -2} 3 = 16 + 3 = 19.$$

Also,

$$\lim_{x \to -2} (x^2 + 1) = \lim_{x \to -2} x^2 + \lim_{x \to -2} 1 = (-2)^2 + 1 = 5.$$

Therefore,

Application of Theorem 2(iii)

$$\lim_{x \to -2} \frac{x^4 + 3}{x^2 + 1} = \frac{\lim_{x \to -2} (x^4 + 3)}{\lim_{x \to -2} (x^2 + 1)} = \frac{19}{5}.$$

Example 4. Evaluate

$$\lim_{x \to 3} \sqrt[3]{\frac{x^3 - 27}{x^2 - 9}}, \qquad x \neq \pm 3.$$

Solution. For $x \neq \pm 3$, we have

Difference of two cubes:

$a^3 - b^3 = (a - b)(a^2 + ab + b^2)$

$$\frac{x^3 - 27}{x^2 - 9} = \frac{(x - 3)(x^2 + 3x + 9)}{(x - 3)(x + 3)} = \frac{x^2 + 3x + 9}{x + 3}.$$

Now,

$$\lim_{x \to 3} x^2 = 9$$

and by Theorem 1

$$\lim_{x \to 3} (3x + 9) = 3(3) + 9 = 18.$$

Therefore,

$$\lim_{x \to 3} [x^2 + (3x + 9)] = \lim_{x \to 3} x^2 + \lim_{x \to 3} (3x + 9) = 9 + 18 = 27.$$

Also by Theorem 1

$$\lim_{x \to 3} (x + 3) = 3 + 3 = 6$$

and application of Theorem 2(iii) yields

$$\lim_{x \to 3} \frac{x^2 + 3x + 9}{x + 3} = \frac{\lim_{x \to 3} (x^2 + 3x + 9)}{\lim_{x \to 3} (x + 3)} = \frac{27}{6} = \frac{9}{2}.$$

Finally by Theorem 2(iv)

$$\lim_{x \to 3} \sqrt[3]{\frac{x^3 - 27}{x^2 - 9}} = \lim_{x \to 3} \sqrt[3]{\frac{x^2 + 3x + 9}{x + 3}} = \sqrt[3]{\frac{9}{2}}.$$

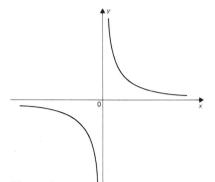

Figure 6

Graph of $f(x) = \dfrac{1}{x}$

We shall now consider some other important limits: the "limit at infinity" and the "infinite limit." Consider the function f where

$$f(x) = \frac{1}{x}, \qquad x \neq 0.$$

The graph of f is given in Figure 6. A few points belonging to the graph of f are given in the following table:

x	-10^{100}	-10^{10}	-10^2	-10	10	10^2	10^{10}	10^{100}
$f(x) = \dfrac{1}{x}$	-10^{-100}	-10^{-10}	-10^{-2}	-10^{-1}	10^{-1}	10^{-2}	10^{-10}	10^{-100}

As x gets larger and larger $f(x)$ gets closer and closer to zero. The expression "x tends to infinity" is used to indicate that x increases without bound. The following symbols are used:

$x \to +\infty$ if x increases without bound through positive values

$x \to -\infty$ if x decreases without bound

From the table above we see that as x increases without bound through positive values $(x \to +\infty)$, $1/x$ approaches 0. Also, $1/x$ approaches 0 as x decreases without bound.

We write: $\displaystyle \lim_{x \to +\infty} \frac{1}{x} = 0$ *or*

$\dfrac{1}{x} \to 0$ *as* $x \to +\infty$

and $\displaystyle \lim_{x \to -\infty} \frac{1}{x} = 0$ *or*

$\dfrac{1}{x} \to 0$ *as* $x \to -\infty$.

Theorem 3. If f is the function defined by

$$f(x) = \frac{1}{x}$$

then

(i) $\displaystyle \lim_{x \to +\infty} f(x) = 0$ (ii) $\displaystyle \lim_{x \to -\infty} f(x) = 0$

The results of Theorem 2 still hold when the number c is replaced by the symbols $+\infty$ or $-\infty$. We apply these results in the following examples.

Example 5. Evaluate

$$\lim_{x \to +\infty} \frac{2x + 5}{7x - 3}.$$

Solution. For $x \neq 0$ we have

Dividing numerator and denominator by x

$$\frac{2x + 5}{7x - 3} = \frac{2 + \dfrac{5}{x}}{7 - \dfrac{3}{x}}.$$

Now

limit of sum = sum of limits

$$\lim_{x \to +\infty} \left(2 + \frac{5}{x}\right) = \lim_{x \to +\infty} 2 + \lim_{x \to +\infty} \frac{5}{x}$$

$\lim_{x \to +\infty} b = b$, *limit of product = product*
of limits

$$= 2 + 5 \lim_{x \to +\infty} \frac{1}{x}$$

$$= 2 + 0 = 2$$

and

$$\lim_{x \to +\infty} \left(7 - \frac{3}{x}\right) = \lim_{x \to +\infty} 7 + \lim_{x \to +\infty} \frac{-3}{x}$$

$$= 7 - 3 \lim_{x \to +\infty} \frac{1}{x}$$

$$= 7 - 0 = 7.$$

Then,

$$\lim_{x \to +\infty} \frac{2x + 5}{7x - 5} = \lim_{x \to +\infty} \frac{2 + \dfrac{5}{x}}{7 - \dfrac{3}{x}}$$

limit of quotient = quotient of limits

$$= \frac{\lim_{x \to +\infty} \left(2 + \dfrac{5}{x}\right)}{\lim_{x \to +\infty} \left(7 - \dfrac{3}{x}\right)}$$

$$= \frac{2}{7}.$$

Example 6. Evaluate

$$\lim_{x \to +\infty} \frac{\sqrt[3]{x^3 + 1}}{3x - 5}, \qquad x \neq \frac{5}{3}.$$

Solution. We have

$$\frac{\sqrt[3]{x^3 + 1}}{3x - 5} = \frac{\dfrac{1}{x} \sqrt[3]{x^3 + 1}}{\dfrac{1}{x}(3x - 5)} = \frac{\sqrt[3]{1 + \dfrac{1}{x^3}}}{3 - \dfrac{5}{x}}.$$

Now

$$\lim_{x \to +\infty} \frac{1}{x^2} = \lim_{x \to +\infty} \left(\frac{1}{x} \cdot \frac{1}{x} \right) = \lim_{x \to +\infty} \frac{1}{x} \cdot \lim_{x \to +\infty} \frac{1}{x} = 0.$$

Hence,

$$\lim_{x \to +\infty} \frac{1}{x^3} = \lim_{x \to +\infty} \left(\frac{1}{x^2} \cdot \frac{1}{x} \right) = 0$$

and

$$\lim_{x \to +\infty} \left(1 + \frac{1}{x^3} \right) = \lim_{x \to +\infty} 1 + \lim_{x \to +\infty} \frac{1}{x^3} = 1 + 0 = 1.$$

Also,

$$\lim_{x \to +\infty} \left(3 - \frac{5}{x} \right) = \lim_{x \to +\infty} 3 + \lim_{x \to +\infty} \frac{-5}{x}$$

$$= 3 - 5 \lim_{x \to +\infty} \frac{1}{x} = 3 - 0 = 3.$$

Therefore,

$$\lim_{x \to +\infty} \frac{\sqrt[3]{x^3 + 1}}{3x - 5} = \lim_{x \to +\infty} \frac{\sqrt[3]{1 + \dfrac{1}{x^3}}}{3 - \dfrac{5}{x}}$$

$$= \frac{\lim\limits_{x \to +\infty} \sqrt[3]{1 + \dfrac{1}{x^3}}}{\lim\limits_{x \to +\infty} \left(3 - \dfrac{5}{x} \right)}$$

$$= \frac{1}{3}.$$

Returning to the function f defined by

$$f(x) = \frac{1}{x}$$

we find that (see Figure 6) as x approaches 0 from the right, $f(x)$ increases without bound. We write

$$\lim_{x \to 0^+} \frac{1}{x} = +\infty$$

and as x approaches 0 from the left, $f(x)$ decreases without bound.

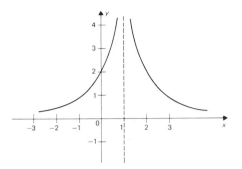

Figure 7

Graph of $f(x) = \dfrac{2}{(x-1)^2}$

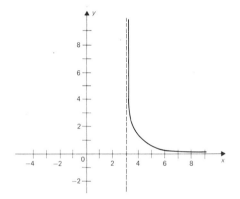

Figure 8

Graph of $f(x) = \dfrac{1}{\sqrt{x-3}}$

We write

$$\lim_{x \to 0^-} \frac{1}{x} = -\infty.$$

For the function whose graph is shown in Figure 7 we have

$$\lim_{x \to 1^+} f(x) = +\infty$$

and

$$\lim_{x \to 1^-} f(x) = +\infty.$$

For the function whose graph is shown in Figure 8, we have

$$\lim_{x \to 3^+} f(x) = +\infty$$

and since $f(x)$ is not defined for $x \le 3$, we say

$$\lim_{x \to 3^-} f(x) \text{ does not exist.}$$

The student should be careful when operating with limits involving the symbol ∞. It should be remembered that ∞ is *not* a number and cannot be treated as such.

We have the following rules of operation.

Rule 1. If $\lim\limits_{x \to c} f(x) = +\infty$ (or $-\infty$) and $\lim\limits_{x \to c} g(x) = b$, then

$$\lim_{x \to c} (f + g)(x) = +\infty \text{ (or } -\infty).$$

If $\lim\limits_{x \to c} g(x) = -\infty$ (or $+\infty$), nothing in general can be said about

$$\lim_{x \to c} (f + g)(x).$$

Further investigation of the particular problem will be necessary.

Rule 2. If $\lim\limits_{x \to c} f(x) = +\infty \; (-\infty)$ and $\lim\limits_{x \to c} g(x) = b$, then

(i) $\lim\limits_{x \to c} (fg)(x) = +\infty \; (-\infty)$ if $b > 0$

(ii) $\lim\limits_{x \to c} (fg)(x) = -\infty \; (+\infty)$ if $b < 0$.

If $b = 0$, further investigation is needed.

Rule 3. If $\lim\limits_{x \to c} f(x) = +\infty \; (-\infty)$ and $\lim\limits_{x \to c} g(x) = b$, then

$$\lim_{x \to c} \left(\frac{g}{f}\right)(x) = 0.$$

Exercise 2

In problems 1–14, apply Theorems 1 and 2 to find the indicated limits.

1. $\lim\limits_{x \to 2} (3 - 5x)$

2. $\lim\limits_{x \to -1} (5x^2 + 4)$

3. $\lim\limits_{x \to -3} (2x^3 - 3x^2 + 4x + 1)$

4. $\lim\limits_{x \to 4} \dfrac{x^2}{2x - 1}$

5. $\lim\limits_{x \to 2} \dfrac{x^3 + 5}{x^2 - 3x}$

6. $\lim\limits_{x \to 4} \dfrac{2x^2 - 7}{x^3 + 3}$

7. $\lim\limits_{x \to -2} \dfrac{x + 5}{2x^2 - 7x + 5}$

8. $\lim\limits_{x \to 2} \sqrt{\dfrac{3x + 1}{2x - 2}}$

9. $\lim\limits_{x \to 1} \sqrt{\dfrac{2x^2 + 3x - 1}{x^2 + 1}}$

10. $\lim\limits_{x \to 2} \sqrt{\dfrac{x^2 - 4}{x^2 - 3x + 2}}$

11. $\lim\limits_{x \to -2} \sqrt{\dfrac{x^3 + 8}{2x^2 + x - 1}}$

12. $\lim\limits_{x \to 1} \sqrt{\dfrac{x^3 - 1}{x^2 - 1}}$

13. $\lim\limits_{x \to -4} \sqrt[3]{\dfrac{x^3 - 27}{x^2 + 3x + 9}}$

14. $\lim\limits_{x \to -1} \sqrt[3]{\dfrac{x + 1}{x^3 + 1}}$

15. Prove Theorem 2(i).
16. Prove Theorem 2(iii).
17. Prove Theorem 2(iv).

In problems 18–32, evaluate the indicated limits.

18. $\lim\limits_{x \to +\infty} \dfrac{5x - 1}{-x + 4}$

19. $\lim\limits_{x \to +\infty} \dfrac{x^2 + 3x + 7}{3x^2 - 2x - 1}$

20. $\lim\limits_{x \to +\infty} \dfrac{2x + 5}{x^2 + 1}$

21. $\lim\limits_{x \to +\infty} \dfrac{2x^2 + 4x - 1}{x^3 + 4}$

22. $\lim\limits_{x \to +\infty} \dfrac{\sqrt{x^2 + 6}}{x + 6}$

23. $\lim\limits_{x \to -\infty} \dfrac{\sqrt{x^2 + 6}}{x + 6}$

24. $\lim\limits_{x \to +\infty} \dfrac{2 - x^2}{3x + 5}$

25. $\lim\limits_{x \to +\infty} \dfrac{2x^2 - 3x + 4}{x + 4}$

26. $\lim\limits_{x \to 2} \dfrac{x}{x - 2}$

27. $\lim\limits_{x \to 3} \dfrac{x^2}{9 - x^2}$

28. $\lim\limits_{x \to 1^+} \dfrac{x + 1}{x^2 - 1}$

29. $\lim\limits_{x \to 1^-} \dfrac{x + 1}{x^2 - 1}$

30. $\lim\limits_{x \to 3^-} \dfrac{\sqrt{9 - x^2}}{x - 3}$

31. $\lim\limits_{x \to 0} \left(\dfrac{1}{x} - \dfrac{1}{x^2} \right)$

32. $\lim\limits_{x \to 3} \left(\dfrac{1}{x - 3} - \dfrac{2}{x^2 - 9} \right)$

3. Continuity of a Function

Let us consider the functions whose graphs are illustrated in Figures 9–12. In each case we are interested in what happens to the graph at the point $x = c$. In the first three graphs we note that there is a break in the curve at c. In other words, if you move your pencil along the curve, you will have to lift the pencil at $x = c$. We say that the functions involved are *discontinuous* at $x = c$. On the other hand, you could move your pencil along the curve in Figure 12 near c without lifting the pencil at c. We say that the function whose graph is shown in Figure 12 is *continuous* at $x = c$.

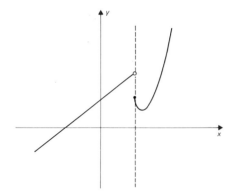

Figure 9

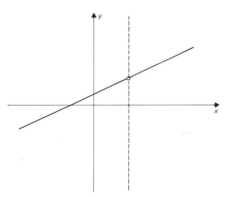

Figure 10

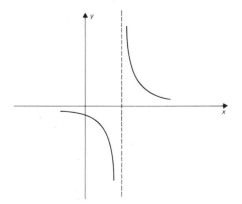

Figure 11

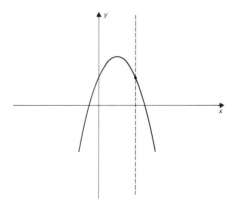

Figure 12

Bernhard Bolzano (1781–1848), an Austrian mathematician and theologian, was a priest and a professor of religious philosophy at the University of Prague. In 1820 he was dismissed from his post by the Austrian government because his sermons were considered subversive. His mathematical work Paradoxes of Infinity *was published after his death.*

Karl W. T. Weierstrass (1815–1897) was born in Ostenfelde, Germany, the eldest in a family of two brothers and two sisters. At nineteen he entered the University of Bonn, where he indulged in fencing and beer drinking; he returned home after four years without a degree. At the age of twenty-six he took his teachers' certificate examinations and then taught in secondary schools until 1853.

Weierstrass is called the father of modern analysis. His early publications in an unknown high school paper went unnoticed. In 1854 he published a memoir on Abelian functions which created a sensation and won him an honorary doctorate from the University of Königsberg. He was elected to the Berlin Academy and was given a professorship at the University of Berlin, where he stayed until his death. His fame as an outstanding mathematician and an excellent teacher spread all over Europe and to America. He emphasized rigor and logic.

The mathematicians Bolzano and Weierstrass offered definitions of the concept of continuity, which as we have seen is connected with the notion of unbroken curves. Bolzano's definition was imprecise. Weierstrass offered the more rigorous definition, which is essentially the one that is used today.

We now see that continuity of a function at *c* is closely related to the limit of a function at $x = c$.

Definition 3. A function *f* is said to be *continuous* at the number *c* if and only if

(i) $f(c)$ **exists, that is, *f* is defined at *c***
(ii) $\lim\limits_{x \to c} f(x)$ **exists**
(iii) $\lim\limits_{x \to c} f(x) = f(c)$.

When a function *f* is not continuous, we say that *f* is *discontinuous* at *c*. If a function is continuous at every point in an open interval *I*, it is said to be *continuous on I*.

Using Definition 3, we shall now consider some examples.

Example 1. Let *f* be the function defined by

$$f(x) = \begin{cases} 1, & x \geq 2 \\ -1, & x < 2. \end{cases}$$

Is *f* continuous at $x = 2$?

Solution. Figure 13 illustrates the graph of *f*. Here $f(2) = 1$ and so condition (i) is satisfied. Now

$$\lim_{x \to 2^+} f(x) = 1, \qquad \lim_{x \to 2^-} f(x) = -1.$$

Hence, $\lim\limits_{x \to 2} f(x)$ does not exist and so condition (ii) is not satisfied. Therefore, *f* is discontinuous at $x = 2$.

Example 2. Let *f* be the function defined by

$$f(x) = \begin{cases} \dfrac{(x + 2)(x - 1)}{x - 1}, & x \neq 1 \\ \\ 2, & x = 1. \end{cases}$$

Is *f* continuous at $x = 1$?

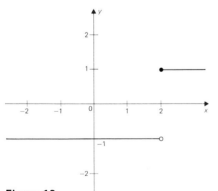

Figure 13

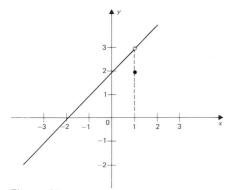

Figure 14

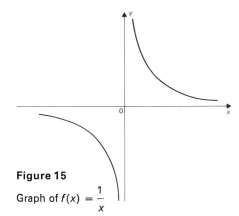

Figure 15

Graph of $f(x) = \dfrac{1}{x}$

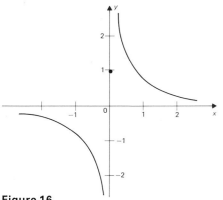

Figure 16

Solution. A graph of f is shown in Figure 14. Here $f(1) = 2$. Furthermore,

$$\lim_{x \to 1^+} f(x) = 3 \text{ and } \lim_{x \to 1^-} f(x) = 3.$$

Hence,

$$\lim_{x \to 1} f(x) = 3.$$

Thus conditions (i) and (ii) are satisfied. However,

$$\lim_{x \to 1} f(x) \neq f(1).$$

Therefore, f is discontinuous at $x = 1$.

Example 3. Let the function f be defined by

$$f(x) = \frac{1}{x}.$$

Is f continuous at $x = 0$?

Solution. Since $f(0)$ is not defined, condition (i) is not satisfied and so f is discontinuous at $x = 0$.

Example 4. Let f be defined by

$$f(x) = \begin{cases} \dfrac{1}{x}, & x \neq 0 \\ 1, & x = 0. \end{cases}$$

Is f continuous at $x = 0$?

Solution. Here $f(0) = 1$ and so condition (i) is satisfied. Now,

$$\lim_{x \to 0^+} f(x) = +\infty \text{ and } \lim_{x \to 0^-} f(x) = -\infty.$$

Therefore, condition (ii) is not satisfied and so f is discontinuous at $x = 0$.

Example 5. Let f be the function defined by

$$f(x) = \sqrt{x + 1}.$$

Is f continuous at $x = -1$?

Solution. A graph of f is shown in Figure 17. We find that $f(-1) = 0$. Now,

$$\lim_{x \to -1^+} f(x) = 0, \text{ but } \lim_{x \to -1^-} f(x) \text{ does not exist.}$$

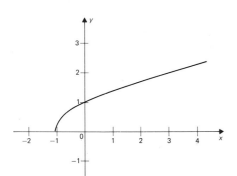

Figure 17

Note that f is continuous at every other point in I_2. But still f is not continuous on I_2.

Sonja Kowalewski (1850–1891) was born into a prosperous family at Moscow. At fifteen she began studying mathematics and later went to the University at Heidelberg, where she attended lectures given by Kirchhoff and Helmholtz. In 1868 after a nominal marriage she left her husband in Russia and returned to Germany to study. At twenty Kowalewski went to Berlin to study under Weierstrass, who tutored her at his home after trying in vain to get the University Senate to admit her to his lectures. She received her degree in absentia from Göttingen (1874) and returned to Russia. After many trials Kowalewski obtained a position at the University of Stockholm where she stayed until her death. In 1888 her memoir "On the rotation of a solid body about a fixed point" earned the Bordin Prize of the French Academy of Sciences.

Hence,

$$\lim_{x \to -1} f(x) \text{ does not exist}$$

and so condition (ii) is not satisfied. Therefore, *f* is discontinuous at $x = -1$.

Example 6. Let *f* be the function defined by

$$f(x) = \frac{1}{x + 1}$$

(a) Is *f* continuous on $I_1 = \{x \mid 0 < x < 4\}$?
(b) Is *f* continuous on $I_2 = \{x \mid -3 < x < 4\}$?

Solution. (a) Let *c* be any number on I_1. Then $f(c) = 1/(c + 1)$. Also,

$$\lim_{x \to c} f(x) = \lim_{x \to c} \frac{1}{x + 1} = \frac{1}{c + 1}, \; c \neq -1.$$

Hence

$$\lim_{x \to c} f(x) = f(c).$$

Conditions (i), (ii), and (iii) are satisfied for every *c* in I_1. Therefore, *f* is continuous on I_1.

(b) Here -1 is in I_2 and $f(-1)$ is not defined. Thus, *f* is not continuous at -1. Therefore, *f* is not continuous on I_2. We have the following theorem.

Theorem 4. **Suppose *f* and *g* are functions both continuous at *c*. Then**

(i)   ***f + g* is continuous at *c***
(ii) ***fg* is continuous at *c***
(iii) ***f/g* is continuous at *c* provided $g(c) \neq 0$.**

Proof. We shall prove part (i) and leave parts (ii) and (iii) as exercises for the student.
By Definition 3 we have

$$\lim_{x \to c} f(x) = f(c) \text{ and } \lim_{x \to c} g(x) = g(c).$$

Applying Theorem 2(i) we have

$$\lim_{x \to c} (f + g)(x) = \lim_{x \to c} f(x) + \lim_{x \to c} g(x)$$
$$= f(c) + g(c)$$
$$= (f + g)(c).$$

Thus Definition 3 is satisfied and the function $f + g$ is continuous at c.

Exercise 3

In problems 1–14, sketch the graphs of each of the functions defined and determine if each is continuous at the given value c.

1. $f(x) = \dfrac{3}{x - 2}$,   at $c = 2$

2. $f(x) = \begin{cases} \dfrac{1}{x + 4}, & x \neq -4 \\ 0, & x = -4 \end{cases}$ at $c = -4$

3. $f(x) = \begin{cases} \dfrac{x^2 - 3x + 2}{x - 1}, & x \neq 1 \\ -5, & x = 1 \end{cases}$ at $c = 1$

4. $f(x) = \sqrt{3x - 1}$,   at $c = \frac{1}{3}$

5. $f(x) = \begin{cases} 2x + 1, & x \leq 1 \\ -x + 4, & x > 1 \end{cases}$ at $c = 1$

6. $f(x) = \begin{cases} 2x + 1, & x \leq 2 \\ -x + 4, & x > 2 \end{cases}$ at $c = 2$

7. $f(x) = |x - 2|$,   at $c = 2$

8. $f(x) = \begin{cases} |x - 2|, & x \neq 2 \\ 1, & x = 2 \end{cases}$ at $c = 2$

9. $f(x) = \begin{cases} \dfrac{x^2 - 9}{x + 3}, & x \neq -3 \\ 3, & x = -3 \end{cases}$ at $c = -3$

10. $f(x) = \begin{cases} \dfrac{x^2 - 4}{x - 2}, & x \neq 2 \\ 4, & x = 2 \end{cases}$ at $c = 2$

11. $f(x) = \sqrt[3]{x}$,   at $c = -1$

12. $f(x) = \sqrt[3]{x - 1}$,   at $c = 1$

13. $f(x) = \begin{cases} \dfrac{x - 4}{x^2 - 16}, & x \neq \pm 4 \\ f(4) = \frac{1}{8}, \\ f(-4) = 0, \end{cases}$ at $c = 4, -4$

14. $f(x) = \begin{cases} x + 2, & x \leq -1 \\ x^2, & -1 < x < 1 \\ 3x + 1, & x \geq 1 \end{cases}$ at $c = -1, 1$

In problems 15–22, a function f is defined on an interval I. Determine whether f is continuous on I.

15. $f(x) = \dfrac{1}{x^2 + 1}$, $I = \{x \mid -1 < x < 5\}$

16. $f(x) = \dfrac{x}{x - 1}$, $I = \{x \mid -3 < x < 1\}$

17. $f(x) = \dfrac{x}{x - 1}$, $I = \{x \mid -1 < x < 4\}$

18. $f(x) = \dfrac{x^2 + 3x - 1}{2x^2 + 3}$, $I = \{x \mid -2 < x < 2\}$

19. $f(x) = \begin{cases} \dfrac{x^2 + x - 2}{x + 2}, & x \neq -2 \\ -3, & x = -2 \end{cases}$ $I = \{x \mid -3 < x < 3\}$

20. $f(x) = \begin{cases} \dfrac{x^3 - 1}{x - 1}, & x \neq 1 \\ f(1) = 2, \end{cases}$ $I = \{x \mid 0 < x < 3/2\}$

21. $f(x) = \begin{cases} \dfrac{4x^2 - 1}{2x + 1}, & x \neq -\frac{1}{2} \\ f(-\frac{1}{2}) = -2, \end{cases}$ $I = \{x \mid -1 < x < 1\}$

22. $f(x) = \dfrac{x + 2}{x^2 - 4}$, $I = \{x \mid 0 < x < 2\}$

In problems 23–26, define $f(c)$ so that the function f is continuous at c.

23. $f(x) = \dfrac{(x - 1)(x + 2)}{x - 1}$, $x \neq 1$, $c = 1$

24. $f(x) = \dfrac{x^2 - 3x + 2}{x^2 - 4}$, $x \neq \pm 2$, $c = 2$

25. $f(x) = \dfrac{x^2 - 1}{x + 1}$, $x \neq -1$, $c = -1$

26. $f(x) = \dfrac{x^3 + 8}{x + 2}$, $x \neq -2$, $c = -2$

27. In problem 24 can $f(-2)$ be defined so that the function is continuous at -2?

28. Prove Theorem 4(ii).

29. Prove Theorem 4(iii).

30. Suppose f is defined on the closed interval $I = \{x \mid a \leq x \leq b\}$. We say that the function f is continuous from the right at a if and only if $f(a)$ is defined and $\lim\limits_{x \to a^+} f(x) = f(a)$.

   (i) Define continuity of the function f from the left at b.

   (ii) Define continuity of the function f on I.

**31. Suppose f is continuous at c and $f(c) > 0$ $(f(c) < 0)$. Show that there exists an open interval I containing c such that $f(x) > 0$ $(f(x) < 0)$ for every x in I.

Karl W. T. Weierstrass

5

The

Derivative

1. The Derivative

In this chapter we shall study one of the most basic concepts in the development of the calculus: the *derivative* of a function. The derivative has become a very valuable tool in the sciences, both physical and social. Before we formulate a definition of this concept, we shall illustrate a few concepts needed to build up the definition. In later sections we shall indicate how the derivative is applied in various disciplines and we will also develop some of its basic properties.

Consider the function f defined by

$$f(x) = x^2.$$

When $x = 10$, we get $f(10) = 10^2 = 100$. Now, suppose we increase the value of x from 10 to 10.1. Then, $f(10.1) = (10.1)^2 = 102.01$. The change in the value of x is

$$h = 10.1 - 10 = 0.1$$

and the corresponding change in $f(x)$ is

$$f(10.1) - f(10) = 102.01 - 100 = 2.01.$$

The ratio of the change in $f(x)$ divided by the change in x is called the *difference quotient:*

$$(1) \quad \text{difference quotient} = \frac{\text{change in } f(x)}{\text{change in } x}$$

$$= \frac{f(x+h) - f(x)}{(x+h) - x} = \frac{f(x+h) - f(x)}{h}.$$

This is the *average rate of change* of f from x to $x + h$. In the above case we have

$$\text{difference quotient} = \frac{f(10.1) - f(10)}{0.1} = 20.1.$$

If the value of x changes from 10 to 9.9, the change in x is $h = 9.9 - 10 = -0.1$, a negative number; and the corresponding change in $f(x)$ is

$$f(9.9) - f(10) = 98.01 - 100 = -1.99,$$

so that

$$\text{difference quotient} = \frac{f(9.9) - f(10)}{-0.1} = \frac{-1.99}{-0.1} = 19.9.$$

Table 1 contains more difference quotients corresponding to different changes in the values of x.

Table 1

h (change in x)	$f(10 + h) - f(10)$	$\dfrac{f(10 + h) - f(10)}{h}$
$h = 10.01 - 10 = 0.01$	0.2001	20.01
$h = 10.001 - 10 = 0.001$	0.020001	20.001
$h = 10.0001 - 10 = 0.0001$	0.00200001	20.0001
$h = 9.99 - 10 = -0.01$	-0.1999	19.99
$h = 9.999 - 10 = -0.001$	-0.019999	19.999
$h = 9.9999 - 10 = -0.0001$	-0.00199999	19.9999

It is clear from the above table that as the change in x approaches zero (through positive or negative values), the difference quotient approaches the number 20.

We observe that

$$\lim_{h \to 0} \frac{f(10 + h) - f(10)}{h} = \lim_{h \to 0} \frac{(10 + h)^2 - 10^2}{h}$$

$$= \lim_{h \to 0} \frac{100 + 20h + h^2 - 100}{h}$$

$$= \lim_{h \to 0} (20 + h) = 20.$$

Let f be a function and x a fixed number in an open interval I

in the domain of *f*. If we choose *h* such that $x + h$ is in *I*, then we have a new function *F* where

F depends on x as well.

$$F(h) = \frac{f(x + h) - f(x)}{h}.$$

Here *F* is not defined for $h = 0$. Furthermore, as we have seen above and in Chapter 4, $\lim_{h \to 0} F(h)$ may not exist. This limit, if it exists, will (in general) depend upon *x* and consequently will be a function of *x*. We call this new function the *derived* function of *f* or simply the *derivative* of *f*.

We now give a formal definition of the derivative of a function.

Definition 1. **Let *f* be a function and *x* be a point in an open interval in the domain of *f*. The *derivative* of *f* at the point *x*, denoted by *f′*(*x*), is defined by**

Note that h ≠ 0.

(2) $$f'(x) = \lim_{h \to 0} \frac{f(x + h) - f(x)}{h}$$

f′ is a new function obtained from f by the process of differentiation.

provided the limit exists.

If the limit (2) exists, the function *f* is said to be differentiable at *x*.

Joseph Louis Lagrange (1736–1813), born in Turin, Italy, was a genius in generalizations and analysis. He did his most important mathematical work in partial differential equations, and he contributed immensely to the development of the calculus of variations.

At nineteen Lagrange became professor of mathematics at the Royal Artillery School of Turin. At twenty-eight he won the coveted prize of mathematics of the Academy of Sciences of Paris, a prize he was to win several times in the succeeding years. In 1766 at the invitation of Frederick II of Prussia he replaced Euler as director of the mathematics department of the Academy of Berlin. In 1787 at the invitation of Louis XVI Lagrange went to Paris to work at the French Academy. Napoleon treated Lagrange as a friend, giving him the title of count and making him a senator and a high officer of the Legion of Honor.

Notations. The derivative notation *f′* was introduced by Lagrange. For a function *f* defined by the equation $y = f(x)$, Leibniz denoted the derivative of *f* by

$$\frac{dy}{dx} \quad \text{and} \quad \frac{df}{dx},$$

read "the derivative of *y* (or *f*) with respect to *x*." Other notations used are

$$f', \; Dy, \; Df, \; y'.$$

When we wish to evaluate the derivative at a point *c* we write

$$f'(c) = \frac{dy}{dx}\bigg|_{x=c} = \frac{df}{dx}\bigg|_{x=c} = y'(c) = Dy|_{x=c} = Df|_{x=c}.$$

Example 1. Let *f* be the function defined by

$$f(x) = x^3.$$

Find $f'(x)$, $f'(2)$, $f'(-4)$, and $f'(c)$.

Solution. For $h \neq 0$ we set up the difference quotient

Note. $(a + b)^3 = a^3 + 3a^2b + 3ab^2 + b^3$

$$\frac{f(x + h) - f(x)}{h} = \frac{1}{h}[(x + h)^3 - x^3]$$

$$= \frac{1}{h}[x^3 + 3x^2h + 3xh^2 + h^3 - x^3]$$

$$= \frac{1}{h}[3x^2h + 3xh^2 + h^3]$$

$$= 3x^2 + 3xh + h^2.$$

Thus,

Using Definition 1

$$f'(x) = \lim_{h \to 0} \frac{f(x + h) - f(x)}{h}$$

$$= \lim_{h \to 0} (3x^2 + 3xh + h^2)$$

Application of Theorems 1 and 2 of Chapter 4

$$= 3x^2,$$

and so,

$$f'(2) = 3(2)^2 = 12$$
$$f'(-4) = 3(-4)^2 = 48$$
$$f'(c) = 3(c)^2 = 3c^2.$$

Example 2. Let f be the function defined by

$$f(x) = \frac{1}{x}, x > 0.$$

Find $f'(x)$.

Solution. Using Definition 1, we have for $h \neq 0$,

Since $x + h$ must be in the domain of f, $x + h \neq 0$. (Why?)

$$f'(x) = \lim_{h \to 0} \frac{f(x + h) - f(x)}{h}$$

Note. $\dfrac{1}{x + h} - \dfrac{1}{x} = \dfrac{x - (x + h)}{x(x + h)} = \dfrac{-h}{x(x + h)}$

$$= \lim_{h \to 0} \frac{1}{h}\left[\frac{1}{x + h} - \frac{1}{x}\right]$$

$$= \lim_{h \to 0} \frac{-1}{x(x + h)} = -\frac{1}{x^2}.$$

Example 3. Let f be the function defined by $f(x) = \sqrt{x}$. Find $f'(x)$.

Solution. We have

$$f'(x) = \lim_{h \to 0} \frac{f(x + h) - f(x)}{h}$$

We observe that the indicated limit yields the indeterminate form 0/0. A useful "trick" at this point is to rationalize the numerator:

$$\sqrt{x+h} - \sqrt{x} = (\sqrt{x+h} - \sqrt{x}) \times$$

$$\left(\frac{\sqrt{x+h} + \sqrt{x}}{\sqrt{x+h} + \sqrt{x}}\right) = \frac{(\sqrt{x+h})^2 - (\sqrt{x})^2}{\sqrt{x+h} + \sqrt{x}}$$

$$= \frac{h}{\sqrt{x+h} + \sqrt{x}}$$

$$= \lim_{h \to 0} \frac{1}{h} [\sqrt{x+h} - \sqrt{x}]$$

$$= \lim_{h \to 0} \frac{1}{h} \left[\frac{h}{\sqrt{x+h} + \sqrt{x}}\right]$$

$$= \lim_{h \to 0} \frac{1}{\sqrt{x+h} + \sqrt{x}} = \frac{1}{2\sqrt{x}}.$$

Example 4. Find $f'(x)$ if $f(x) = 3x^2 + 4x + 8$.

Solution. We have

$$f'(x) = \lim_{h \to 0} \frac{f(x+h) - f(x)}{h}, \; h \neq 0$$

$$= \lim_{h \to 0} \frac{1}{h} \{[3(x+h)^2 + 4(x+h) + 8] - [3x^2 + 4x + 8]\}$$

$$= \lim_{h \to 0} \frac{1}{h} [3x^2 + 6xh + 3h^2 + 4x + 4h + 8 - 3x^2 - 4x - 8]$$

$$= \lim_{h \to 0} \frac{1}{h} [6xh + 3h^2 + 4h]$$

$$= \lim_{h \to 0} [6x + 3h + 4] = 6x + 4.$$

Example 5. Let $f(x) = \sqrt[3]{x}$. Show that $f'(0)$ does not exist.

Solution. We point out that f is defined for all real numbers. Now, for $h \neq 0$ we have

$$\lim_{h \to 0} \frac{f(0+h) - f(0)}{h} = \lim_{h \to 0} \frac{1}{h} [\sqrt[3]{0+h} - \sqrt[3]{0}]$$

$$= \lim_{h \to 0} \frac{\sqrt[3]{h}}{h}$$

$$= \lim_{h \to 0} \frac{1}{\sqrt[3]{h^2}}.$$

Clearly, this limit does not exist and so $f(x) = \sqrt[3]{x}$ has no derivative at $x = 0$.

Exercise 1

In problems 1–14, find $f'(x)$ and indicate the domain in which the function f is differentiable.

1. $f(x) = x$ **2.** $f(x) = x^2$

3. $f(x) = 5x^2$ **4.** $f(x) = 7$

5. $f(x) = 3x^2 + 2$

6. $f(x) = x^2 + 3x + 6$

7. $f(x) = \dfrac{2}{x}$

8. $f(x) = \dfrac{1}{x + 3}$

9. $f(x) = x + \dfrac{1}{x}$

10. $f(x) = \dfrac{x}{x^2 + 1}$

11. $f(x) = \dfrac{1}{\sqrt{x}}$

12. $f(x) = x - \sqrt{x}$

13. $f(x) = \sqrt{x^2 + 1}$

14. $f(x) = x^3 + \dfrac{1}{\sqrt{x}}$

In problems 15–20, find the values of x for which $f'(x) = 0$, if any. Sketch a graph of f in each case.

15. $f(x) = 3x^2$

16. $f(x) = 2x^2 - 4x$

17. $f(x) = 2x^2 + 4x$

18. $f(x) = \sqrt{x + 1}$

19. $f(x) = x^3$

20. $f(x) = \dfrac{1}{x}$

*21. The limits

$$\lim_{h \to 0^+} \frac{f(c + h) - f(c)}{h} = f'_+(c)$$

$$\lim_{h \to 0^-} \frac{f(c + h) - f(c)}{h} = f'_-(c),$$

if they exist, are called the *right derivative* and the *left derivative* of f at a point $x = c$.

(a) Show that f is differentiable at $x = c$ if and only if $f'_+(c)$ and $f'_-(c)$ exist and $f'_+(c) = f'_-(c)$.

(b) Prove that $f(x) = |x|$ does not have a derivative at $x = 0$. *Hint:* Show that $f'_+(0) \neq f'_-(0)$.

*22. Prove that if f has a derivative at x, then f is continuous at x.

2. Some Applications

The notion of the derivative leads to the definition of the "rate of change of a function." There are numerous applications of this notion in a wide variety of disciplines: the biological, social, and physical sciences, and economic theory, among others. Here we shall illustrate some of these applications.

We assume that no liquid is leaving the lake. We disregard loss of water due to evaporation, or the addition of water due to rain, etc.

Example 1. *The derivative in ecology.* Suppose a lake is being polluted by a liquid chemical. At any time t (minutes), the quantity of chemical in the lake is $Q(t)$ (gallons). Hence Q is a function of time. As the time changes from t to $t + h$ ($h > 0$), the quantity of chemical in the lake changes by an amount

$$(3) \qquad Q(t + h) - Q(t)$$

and the average rate of change of Q from t to $t + h$ is then

$$(4) \qquad \frac{Q(t + h) - Q(t)}{(t + h) - t} = \frac{Q(t + h) - Q(t)}{h}, h \neq 0.$$

The student should recognize this as the difference quotient. Now, we define the *instantaneous rate of change* (or simply the rate of change) of Q at time t to be

$$(5) \qquad Q'(t) = \lim_{h \to 0} \frac{Q(t + h) - Q(t)}{h},$$

which is simply the derivative of Q with respect to t. At a given time t^*, the number $Q'(t^*)$ represents the rate at which the amount of undesirable chemical in the lake is increasing.

Example 2. *The derivative in economic theory.* The total cost C of producing a commodity is often found to be dependent on the number x of the item produced. That is, C is a function of x, $C(x)$. If the number of the item produced is changed from x to $x + h$, then the average rate of change of cost is given by

In reality $x \in N$. However, in analysis of this type the function C is approximated by a continuous function of x.

$$(6) \qquad \frac{C(x + h) - C(x)}{(x + h) - x} = \frac{C(x + h) - C(x)}{h}, h \neq 0$$

and the rate of change of the cost is given by

$$(7) \qquad C'(x) = \lim_{h \to 0} \frac{C(x + h) - C(x)}{h}.$$

This is called the *marginal cost*, which is simply the derivative of the cost C with respect to x.

marginal cost $= C'(x)$

$$= \lim_{h \to 0} \frac{C(x + h) - C(x)}{h}, h \neq 0$$

We shall assume that the manufacturer is interested in revenue and profit. To sell his output of x items he charges a price p (dollars) per item. It is obvious that the price will depend on the number of items that can be sold. His *total revenue* is given by

Note that in general the higher the price the smaller the number of items sold.

$$(8) \qquad R(x) = xp(x)$$

and the *marginal revenue* (or the rate of increase of revenue per unit increase in output) is given by

$$(9) \qquad \text{marginal revenue} = \lim_{h \to 0} \frac{R(x + h) - R(x)}{h}, h \neq 0$$

$$= R'(x).$$

It is easy to see that the profit P is given by

(10)
$$P(x) = R(x) - C(x).$$

In the next chapter, the reader will be shown how the manufacturer who has had an elementary course in calculus can then find out how to maximize his profit by using certain properties of the derivative.

Example 3. *The derivative in the physical sciences.* Suppose the distance s traveled by a moving object is a function of time. Say

(11)
$$s = f(t)$$

where t is time measured in seconds and s is the distance traveled in feet. The average rate of change of s from t to $t + h$ is

(12)
$$V_{ave}(h) = \frac{f(t + h) - f(t)}{h}, h \neq 0.$$

This is called the *average velocity* V_{ave}. The instantaneous velocity (or the rate of change of s at time t) is

(13)
$$V = \lim_{h \to 0} V_{ave}(h)$$
$$= \lim_{h \to 0} \frac{f(t + h) - f(t)}{h}.$$

For example, in an experiment conducted above the moon's surface it is found that an object thrown from a spacecraft travels according to the law

(14)
$$s = kt^2 + t$$

where k (ft/sec^2) is a constant, s (ft) is the distance of the object from the spacecraft, and t (sec) is the time. As the time changes from t to $t + h$ we find that

$$V_{ave} = \frac{f(t + h) - f(t)}{h}$$
$$= \frac{[k(t + h)^2 + (t + h)] - [kt^2 + t]}{h}$$

(15)
$$= 2kt + kh + 1.$$

Starting at $t = 0$ with $h = 0.5$ we obtain

(16)
$$V_{ave} = 2k(0) + k(0.5) + 1$$
$$= 0.5k + 1.$$

The following table illustrates the changes in the average velocity over shorter and shorter time intervals.

Table 2

$t + h$	h	$V_{ave} = 2kt + kh + 1$
$0 + 0.5$	0.5	$k(0.5) + 1$
$0 + 0.1$	0.1	$k(0.1) + 1$
$0 + 0.05$	0.05	$k(0.05) + 1$
$0 + 0.01$	0.01	$k(0.01) + 1$
$0 + 0.001$	0.001	$k(0.001) + 1$

It is clear that as h approaches zero the average velocity approaches 1. This, then, is the instantaneous velocity at $t = 0$. That is,

$$(17) \qquad V(t) = \lim_{h \to 0} V_{ave}(h)$$
$$= \lim_{h \to 0} [2kt + kh + 1]$$
$$= 2kt + 1.$$

At $t = 0$ we have

$$V(0) = 2k(0) + 1 = 1.$$

This implies that the object left the spacecraft with an initial instantaneous velocity of 1 ft/sec.

Example 4. *The derivative in the biological sciences.* Consider a colony of bacteria introduced to a food supply in a jar. It is found that at the beginning there is no change in the number of bacteria (period of adaptation). Then there is a rapid increase in the number of bacteria (period of reproduction), and once the food supply is exhausted the bacteria begin to die. In such a model the number N of live bacteria is dependent on time t.

Now, the average rate of change of N from t to $t + h$ is

$$(18) \qquad \frac{N(t + h) - N(t)}{h}, h \neq 0$$

and the instantaneous rate of change of N at the point t is

$$(19) \qquad N'(t) = \lim_{h \to 0} \frac{N(t + h) - N(t)}{h}.$$

For studies of this type, the change in N is assumed to be continuous. In reality, the change in N occurs in jumps, and the function describing the model is usually approximated by a continuous function.

In practice it is often deduced from experimental data that at time t the instantaneous rate of increase is directly proportional to the number present. Thus we have the equation

$$N'(t) = k N(t)$$

where k is a constant of proportionality. Equations of this type, called differential equations, will be discussed in later chapters.

Exercise 2

8 miles

1. A factory at the east end of the city of Polluteville starts operation at 7:00 A.M. Smoke emitted from the factory travels westward according to the law

$$s(t) = 100t + 40$$

where s (ft) is the distance from the edge of the smoke cloud to the factory at t minutes after 7:00 A.M.
   (a) Find the average velocity of the edge of the smoke cloud during the time interval h.
   (b) Find the instantaneous velocity at $t = 0$.
   (c) At what time will the smoke cloud reach the west end of town?

2. If in the problem above we have

$$s(t) = 3t^2 + 50t$$

   (a) At what velocity is the smoke moving when $t = 0$? $t = 10$ min? $t = 20$ min?
   (b) When the cloud reaches point A it is found that the instantaneous velocity is 650 ft/min. At what time will this occur?
   (c) How far is A from the factory?

3. Consider Example 3 and let

$$s(t) = 3t^2 + t.$$

   (a) Starting at $t = 1$ sec construct a table similar to Table 2.
   (b) Find the instantaneous velocity of the object at $t = 1$ sec.

4. If the total cost C (dollars) of producing x units of a product is given by

$$C(x) = 2x + 160$$

   (a) What is the average cost per unit if 80 units are made?
   (b) What is the average rate of change of cost?
   (c) What is the marginal cost of the 50th unit?   80th unit?

5. Repeat problem 4 if $C(x) = 2x^2$.

6. If the total cost C of making x gallons of oil is $C(x) = 20\sqrt{x} + 5$
   (a) Find the marginal cost at 160-gallon output.
   (b) At what level of output is the marginal cost $0.20? $0.40? *Hint:* Solve for x such that $C'(x) = \$0.20$.

7. A manufacturer sells x units of a product at a price of p dollars per unit where

$$p(x) = 100 + \frac{x}{50}.$$

(a) Find the total revenue if his total output is 2500 units.

(b) Find the marginal revenue when the output is 500 units, 1000 units, 2500 units.

(c) At what level of output is the marginal revenue 150 dollars per unit?

8. A colony of bacteria is sprayed with a bactericidal agent. Suppose N represents the number of viable bacteria per milliliter remaining t minutes after spraying, and $N(t) = 10^6(40 - t)^2$.

(a) What is the average rate of decrease in the number of viable bacteria during the first 10 minutes?   20 minutes?

(b) What is the instantaneous rate of decrease in N at the end of 10 minutes?   20 minutes?

(c) At what time t^* would all the bacteria be dead?

(d) Find $N'(t^*)$.

3. Rules of Differentiation

We have introduced the notion of the derivative of a function and indicated some of its applications. We also calculated the derivative of some simple functions by using the definition. Evidently finding the derivative of a function by direct use of the definition can be tedious. We will therefore develop standard formulas to simplify the task of finding derivatives.

If $f(x) = c$ for all x, then $f'(x) = 0$ for all x.

Theorem 1. **The derivative of a constant function is the constant function 0.**

Proof. A constant function is defined by

$$f(x) = c$$

for all $x \in R$ where c is a constant. Application of Definition 1 yields

Note. Since $f(x) = c$ for every x, $f(x + h) = c$.

$$f'(x) = \lim_{h \to 0} \frac{f(x + h) - f(x)}{h}, \, h \neq 0$$

$$= \lim_{h \to 0} \frac{c - c}{h}$$

$$= \lim_{h \to 0} \frac{0}{h} = 0.$$

This proves Theorem 1.

Theorem 2. **If *f* is the function defined by**

(20) $$f(x) = x^n$$

where *n* is a positive integer, then $f'(x) = nx^{n-1}$.

Proof. First we shall consider specific values of *n*. For $n = 1$, we have $f(x) = x$ and by Definition 1,

Here we have $f(x) = x^1$ and

$f'(x) = 1 \cdot x^{1-1} = x^0 = 1.$

$$f'(x) = \lim_{h \to 0} \frac{(x+h) - x}{h} = \lim_{h \to 0} \frac{h}{h} = \lim_{h \to 0} 1 = 1.$$

For $n = 2$, we have $f(x) = x^2$ and by Definition 1,

$$f'(x) = \lim_{h \to 0} \frac{(x+h)^2 - x^2}{h} = \lim_{h \to 0} \frac{x^2 + 2xh + h^2 - x^2}{h}$$

$$= \lim_{h \to 0} (2x + h) = 2x.$$

For $n = 3$, we have $f(x) = x^3$ and by Definition 1,

$$f'(x) = \lim_{h \to 0} \frac{(x+h)^3 - x^3}{h} = \lim_{h \to 0} \frac{x^3 + 3x^2h + 3xh^2 + h^3 - x^3}{h}$$

$$= \lim_{h \to 0} (3x^2 + 3xh + h^2) = 3x^2.$$

For *n* we have

This is an application of the binomial theorem (see Appendix 1).

$(x + h)^n = x^n + nx^{n-1}h$

$+ \dfrac{n(n-1)}{2} x^{n-2}h^2 + \cdots$

$+ \dfrac{n!}{(n-k)!\,k!} x^{n-k}h^k + \cdots + h^n$

where for any positive integer k, k! (read k factorial) is defined as

$k! = k \cdot (k-1) \cdot (k-2) \cdots 3 \cdot 2 \cdot 1.$

Thus,

$5! = 5 \cdot 4 \cdot 3 \cdot 2 \cdot 1 = 120.$

The binomial theorem was first proved by Isaac Newton.

$$f'(x) = \lim_{h \to 0} \frac{(x+h)^n - x^n}{h}$$

$$= \lim_{h \to 0} \frac{1}{h} \left[x^n + nx^{n-1}h + \frac{n(n-1)}{2} x^{n-2}h^2 + \cdots + h^n - x^n \right]$$

$$= \lim_{h \to 0} \left[nx^{n-1} + \frac{n(n-1)}{2} x^{n-2}h + \cdots + h^{n-1} \right].$$

Since all terms but the first contain a power of *h*, we obtain

(21) $$f'(x) = nx^{n-1}.$$

This proves Theorem 2.

It can be shown that Theorem 2 holds for any real number *n*. We shall assume this result. The following generalization of Theorem 2 can be obtained:

Theorem 3. **If *f* is the function defined by**

(22) $$f(x) = cx^n$$

Isaac Newton (1642–1727) was born on Christmas Day at Woolsthorpe, England. Newton's father died before the frail child's birth. He received his A.B. degree in 1664 at Trinity College in Cambridge. The Great Plague (1664–65) forced Newton to return to Woolsthorpe to meditate for two years, during which time he invented the calculus, proved that light is composed of different colors, and discovered the universal law of gravitation. In 1669 he became Professor of Mathematics at Trinity. He constructed a telescope and observed the satellites of Jupiter, and in 1686 he published (at Halley's expense) his masterpiece The Mathematical Principles of Natural Philosophy.

Newton had a strong interest in theology and in alchemy. In addition, he spent some time in Parliament and was made Master of the Mint in 1699. Four years later he was elected President of the Royal Society and was knighted by Queen Anne.

where n is a positive integer and c is a constant, then

$$(23) \qquad f'(x) = cnx^{n-1}.$$

The proof of this theorem is left as an exercise for the student. We also remark that Theorem 3 holds for any real number n.

Example 1. Find $f'(x)$ for each of the functions:

(a) $f(x) = x^{13}$ (b) $f(x) = 2x^{-4}$

(c) $f(x) = x^{2/3}$ (d) $f(x) = 5^3$.

Solution. By direct application of the above theorems, we have

(a) $f'(x) = 13x^{13-1} = 13x^{12}$

(b) $f'(x) = 2(-4)x^{-4-1} = -8x^{-5}$

(c) $f'(x) = \frac{2}{3}x^{2/3-1} = \frac{2}{3}x^{-1/3}$

(d) $f'(x) = 0.$

Note that in (d) f is a constant function and so $f'(x) = 0$ (*not* $3 \cdot 5^2$).

Another useful result is

Theorem 4. If u and v are functions differentiable at x and if the function f is defined by

$$(24) \qquad f(x) = u(x) + v(x),$$

then f is differentiable at x and

$$(25) \qquad f'(x) = u'(x) + v'(x).$$

Proof. We have

$$f'(x) = \lim_{h \to 0} \frac{f(x + h) - f(x)}{h}$$

$$= \lim_{h \to 0} \frac{1}{h} [(u + v)(x + h) - (u + v)(x)]$$

$$= \lim_{h \to 0} \frac{1}{h} [u(x + h) + v(x + h) - u(x) - v(x)]$$

$$= \lim_{h \to 0} \frac{1}{h} [u(x + h) - u(x) + v(x + h) - v(x)]$$

Here we are using Theorem 2 of Chapter 4.

$$= \lim_{h \to 0} \left[\frac{u(x + h) - u(x)}{h} + \frac{v(x + h) - v(x)}{h} \right]$$

$$= u'(x) + v'(x).$$

Example 2. Find the derivative of $f(x) = 5x^2 + 4x^7$.

Solution. Application of Theorems 3 and 4 yields

$$f'(x) = (5)(2)x + (4)(7)x^6$$
$$= 10x + 28x^6.$$

By an extension of Theorem 4, the derivative of the sum of any finite number of differentiable functions at the point x is the sum of the derivatives of the functions. That is, if

(26) $$f(x) = (u_1 + u_2 + u_3 + \cdots + u_n)(x),$$

then

(27) $$f'(x) = u_1'(x) + u_2'(x) + \cdots + u_n'(x).$$

Example 3. Suppose the price per unit of a product with demand x is $p(x) = 3x^2 + 5x + 7$. Find the total revenue and the marginal revenue.

Solution. We recall that the total revenue is given by

See equation (8).

$$R(x) = xp(x)$$
$$= x(3x^2 + 5x + 7)$$
$$= 3x^3 + 5x^2 + 7x$$

and the marginal revenue is

See equation (9).
Note. If $f(x) = 7x$, then $f'(x) = (7)(1)x^0 = 7$.

$$R'(x) = (3)(3)x^2 + (5)(2)x + 7$$
$$= 9x^2 + 10x + 7.$$

Example 4. A particle moves along a straight line according to the law

(28) $$s(t) = t^2 - 5t + 50$$

where s is the distance measured in feet and t is the time measured in seconds.
   (a) Find the expression for the velocity of the particle at any instant t.
   (b) At what distance from the starting point does the particle come to instantaneous rest?

Solution. We recall that $s'(t)$ gives the velocity of the particle at any instant t. From equation (28) we have

Here we are employing the extension of Theorem 4.

(29) $$s'(t) = 2t - 5.$$

Now the particle comes to instantaneous rest when the velocity is zero. Thus,

$$s'(t) = 0 \Rightarrow 2t - 5 = 0 \Rightarrow t = \tfrac{5}{2}.$$

Hence the particle comes to rest $\frac{5}{2}$ seconds after the start and then reverses its direction. The distance traveled in $\frac{5}{2}$ seconds from the starting point may be obtained from equation (20):

$$s(\tfrac{5}{2}) = (\tfrac{5}{2})^2 - 5(\tfrac{5}{2}) + 50 = \tfrac{25}{4} - \tfrac{25}{2} + 50 = \tfrac{175}{4} \text{ ft.}$$

Exercise 3

In problems 1 through 12, find the derivative of the given functions.

1. $f(x) = x^3 + 7x^2 - 9x$

2. $s(t) = t^{7/2} - 9t^3 + 3t + 8$

3. $f(x) = \dfrac{x^2 + 10x + 3}{5}$

4. $v(x) = -3x + 10{,}000$

5. $f(x) = 5x^{101} + 17x^2 + 8$

6. $f(x) = (x + 3)^2$

7. $w(s) = \left(s + \dfrac{1}{s}\right)^2$

8. $f(x) = \left(x + \dfrac{1}{\sqrt{x}}\right)^3$

9. $f(x) = (x + a)(x + b)(x + c)$ where a, b, and c are constants

10. $f(x) = \dfrac{5x^3 + 3x^2 + 4x - 2}{\sqrt{x}}$

11. $f(x) = \dfrac{(x^2 + 3)^2}{x^{3/2}}$

12. $f(x) = (x^2 + 3x + 5)(3x^2 - 2x + 4)$

13. Using the method of proof of Theorem 2, prove Theorem 3.

14. A brush fire in a valley spreads along a straight line according to the law

$$s(t) = \tfrac{3}{2}t - \tfrac{1}{4}t^2$$

where the distance s is measured in miles from a starting point and t is the time measured in hours.

(a) What is the instantaneous rate of spreading (velocity) of the fire at $t = 1$ hr? $t = 1.5$ hr? $t = 2$ hr?

(b) At what distance from the starting point does the fire stop spreading?

(c) If the fire spreads radially at the rate given above, find the area of the circular region devastated by the fire.

15. Suppose the total cost of making x units of a product is C dollars where

$$C(x) = 20 + 18x^{2/3}.$$

For what value of x will the marginal cost be \$2?

16. If the total cost C of making x units of a product is

$$C(x) = 6x - 0.03x^2 + 0.00005x^3,$$

construct a table showing the marginal cost at 50, 100, 150, 200, 250, and 300 units.

17. If the cost of making x units is C dollars, where

$$C(x) = 12x - 0.045x^2 + 0.0001 \frac{x^3}{3},$$

for what output will the marginal cost be \$22?

18. Suppose W is the number of gallons of water in a pool being drained for cleaning, and t minutes after the draining starts

$$W(t) = 300(40 - t)^2.$$

(a) Find the instantaneous rate at which the water is running out 5 min, 10 min, 20 min, and 30 min after the pool has started to drain.

(b) How many minutes after the pool has started to drain will the pool be empty?

19. Applying Newton's law of universal gravitation, it is found that the force F exerted by the Earth on a spacecraft at a distance r from the center of the Earth is given by

$$F(r) = \frac{-k}{r^2}$$

where k is a positive number. Construct a table to show the instantaneous rate of change of F with respect to r when r is 4000 mi, 5000 mi, 10,000 mi, 20,000 mi. (An approximate value for the radius of the Earth is 3959 miles.)

4. The Derivative of the Product and Quotient

In this section, we develop formulas for the derivative of the product of two functions and for the quotient of two functions.

Theorem 5. Let $f = uv$ and suppose u and v are functions differentiable at x. Then f is differentiable at x and its derivative at x is given by

(30) $$f'(x) = u(x)\, v'(x) + u'(x)\, v(x).$$

Proof. Setting up the difference quotient we have

$$\frac{f(x + h) - f(x)}{h} = \frac{u(x + h)\, v(x + h) - u(x)\, v(x)}{h}$$

Here we have added and subtracted the term

$$\frac{v(x)\,u(x+h)}{h}.$$

$$= \frac{u(x+h)\,v(x+h) - v(x)\,u(x+h) + v(x)\,u(x+h) - u(x)\,v(x)}{h}$$

$$= u(x+h)\left[\frac{v(x+h)-v(x)}{h}\right] + \left[\frac{u(x+h)-u(x)}{h}\right]v(x).$$

Now, we are given that

$$\lim_{h\to 0}\frac{v(x+h)-v(x)}{h} = v'(x)$$

and

$$\lim_{h\to 0}\frac{u(x+h)-u(x)}{h} = u'(x).$$

Also, it can be shown that

$$\lim_{h\to 0} u(x+h) = u(x).$$

Since u is differentiable at x, u is continuous at x (see problem 22, Exercise 1). Therefore,

$$\lim_{z\to a} u(z) = u(a).$$

Therefore,

$$f'(x) = \lim_{h\to 0}\frac{f(x+h)-f(x)}{h}$$

$$= \lim_{h\to 0}\left\{u(x+h)\left[\frac{v(x+h)-v(x)}{h}\right] + \left[\frac{u(x+h)-u(x)}{h}\right]v(x)\right\}$$

Here we have applied Theorem 2 of Chapter 4.

$$= u(x)\,v'(x) + u'(x)\,v(x).$$

The above theorem can be used to find the derivative of the product of three or more functions. For example, let

(31) $$f = uvw$$

and suppose u, v, and w are functions differentiable at x. Then

(32) $$f'(x) = u(x)\,v(x)\,w'(x) + u(x)\,v'(x)\,w(x) + u'(x)\,v(x)\,w(x).$$

Example 1. It is found experimentally that the change in the temperature T of the body due to a dose of a drug is a function of the amount x of the drug administered. Suppose for a given drug

(33) $$T(x) = x^2(k_1 - k_2 x)$$

where k_1 and k_2 are positive constants.
  (a) Find the rate of change of T with respect to the amount of the drug administered.
  (b) For what values of x is the rate of change in T zero?

Analysis of this type is used to determine the exact dosage of a drug to be administered to give the greatest change in the reaction of the body with respect to a change in dosage.

Solution. (a) Let $u(x) = x^2$ and $v(x) = k_1 - k_2 x$. Then

$$T(x) = u(x)\,v(x).$$

Application of Theorem 5 yields

$$T'(x) = u(x) v'(x) + u'(x) v(x)$$
$$= (x^2)(-k_2) + (2x)(k_1 - k_2x)$$
$$= 2k_1x - 3k_2x^2.$$

(b) The rate of change of T with respect to x is zero when

$$2k_1x - 3k_2x^2 = 0.$$

This occurs when $x = 0$ or $x = 2k_1/3k_2$.

Example 2. If $f(x) = (3x^3 + 5x^2 + 7)(2x^5 - 8x^3 + 6x)$, find $f'(x)$.

Solution. Let $u(x) = 3x^3 + 5x^2 + 7$ and $v(x) = 2x^5 - 8x^3 + 6x$.
Then

$$u'(x) = 9x^2 + 10x$$
$$v'(x) = 10x^4 - 24x^2 + 6.$$

Now,

$$f = uv$$

and by Theorem 5, we have

$$f'(x) = u(x) v'(x) + u'(x) v(x).$$

Substitution then yields

$$f'(x) = (3x^3 + 5x^2 + 7)(10x^4 - 24x^2 + 6)$$
$$+ (9x^2 + 10x)(2x^5 - 8x^3 + 6x).$$

Theorem 6. Let $f = u/v$ and suppose u and v are differentiable at x. Then f is differentiable at x and

(34)
$$f'(x) = \frac{u'(x) v(x) - u(x) v'(x)}{[v(x)]^2}$$

provided $v(x) \neq 0$.

Proof. Setting up the difference quotient we have

$$\frac{f(x + h) - f(x)}{h}$$

$$= \frac{1}{h}\left[\frac{u(x + h)}{v(x + h)} - \frac{u(x)}{v(x)}\right]$$

$$= \frac{1}{h}\left[\frac{u(x + h) v(x) - u(x) v(x + h)}{v(x + h) v(x)}\right]$$

Here we have added and subtracted the
term u(x) v(x)/h.

$$= \left[\frac{u(x + h)\, v(x) - u(x)\, v(x) + u(x)\, v(x) - u(x)\, v(x + h)}{h} \right]$$

$$\times \frac{1}{v(x + h)\, v(x)}$$

$$= \left\{ \left[\frac{u(x + h) - u(x)}{h} \right] v(x) - u(x) \left[\frac{v(x + h) - v(x)}{h} \right] \right\}$$

$$\times \frac{1}{v(x + h)\, v(x)} .$$

And so, we have

$$f'(x) = \lim_{h \to 0} \frac{f(x + h) - f(x)}{h}$$

Again we have used the property that
$$\lim_{h \to 0} v(x + h) = v(x).$$

$$= \lim_{h \to 0} \left\{ \left(\frac{u(x + h) - u(x)}{h} \right) v(x) - u(x) \left(\frac{v(x + h) - v(x)}{h} \right) \right.$$

$$\left. \times \frac{1}{v(x + h)\, v(x)} \right\}$$

$$= \frac{u'(x)\, v(x) - u(x)\, v'(x)}{[v(x)]^2} .$$

This proves Theorem 6.

Example 3. If $f(x) = (x^2 + 1)/(2x + 1)$, find $f'(x)$.

Solution. Let $u(x) = x^2 + 1$ and $v(x) = 2x + 1$. Then,

$$u'(x) = 2x \quad \text{and} \quad v'(x) = 2$$

and for $f = u/v$ we have

Application of Theorem 6

$$f'(x) = \frac{u'(x)\, v(x) - u(x)\, v'(x)}{[v(x)]^2}$$

This holds for $x \neq -1/2$.

$$= \frac{(2x)(2x + 1) - (x^2 + 1)(2)}{(2x + 1)^2}$$

$$= \frac{4x^2 + 2x - 2x^2 - 2}{(2x + 1)^2}$$

$$= \frac{2x^2 + 2x - 2}{(2x + 1)^2}$$

$$= \frac{2(x^2 + x - 1)}{(2x + 1)^2} .$$

Example 4. If $f(x) = (x^2 + 2)(x^3 + 1)/(x^2 - 2)$ and $x^2 \neq 2$, find $f'(x)$.

Solution. Set $f = u/v$ where

$$u(x) = (x^2 + 2)(x^3 + 1)$$

and

$$v(x) = x^2 - 2.$$

It is easily seen that $v'(x) = 2x$ and application of Theorem 5 yields

$$u'(x) = (x^2 + 2)(3x^2) + (x^3 + 1)(2x)$$
$$= 5x^4 + 6x^2 + 2x.$$

Now,

$$f'(x) = \frac{u'(x)\, v(x) - u(x)\, v'(x)}{v^2(x)}.$$

Hence,

$$f'(x) = \frac{(5x^4 + 6x^2 + 2x)(x^2 - 2) - (x^2 + 2)(x^3 + 1)(2x)}{(x^2 - 2)^2}$$

$$= \frac{3x^6 - 8x^4 - 12x^2 - 8x}{(x^2 - 2)^2}.$$

Exercise 4

In problems 1 through 18, find $f'(x)$ and simplify the result as much as possible (a, b, and c are constants).

1. $f(x) = (x + a)(x + b)$

2. $f(x) = x(ax^2 + bx + c)$

3. $f(x) = (ax^2 + bx + c)(3x^3 + 7x^2 + 8)$

4. $f(x) = (x + 2x^3)(x^4 - 2x^5)$

5. $f(x) = \dfrac{ax + b}{cx + d}$

6. $f(x) = \dfrac{3 + 4x + 5x^2}{7}$

7. $f(x) = \dfrac{9x^{-1} + 5x^{-2}}{x^3}$

8. $f(x) = \dfrac{5x^2 + 7x^3}{x^{-4}}$

9. $f(x) = \dfrac{1}{x^2 - 5}$

10. $f(x) = \dfrac{5}{x - 7}$

11. $f(x) = \dfrac{4}{2x^3 + 3x^2 + 5}$

12. $f(x) = \dfrac{3x^2 + 5x}{7x + 4}$

13. $f(x) = \dfrac{ax^2 + bx}{1 + x + x^{-1}}$

14. $f(x) = \dfrac{(x + a)(x + b)}{(x + c)(x + 1)}$

15. $f(x) = \dfrac{x^3 + x^{-3}}{x^3 - x^{-3}}$

16. $f(x) = \dfrac{x^2 + 5}{x(x + 7)}$

17. $f(x) = \dfrac{(2x + 3)^2}{(3x + 1)^3}$

18. $f(x) = \dfrac{3x^4 + 5x^2 + 9x - 7}{x^3 - 5x + 6}$

19. Apply Theorem 5 to obtain equation (32).

5. The Derivative: A Geometric Interpretation

The problem already discussed of computing the velocity of a moving object led to the development of the derivative concept. A second problem that led to the development of the differential calculus was that of finding the direction of the tangent line at a specified point on a given curve. This problem was successfully solved by the French mathematician Fermat.

Here, we shall illustrate how the derivative concept is used to define the tangent line to the graph of a given function at a given point.

Consider a function f whose graph is given in Figure 1. Let P be a point on the graph of f with coordinates $(c, f(c))$. Let h be a number ($h \neq 0$) such that $c + h$ is in the domain of f. Then there is a point Q on the graph of f with coordinates $(c + h, f(c + h))$. The line L containing the points P and Q is called a *secant line* (in geometry, a line which crosses a circle at two points is called a secant line). Now the slope of L (see Chapter 2) is given by

(35) $$\text{slope of } PQ = \frac{f(c + h) - f(c)}{h}, \; h \neq 0.$$

Pierre de Fermat (1601–1665) was born at Beaumont-de-Lomagne, France. His father, Dominique Fermat, was a leather merchant. Fermat led an uneventful life. He received his education in his home town and later at Toulouse. At the age of thirty he married and fathered three children, a boy and two girls. The latter became nuns. In 1648 he became Counselor of the Parliament of Toulouse, a position which gave him ample time to do mathematics.

Fermat is considered one of the creators of the calculus (along with Newton and Leibniz). He solved the problem of finding the extrema of the graph of a function and deduced the laws of reflection and refraction. He extended Descartes' analytic geometry to three dimensions, and was a cofounder (with Pascal) of the theory of probability. His greatest work, however, was done in the theory of numbers. He stated and claimed to have proved a famous problem (Fermat's Last Theorem): If n is an integer, $n > 2$, then there exist no rational numbers x, y, and z such that $x^n + y^n = z^n$. Today this remains as an unproved theorem.

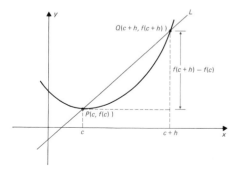

Figure 1

We shall assume that the function f is continuous on an interval containing the points c and $c + h$. Now let the point Q approach P along the curve. Clearly, as Q approaches P, h approaches zero. Taking the limit in equation (35) we have

$$(36) \qquad \lim_{Q \to P} (\text{slope of } PQ) = \lim_{h \to 0} \frac{f(c + h) - f(c)}{h}.$$

If f has a derivative at the point P, then from equation (36) we have

$$(37) \qquad \lim_{Q \to P} (\text{slope of } PQ) = f'(c).$$

Note that in Figure 1, h is a positive number. However, h approaches zero through positive and negative values.

We define the *tangent line* to the graph of f at the point P as the line L_{PT} having slope number $f'(c)$ and containing the point P. Thus, using the point-slope form of the equation of a line, we obtain

$$(38) \qquad L_{PT} = \{(x, y) \mid y - f(c) = f'(c)(x - c)\}.$$

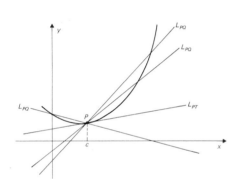

Figure 2

Example 1. Let f be the function defined by

$$f(x) = 2x - x^2.$$

(a) Find the equation of the tangent line to the graph of f at the point $(\frac{1}{2}, \frac{3}{4})$.

(b) At what point on the curve is the tangent line parallel to the x-axis?

Solution.

(a) Since $f(x) = 2x - x^2$, application of the rules of differentiation yields

$$(39) \qquad f'(x) = 2 - 2x.$$

Thus the slope of the tangent line to the graph of f at a point (x, y) is $2 - 2x$. Hence, the slope of the tangent line at the point $(\frac{1}{2}, \frac{3}{4})$ is

$$(40) \qquad f'(\tfrac{1}{2}) = 2 - 2(\tfrac{1}{2}) = 1.$$

The tangent line is then given by

$$(41) \qquad \begin{aligned} L_{PT} &= \{(x, y) \mid y - \tfrac{3}{4} = 1(x - \tfrac{1}{2})\} \\ &= \{(x, y) \mid y = x + \tfrac{1}{4}\} \end{aligned}$$

(b) To find the point at which the tangent line is horizontal, or parallel to the x-axis, we write

$$f'(x) = 0,$$

and so

$$2 - 2x = 0$$

or

$$x = 1.$$

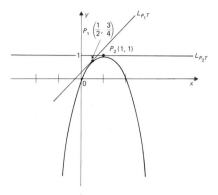

Figure 3

To find the corresponding value of *f* we have

$$f(1) = 2(1) - (1)^2 = 1.$$

Therefore, at the point $P_2(1, 1)$ the tangent line is parallel to the *x*-axis (see Figure 3).

Example 2. Find the point at which the slope of the tangent line to the curve

(42) $$y = 3x - x^3 + 5$$

is 3.

Solution. From equation (42) we obtain

$$\frac{dy}{dx} = 3 - 3x^2.$$

To find the point at which the tangent line has slope 3, we set

$$\frac{dy}{dx} = 3.$$

Hence

$$3 - 3x^2 = 3,$$

which implies

$$x = 0.$$

To find the corresponding value of *y* we substitute *x* into equation (42) to obtain

$$y = 3(0) - (0)^3 + 5$$
$$= 5.$$

Therefore, the required point is (0, 5).

We shall now consider some important results that are crucial to the next chapter with its applications. First, for many curves which cross the *x*-axis at two points, there is at least one point between the crossings where the tangent line is parallel to the *x*-axis. (See Figure 4.)

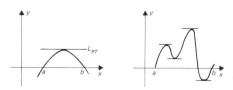

Figure 4

Michel Rolle (1652–1719), a French mathematician, was born on April 21, 1652 at Ambert, and died on October 17, 1719 in Paris. Rolle first attracted the attention of the scientific community when he solved an algebraic problem proposed by Ozanam. His remarkable work in algebra and geometry gained him membership in the French Academy of Sciences. His works include A Study of Descartes' Geometry *and* A Treatise of Algebra.

Remark. Theorem 7 (ii) gives the mathematical meaning of smooth.

Theorem 7. **Rolle's Theorem. Suppose**

(i) **f is a function continuous on $[a, b]$**

(ii) **$f'(x)$ exists for each x on (a, b)**

(iii) **$f(a) = f(b) = 0$.**

Then there is at least one point c between a and b such that

$$f'(c) = 0.$$

It is important to note that hypothesis (ii) is essential. If the function has no derivative at some point between a and b, Rolle's Theorem does not necessarily apply.

For example, the function f where

$$f(x) = 1 - x^{2/3}$$

is continuous on $[-1, 1]$ and $f(-1) = f(1) = 0$. However,

$$f'(x) = -\tfrac{2}{3}x^{-1/3}$$

is zero nowhere. Here we find that f' is not defined at $x = 0$ and so Rolle's Theorem does not apply.

Example 3. If $f(x) = x^3 - x$, find all numbers c between 0 and 1 such that $f'(c) = 0$.

Solution. We find that

$$f(0) = (0)^3 - 0 = 0$$
$$f(1) = 1^3 - 1 = 0.$$

Computing the derivative we have

$$f'(x) = 3x^2 - 1.$$

Setting $f'(x)$ equal to zero for some value $x = c$ we obtain

$$3c^2 - 1 = 0$$

or

$$c = \sqrt{\tfrac{1}{3}} \quad \text{or} \quad c = -\sqrt{\tfrac{1}{3}}.$$

Now $-\sqrt{\tfrac{1}{3}}$ is not between 0 and 1. Therefore, the only answer is $c = \sqrt{\tfrac{1}{3}}$. (See Figure 5.) This is the point called for by Rolle's Theorem.

A generalization of Rolle's Theorem is

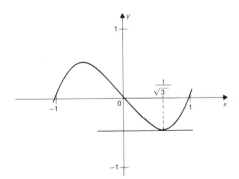

Figure 5
Graph of $f(x) = x^3 - x$

The Mean Value Theorem

Theorem 8. **The Mean Value Theorem. Suppose**

(i) **f is a continuous function on $[a, b]$**

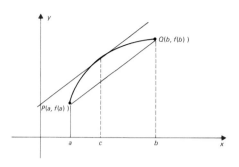

Figure 6

(ii) $f'(x)$ exists for each x on (a, b).

Then there is a number c between a and b such that

(43) $$f'(c) = \frac{f(b) - f(a)}{b - a}.$$

The proof of this theorem is left for the student as an exercise. Here we shall give a geometrical interpretation of the theorem. In Figure 6 we have the graph of a smooth function f between the points a and b. The points P and Q have coordinates $(a, f(a))$ and $(b, f(b))$ respectively. We find that the slope of the line segment PQ is

$$\frac{f(b) - f(a)}{b - a},$$

which is the right-hand side of equation (43). The theorem says that the tangent line to the graph of f at c is parallel to the line segment PQ, where c is some point in the open interval (a, b).

Example 4. If $f(x) = x^2 - x - 2$, find all numbers c between 1 and 3 which satisfy the equation

$$f'(c) = \frac{f(3) - f(1)}{3 - 1}.$$

Solution. Computing the derivative of f we find that

$$f'(x) = 2x - 1.$$

Now,

$$\frac{f(3) - f(1)}{3 - 1} = \frac{(3^2 - 3 - 2) - (1^2 - 1 - 2)}{2}$$
$$= 3.$$

Thus for some value c we have

$$f'(c) = 3$$

or

$$2c - 1 = 3,$$

that is,

$$c = 2.$$

The student can check that $f'(2) = 3$. In Figure 7, the slope of the line segment PQ is equal to the slope of the tangent line to the graph of f at $(2, 0)$.

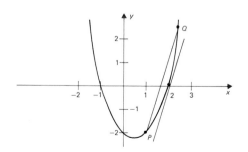

Figure 7

Exercise 5

In problems 1 through 5, find the equation of the tangent line to the graph of the given function at the given point.

1. $f(x) = x^3$ at $(2, 8)$
2. $f(x) = \sqrt{x}$ at $(4, 2)$
3. $f(x) = 1/x$ at $(2, \frac{1}{2})$
4. $f(x) = \frac{1}{4}x^2 + \frac{3}{4}x + \frac{1}{2}$ at $(2, 3)$
5. $f(x) = x^2 + 2x - 3$ at points where the x coordinates are -1 and 1.
6. At what point on the curve $y = 2x - x^2$ is the slope of the tangent line 5?
7. If $f(x) = 3x^2$, show that the slope of the tangent line to the graph of f at $(4, 48)$ is twice the slope at $(2, 12)$.
8. At what point on the curve $y = x + (4/x)$ is the tangent line parallel to the x-axis?
9. Let $y = 6x - x^2$. Find the point at which the slope is -4.
10. Find the coordinates of the point on the curve $y = 5x^2 + (80/x)$ at which the tangent line is parallel to the x-axis.
11. $A(1, -3)$ and $B(4, 3)$ are two points on the curve $y = x - (4/x)$. Find the coordinates of the point on the arc AB of the curve at which the tangent to the curve is parallel to the line through A and B.

In problems 12 through 16, find all possible values c between a and b which satisfy the equation

$$f'(c) = \frac{f(b) - f(a)}{b - a}.$$

12. $f(x) = -3x + 1$, $a = 0$, $b = 2$
13. $f(x) = x^2 - 3x - 4$, $a = -1$, $b = 3$
14. $f(x) = x^3 - 2x^2 + 3x - 2$, $a = 0$, $b = 2$
15. $f(x) = x^2 + 1$, $a = 1$, $b = 2$
16. $f(x) = \dfrac{x - 2}{x + 2}$, $a = 0$, $b = 3$
17. Given that $f(x) = (x + 3)/(2x + 1)$, $a = -1$, and $b = 2$, discuss the validity of the Mean Value Theorem.
18. Given that $f(x) = (x^2 + 4x + 3)/(x - 1)$, $a = 0$, $b = 2$, show that there is no number c between a and b which satisfies the Mean Value Theorem.
*19. Prove Theorem 8. *Hint:* Use the function

$$g(x) = f(x) - \frac{f(b) - f(a)}{b - a}(x - a) - f(a)$$

and apply Rolle's Theorem.

6. The Chain Rule

Suppose we wish to find the derivative of f where

(44) $$f(x) = (3x^3 + 7x^2 + 8)^{70}.$$

From what we have learned so far, the only way to find $f'(x)$ is to compute the seventieth power of $3x^3 + 7x^2 + 8$ and then differentiate the resulting polynomial. To avoid this unpleasant task, we can fortunately use what is called the *chain rule*. By using the chain rule, we shall see that from (44) we obtain

(45) $$f'(x) = 70(3x^3 + 7x^2 + 8)^{69}(9x^2 + 14x).$$

Before studying the chain rule the student should review Chapter 2, Section 4 on the composition of functions. We recall that if g and u are functions, then the composition of g with u is a function f defined by

$$f(x) = g(u(x)).$$

We now state the chain rule (without proof).

The chain rule

Theorem 9. Suppose the function u is differentiable at x and the function g is differentiable at $u(x)$. Then $f = g \circ u$ is differentiable at x and

Suppose $f(x) = g(u(x))$, then using Leibniz'

notation

$$\frac{df}{dx} = \frac{dg}{du}\frac{du}{dx}.$$

(46) $$\begin{aligned} f'(x) &= (g \circ u)'(x) \\ &= g'(u(x)) \cdot u'(x). \end{aligned}$$

Example 1. Let us return to the function f defined by

$$f(x) = (3x^3 + 7x^2 + 8)^{70}.$$

Find $f'(x)$.

Solution. We can write $f(x)$ as

$$f(x) = g(u(x))$$

where

$$u(x) = 3x^3 + 7x^2 + 8$$

and

$$g(u) = u^{70}.$$

Now,

$$g'(u) = 70u^{69}$$

and

$$u'(x) = 9x^2 + 14x.$$

Therefore, application of Theorem 9 yields

$$f'(x) = g'(u(x)) \cdot u'(x)$$

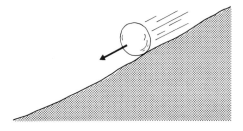

$$= 70[u(x)]^{69}(9x^2 + 14x)$$
$$= 70(3x^3 + 7x^2 + 8)^{69}(9x^2 + 14x).$$

Example 2. A snowball is rolling down a mountainside. At a certain instant, the surface area is increasing at the rate of 50 square feet per second. If the radius of the snowball at that instant is 8 feet, how fast is the volume increasing?

Solution. In this model there are three functions of time: the volume $V(t)$, the surface area $S(t)$, and the radius $r(t)$. We assume that the snowball is a perfect sphere at all times and recall that the surface area of a sphere is

(47) $$S(t) = 4\pi[r(t)]^2$$

and that the volume of the sphere is

(48) $$V(t) = \tfrac{4}{3}\pi[r(t)]^3.$$

Applying the chain rule, we find from equation (47) that

(49) $$S'(t) = 4\pi[2r(t)]r'(t).$$

Now, when $r = 8$, $S'(t) = 50$. Thus from equation (49) we have

$$r'(t) = \frac{50}{8\pi(8)}$$
$$= \frac{25}{32\pi} \text{ ft per second.}$$

From equation (48)

$$V'(t) = \frac{4}{3}\pi \cdot 3[r(t)]^2 r'(t)$$
$$= \frac{4}{3}\pi(3)(8)^2 \frac{25}{32\pi} = 200.$$

Hence at the instant in question the volume of the snowball is increasing at the rate of 200 cubic feet per second.

Example 3. A rocket is fired vertically upward. The equation of motion is given by

(50) $$s(t) = R\left(1 + \frac{3}{2}\frac{v_0}{R}t\right)^{2/3}$$

where R is the radius of the Earth, s is the distance of the rocket from the Earth's center, v_0 is a constant $= \sqrt{2gR}$ called the escape velocity, and t is time (see Figure 8). Find the instantaneous velocity of the rocket.

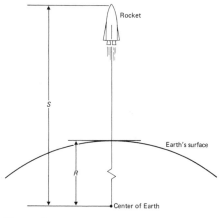

Figure 8

Solution. Let

$$u(t) = 1 + \frac{3}{2} \frac{v_0}{R} t.$$

Then

(51)
$$u'(t) = \frac{3}{2} \frac{v_0}{R}.$$

Now from equation (50) we have

$$s(t) = Ru^{2/3}$$

and by the chain rule

(52)
$$s'(t) = \left[R \frac{2}{3} u^{-1/3} \right] u'(t) = \left(\frac{2}{3} Ru^{-1/3} \right) \left(\frac{3}{2} \frac{v_0}{R} \right)$$

$$= \frac{v_0}{\left[1 + \dfrac{3}{2} \dfrac{v_0}{R} t \right]^{1/3}},$$

which is the instantaneous velocity of the rocket.

Example 4. For each of the following functions find $f'(x)$:

  (a) $f(x) = (x^3 + 5x^2 + 7)^{9/2}$

  (b) $f(x) = \dfrac{1}{\sqrt[3]{4x^3 + 6x + 7}}.$

Solution. (a) Let

$$u(x) = x^3 + 5x^2 + 7$$

and

$$g(u) = u^{9/2}.$$

Then,

$$g'(u(x)) = \tfrac{9}{2}u(x)^{9/2-1} = \tfrac{9}{2}u(x)^{7/2}$$
$$= \tfrac{9}{2}(x^3 + 5x^2 + 7)^{7/2}$$

and

$$u'(x) = 3x^2 + 10x.$$

Hence by the chain rule

$$f'(x) = g'(u(x)) \cdot u'(x)$$
$$= \tfrac{9}{2}(x^3 + 5x^2 + 7)^{7/2}(3x^2 + 10x)$$

  (b) Let

$$u(x) = 4x^3 + 6x + 7$$

and

$$g(u) = u^{-1/3}.$$

Then,

$$g'(u(x)) = -\tfrac{1}{3}u(x)^{-1/3-1} = -\tfrac{1}{3}u^{-4/3}$$

$$= -\tfrac{1}{3}(4x^3 + 6x + 7)^{-4/3}$$

and

$$u'(x) = 12x^2 + 6.$$

Hence by the chain rule, since $f = g \circ u$,

$$\begin{aligned}
f'(x) &= g'(u(x)) \cdot u'(x) \\
&= -\tfrac{1}{3}(4x^3 + 6x + 7)^{-4/3} \cdot (12x^2 + 6) \\
&= -2(4x^3 + 6x + 7)^{-4/3}(2x^2 + 1)
\end{aligned}$$

Example 5. If $y = x^2\sqrt{x^2 - 1}$, find dy/dx.

Solution. Applying Theorem 5 we have

$$\frac{dy}{dx} = x^2 \frac{d}{dx}\left((x^2 - 1)^{1/2}\right) + (x^2 - 1)^{1/2}\frac{d}{dx}(x^2)$$

and by the chain rule

$$\begin{aligned}
\frac{d}{dx}\left((x^2 - 1)^{1/2}\right) &= \tfrac{1}{2}(x^2 - 1)^{1/2-1}\frac{d}{dx}(x^2 - 1) \\
&= \tfrac{1}{2}(x^2 - 1)^{-1/2}(2x) \\
&= x(x^2 - 1)^{-1/2}.
\end{aligned}$$

Hence

$$\begin{aligned}
\frac{dy}{dx} &= x^2[x(x^2 - 1)^{-1/2}] + (x^2 - 1)^{1/2}(2x) \\[2mm]
&= \frac{x^3}{\sqrt{x^2 - 1}} + 2x\sqrt{x^2 - 1} \\[2mm]
&= \frac{x^3 + 2x(x^2 - 1)}{\sqrt{x^2 - 1}} = \frac{3x^3 - 2x}{\sqrt{x^2 - 1}}.
\end{aligned}$$

Exercise 6

1. Use two methods to differentiate each of the following functions and compare your results.

(a) $f(x) = (3x + 1)^2$ (b) $f(x) = \left(x - \dfrac{1}{x}\right)^2$

(c) $f(x) = (x + 1)^{-1}$ (d) $f(x) = (x + 1)^3$

In problems 2 through 15, compute the derivative of the given function (a, b, and c are constants).

2. $f(x) = (2x + 4)^8$

3. $f(x) = (5x^3 + 6x^2 + 8x + 9)^{15}$

4. $y = (3 - 8x^2 + 9x^3)^{5/2}$

5. $f(x) = \sqrt[3]{ax^2 + bx + c}$

6. $f(x) = (2x + 1)\sqrt{x^2 + 5}$

7. $y = \dfrac{1}{\sqrt[5]{x^3 + ax^2 + bx + c}}$

8. $y = x^2 + 3 + \dfrac{1}{\sqrt{x^2 + 3}}$

9. $y = \dfrac{x}{\sqrt{x^2 + 1}}$

10. $y = (x^2 + a^2)^3 (x^2 + b^2)^5$

11. $y = \sqrt{\dfrac{3x + 4}{5x + 11}}$

12. $y = \sqrt[3]{\dfrac{x^2 + 2x + 4}{5x^2 + 7x + 3}}$

13. $f(x) = \sqrt{x + x^2 + 1}$

14. $y = \dfrac{1}{\sqrt{x + a} + \sqrt{x + b}}$

15. $y = \dfrac{1}{\sqrt[3]{x - 3} - \sqrt[3]{x + 3}}$

16. An offshore oil well is leaking. The oil spreads on the surface of the water in a circle. If the radius r of the circle is increasing at the rate of 2 ft per second, at what rate is the area A of the oil spill increasing when $r = 20$ ft?   50 ft?   200 ft?

17. A spherical weather balloon is being inflated. At a certain instant, the volume of the balloon is increasing at the rate of 500 cubic inches per minute. If at that instant the radius of the balloon is 40 inches, how fast is the surface area increasing?

18. A spherical snowball is melting. At a certain instant the volume is decreasing ($V'(t) < 0$) at the rate of 10 cubic inches per hour. At that instant, the radius of the snowball is 25 inches.
(a) How fast is the diameter decreasing?
(b) How fast is the surface area decreasing?

19. Suppose in a model the pollution index $I(t)$ is related to a dose $A(t)$ of an antipollution agent by the equation

$$I \cdot A^{3/2} = k \text{ (a constant)}.$$

At a given instant, the index $I = 30$ per cent, and the dose, $A = 9$ units. If at that moment the dose of the antipollution agent is decreasing at the rate of 3 units per minute, how fast is the index I changing?

20. Smoke is being emitted into the atmosphere in the form of an

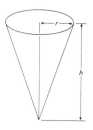

inverted right circular cone. At any instant the diameter is equal to the height *h*. If the rate of smoke emission is 200 cu ft per minute, how fast is the height increasing when $h = 50$ ft? *Hint:* Volume of cone $= \frac{1}{3}\pi r^2 h$.

7. Implicit Differentiation

Note. It is possible that the relation g(x, y) = 0 does not define any function. For example,

$$2x^2 + y^2 + 1 = 0$$

does not define a function, since there are no values of x and y which satisfy this equation.

Suppose *f* is a function given by the equation $y = f(x)$. Then we say that *y* is defined *explicitly* as a function of *x*. The functions we have encountered so far have been of this form. However, a relation between *x* and *y* expressed in the form

$$g(x, y) = 0$$

may also define one or more functions of *x*. In such a case the function or functions are said to be defined *implicitly*.

Thus in the equations

(53) $$3x^3 + y^2 - 8y = 0$$

(54) $$x^2 - 4 - 3xy = 0$$

and

(55) $$x^2 + y^2 - 1 = 0$$

y is defined implicitly as a function of *x*. Sometimes, it is possible to solve an equation for *y* as a function of *x*. For example, from equation (54) we have

$$y = \frac{x^2 - 4}{3x}.$$

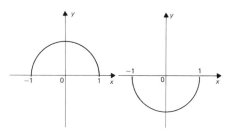

Figure 9
(a) $y = f_1(x) = \sqrt{1 - x^2}$
(b) $y = f_2(x) = -\sqrt{1 - x^2}$

Equation (55) can be solved for *y*, yielding two functions

$$y = f_1(x) = \sqrt{1 - x^2}$$

and

$$y = f_2(x) = -\sqrt{1 - x^2}.$$

Therefore, equation (54) implicitly defines two functions of *x* (see Figure 9).

We shall now indicate how to calculate the derivative of a function defined implicitly by an equation. The process involved is as follows:

Here we are assuming that y is a differentiable function of x defined implicitly by the equation

$$g(x, y) = 0.$$

(a) Using the chain rule, differentiate both sides of the equation $g(x, y) = 0$.

(b) Solve the resulting equation for *dy/dx*.

Observations:

(1) If we set $y = f(x)$, then

$$\frac{d}{dx}(y^3) = \frac{d}{dx}[f(x)]^3$$

$$= 3\,[f(x)]^2\,f'(x) = 3y^2\,\frac{dy}{dx}.$$

(2) $\dfrac{d}{dx}(xy) = \dfrac{d}{dx}[xf(x)] = xf'(x) + 1 \cdot f(x).$

Here we have used the product rule for differentiation.

Example 1. Find dy/dx if

(56) $\qquad\qquad x^3 + y^3 - 3xy = 0.$

Solution. Assuming that y is a differentiable function of x, we differentiate both sides of equation (55) to obtain

$$\frac{d}{dx}(x^3 + y^3 - 3xy) = \frac{d}{dx}(0)$$

or

$$\frac{d}{dx}(x^3) + \frac{d}{dx}(y^3) + \frac{d}{dx}(-3xy) = 0.$$

This implies that

$$3x^2 + 3y^2\,\frac{dy}{dx} - 3\left(x\,\frac{dy}{dx} + y\right) = 0.$$

Collecting terms involving dy/dx and factoring, we obtain

$$(y^2 - x)\,\frac{dy}{dx} = y - x^2.$$

Hence,

$$\frac{dy}{dx} = \frac{y - x^2}{y^2 - x}, \text{ for } y^2 \neq x.$$

Example 2. Find dy/dx if

(57) $\qquad\qquad 2x^2 + 3y^2 = 4.$

Solution 1. Differentiating both sides of equation (57) with respect to x, we obtain

$$\frac{d}{dx}(2x^2 + 3y^2) = \frac{d}{dx}(4)$$

$$\frac{d}{dx}(2x^2) + \frac{d}{dx}(3y^2) = \frac{d}{dx}(4)$$

$$4x + 6y\,\frac{dy}{dx} = 0$$

and so

$$\frac{dy}{dx} = -\frac{4x}{6y} = -\frac{2x}{3y}.$$

Note again:

$$\frac{d}{dx}(y^2) = \frac{d}{dx}[f(x)]^2 = 2f(x) \cdot f'(x) = 2y\,\frac{dy}{dx}.$$

Solution 2. Suppose we solve equation (57) for y. We have

(58) $\qquad y = \dfrac{1}{\sqrt{3}}\sqrt{4 - 2x^2}, \qquad (2 - x^2 \geq 0)$

or

(59) $$y = -\frac{1}{\sqrt{3}}\sqrt{4 - 2x^2}, \qquad (2 - x^2 \geq 0).$$

From (58) we obtain

$$\frac{dy}{dx} = \frac{1}{\sqrt{3}}\frac{1}{2}(4 - 2x^2)^{-1/2}\frac{d}{dx}(4 - 2x^2)$$

$$= \frac{1}{2\sqrt{3}}(4 - 2x^2)^{-1/2}(-4x)$$

$$= -\frac{2x}{\sqrt{3}(4 - 2x^2)^{1/2}}$$

$$= -\frac{\sqrt{3}(2x)}{\sqrt{3}\sqrt{3}(4 - 2x^2)^{1/2}}$$

$$= -\frac{2x}{3y},$$

which agrees with the result above. The student is encouraged to repeat the above starting with equation (59) and show that the same result is obtained.

Example 3. Find the equation of the tangent line to the ellipse

(60) $$9x^2 + 16y^2 = 144$$

at the point $(\sqrt{7}, \frac{9}{4})$.

Solution. Differentiating implicitly we obtain

$$\frac{d}{dx}(9x^2 + 16y^2) = \frac{d}{dx}(144)$$

or

$$\frac{d}{dx}(9x^2) + \frac{d}{dx}(16y^2) = 0$$

$$18x + 32y\frac{dy}{dx} = 0.$$

Hence

$$\frac{dy}{dx} = -\frac{9x}{16y}$$

and so the slope of the tangent line at the point $(\sqrt{7}, \frac{9}{4})$ is

$$\left.\frac{dy}{dx}\right|_{(\sqrt{7}, \frac{9}{4})} = -\frac{9(\sqrt{7})}{16(\frac{9}{4})} = -\frac{\sqrt{7}}{4}.$$

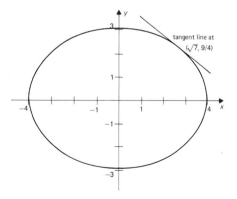

tangent line at $(\sqrt{7}, 9/4)$

Figure 10

Therefore the equation of the tangent line at $(\sqrt{7}, \frac{9}{4})$ is given by

$$y - \frac{9}{4} = -\frac{\sqrt{7}}{4}(x - \sqrt{7}).$$

Simplifying, we have

$$4y + \sqrt{7}x - 16 = 0.$$

Example 4. Two cars C_1 and C_2 are racing on two highways intersecting at right angles (see Figure 11). Car C_1 is traveling south at the rate of 100 ft per second and C_2 is traveling west at the rate of 120 ft per second. At a certain instant, C_1 is 300 ft from point A and C_2 is 400 ft from A. (a) How fast is the distance between the two cars changing at that instant? (b) If the cars maintain the same speed and ignore the stop signs, will they collide?

Solution. (a) Let x, y, and z be the distances from C_2 to A, C_1 to A, and C_1 to C_2 respectively. Here x, y, and z are functions of time t. Using the Pythagorean theorem we have the relation

(61) $$z^2 = x^2 + y^2.$$

Differentiating both sides of equation (61) with respect to t, we obtain

$$\frac{d}{dt}(z^2) = \frac{d}{dt}(x^2 + y^2)$$

and

$$2z\frac{dz}{dt} = 2x\frac{dx}{dt} + 2y\frac{dy}{dt}.$$

Dividing by 2 we have

(62) $$z\frac{dz}{dt} = x\frac{dx}{dt} + y\frac{dy}{dt}.$$

At the instant in question, $x = 400$ ft and $y = 300$ ft, and so

$$z = \sqrt{(400)^2 + (300)^2}$$
$$= 500 \text{ ft}.$$

Therefore, from equation (62)

$$\frac{dz}{dt} = \frac{1}{500}[(400)(120) + (300)(100)]$$
$$= \frac{780}{5}$$
$$= 156.$$

Using the point-slope form: $y - y_1 = m(x - x_1)$

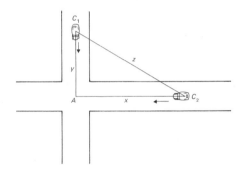

Figure 11

Note that $\dfrac{dx}{dt} = 120\ ft/sec$ = rate of change

(velocity) of C_2

$\dfrac{dy}{dt} = 100\ ft/sec$ = rate of change

(velocity) of C_1.

Therefore the distance between the two cars is changing at the rate of 156 ft per second.

(b) Since car C_2 is 400 ft from point A and is traveling at 120 ft per second, it will arrive at point A in $\frac{400}{120} = \frac{10}{3}$ seconds later, while car C_1 will arrive at point A in $\frac{300}{100} = 3$ seconds. In other words, the cars (we hope!) will miss one another by $\frac{1}{3}$ of a second.

Exercise 7

In problems 1 through 4, find dy/dx using two methods (assume y is a differentiable function of x).

1. $x^2 + y^2 = 25$ **2.** $y^2 = 3x^3$

3. $y^2 = \dfrac{x + 2}{x - 2}$ **4.** $x^3 + y^3 = 1$

In problems 5 through 14, find dy/dx.

5. $y^2 = 3x^3 + 5x^2 + 7$

6. $y^3 + 3xy = x$

7. $x^2y^2 = a^2(x^2 - y^2)$ (a is a constant)

8. $\dfrac{x^2}{a^2} + \dfrac{y^2}{b^2} = 1$ (a, b are constants)

9. $x^5 + y^5 = 5(x^3y^2 + x^2y^3)$

10. $x^{2/3} + y^{2/3} = a^{2/3}$ (a is a constant)

11. $\sqrt{1 - x^2} + \sqrt{1 - y^2} = 8(x - y)$

12. $(x + 4)^5 = 2y^3$

13. $2x^2 = \dfrac{x + y}{x - y}$

14. $y^2 = \dfrac{\sqrt[3]{x^2 + 7}}{3x}$

In problems 15 through 18, find the equation of the tangent line to the given curve at the given point.

15. $2y^2 - 3x^2 = 6$ at $(2, 3)$

16. $x^2 + y^2 = 4$ at $(-\sqrt{3}, 1)$

17. $x - 1 - 2\sqrt{x} + \sqrt{y} = 0$ at $(4, 1)$

18. $\dfrac{y}{x} + \dfrac{x}{y} = 5xy$ at $(1, \frac{1}{2})$

19. Show that the two circles

$$(x - 6)^2 + (y - 3)^2 = 20$$

and

$$x^2 + y^2 + 2x + y = 10$$

have a common tangent line at $(2, 1)$.

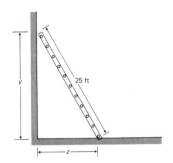

25 ft

z

20. A painter is on top of a ladder 25 ft long which is leaning against a vertical wall. If a prankster pulls the bottom of the ladder away from the wall at the rate of 6 inches per second, how fast is the painter sliding down when the bottom is 20 ft from the wall?

21. Ship *A* leaves port at 2:00 P.M. and travels due east at 15 knots. At 6:00 P.M. on the same day ship *B* leaves the same port and travels south at 20 knots. How fast is the distance changing between *A* and *B* when *B* has traveled 80 nautical miles?

8. Higher Derivatives

Consider the function *f* where $f(x) = x^3$. The derivative of *f* is a new function *f′*, where $f'(x) = 3x^2$. We can again differentiate the function *f′*. If we do so we get a new function *f″* where $f''(x) = 6x$. If we further differentiate the function *f″*, we obtain $f'''(x) = 6$ and so on.

In general, if a function *f* has a derivative *f′* and if *f′* itself is differentiable, we denote the derived function of *f′* by *f″* and call *f″* the *second derivative* of *f*.

Continuing in this manner we obtain the functions f''', $f^{(4)}$, . . . , $f^{(n)}$, each of which is the derivative of the preceding function. Of course here we are assuming that each of the preceding functions is differentiable. There are various notations used for the higher derivatives. If $y = f(x)$, the derivatives can be written as follows:

Note that if f is differentiable at c, f′ is not necessarily differentiable at c. For example, if $f(x) = x^{4/3}$, then $f'(x) = (4/3)x^{1/3}$. Here we find that f is differentiable for all x. In particular, f is differentiable at $x = 0$. However, f′ is not differentiable at $x = 0$, since $f''(x) = (4/9)x^{-2/3}$, which is not defined at $x = 0$.

First derivative: $\dfrac{dy}{dx}, \dfrac{df}{dx}, y', f'$

Second derivative: $\dfrac{d^2y}{dx^2}, \dfrac{d^2f}{dx^2}, y'', f''$

Third derivative: $\dfrac{d^3y}{dx^3}, \dfrac{d^3f}{dx^3}, y''', f'''$

$\vdots$

*n*th derivative: $\dfrac{d^ny}{dx^n}, \dfrac{d^nf}{dx^n}, y^{(n)}, f^{(n)}.$

To avoid confusing powers of f with derivatives of f for $n > 3$ we write

$f^{(n)}$ *to denote the nth derivative of f*

and

f^n *to denote the nth power of f.*

The reason for the peculiar location of the superscripts in the first two notations is that d/dx is a symbol of the operation of differentiation. Thus we have

$$\frac{dy}{dx} = \frac{d}{dx}(y)$$

$$\frac{d^2y}{dx^2} = \frac{d}{dx}\left(\frac{dy}{dx}\right) = \frac{d}{dx}\left(\frac{d}{dx}(y)\right)$$

$$\frac{d^3y}{dx^3} = \frac{d}{dx}\left(\frac{d^2y}{dx^2}\right) = \frac{d}{dx}\left[\frac{d}{dx}\left(\frac{d}{dx}(y)\right)\right]$$

and so on.

Example 1. Find the derivatives of all orders for the function defined by $y = x^5$.

Solution. From Theorem 2 we have

$$\frac{dy}{dx} = 5x^4$$

and Theorem 3 then yields

$$\frac{d^2y}{dx^2} = \frac{d}{dx}\left(\frac{dy}{dx}\right) = \frac{d}{dx}(5x^4) = 20x^3$$

$$\frac{d^3y}{dx^3} = \frac{d}{dx}\left(\frac{d^2y}{dx^2}\right) = \frac{d}{dx}(20x^3) = 60x^2$$

$$\frac{d^4y}{dx^4} = \frac{d}{dx}\left(\frac{d^3y}{dx^3}\right) = \frac{d}{dx}(60x^2) = 120x$$

$$\frac{d^5y}{dx^5} = \frac{d}{dx}\left(\frac{d^4y}{dx^4}\right) = \frac{d}{dx}(120x) = 120$$

$$\frac{d^6y}{dx^6} = \frac{d}{dx}\left(\frac{d^5y}{dx^5}\right) = \frac{d}{dx}(120) = 0$$

and

$$\frac{d^ny}{dx^n} = 0 \text{ for } n > 6.$$

In fact it is not difficult to show (using mathematical induction) that if $y = x^n$, where n is a positive integer, then

(63) $\qquad \dfrac{d^ny}{dx^n} = n! \qquad \text{and} \dfrac{d^ky}{dx^k} = 0 \qquad \text{for } k > n.$

Example 2. If $x^2 + 4y^2 = 25$, show that

$$\frac{d^2y}{dx^2} = -\frac{25}{16y^3}, \, y \neq 0.$$

Here we assume that y and dy/dx define differentiable functions.

Solution. Differentiating with respect to x both sides of the equation

(64) $\qquad\qquad\qquad x^2 + 4y^2 = 25,$

we obtain

$$\frac{d}{dx}(x^2 + 4y^2) = \frac{d}{dx}(25)$$

$$\frac{d}{dx}(x^2) + \frac{d}{dx}(4y^2) = 0$$

or

$$2x + 8y\frac{dy}{dx} = 0.$$

Hence

(65)
$$\frac{dy}{dx} = -\frac{x}{4y}, \, y \neq 0,$$

and so

$$\frac{d^2y}{dx^2} = \frac{d}{dx}\left(-\frac{x}{4y}\right).$$

Theorem 6 then yields

(66)
$$\frac{d^2y}{dx^2} = -\frac{1}{4}\left[\frac{y - x\dfrac{dy}{dx}}{y^2}\right].$$

Substituting (65) into equation (66) we have

$$\frac{d^2y}{dx^2} = -\frac{1}{4}\left[\frac{y - x\left(\dfrac{-x}{4y}\right)}{y^2}\right]$$

$$= -\frac{1}{4}\left[\frac{4y^2 + x^2}{4y^3}\right]$$

and from equation (64) we obtain

$$\frac{d^2y}{dx^2} = -\frac{25}{16y^3}, \, y \neq 0.$$

If the motion of an object is described by distance as a function of time, say

$$s = f(t),$$

then as we have indicated, the first derivative ds/dt gives the instantaneous velocity at time t. In this case, the second derivative d^2s/dt^2 gives what is called the *instantaneous acceleration* of the object at time t. That is, acceleration is the rate at which the velocity is changing with respect to time.

Example 3. A toy rocket is launched vertically upward with an initial velocity of 120 ft per second. The distance s traveled t seconds later is given by

$$s = 120t - 16t^2.$$

Find the velocity and the acceleration. How high will the rocket rise?

Solution. Here we find that the velocity $v(t)$ is

$$v(t) = \frac{ds}{dt}$$
$$= 120 - 32t$$

and the acceleration $a(t)$ is

$$a(t) = \frac{d^2s}{dt^2}$$
$$= -32.$$

This tells us that the velocity of the rocket is changing at the rate of -32 ft per second every second. The negative sign indicates that the velocity is decreasing; it will become zero ($v = 0$), when

$$120 - 32t = 0$$

or when

$$t = \frac{120}{32} = \frac{15}{4} \text{ sec.}$$

The distance traveled during this time interval is then

$$s(\tfrac{15}{4}) = 120(\tfrac{15}{4}) - 16(\tfrac{15}{4})^2$$
$$= 450 - 225$$
$$= 225 \text{ ft.}$$

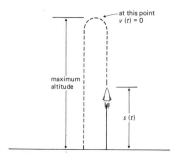

at this point
$v(t) = 0$

maximum
altitude

$s(t)$

Figure 12

Exercise 8

1. Find the second and third derivatives of each of the following functions.
   (a) $f(x) = x^3 - 6$
   (b) $f(x) = x^5 + 3x^2 - 8x + 5$
   (c) $f(x) = x^3 + 2x^{3/2} - \dfrac{4}{x^3}$
   (d) $f(x) = \dfrac{x + 1}{x - 1}$
   (e) $f(x) = \left(\sqrt{x} - \dfrac{2}{\sqrt[3]{x}}\right)^2$
   (f) $f(x) = \dfrac{x^2 - 1}{x^2 + 1}$

2. Find d^2y/dx^2 when
   (a) $x + y = 6y^2$
   (b) $x^{2/3} + y^{2/3} = a^{2/3}$
   (c) $b^2x^2 + a^2y^2 = a^2b^2$
   (d) $4x^2 + 9y^2 - 16x - 18y - 11 = 0$.

3. If $y^3 + 3x^2y + 1 = 0$, show that

$$(x^2 + y^2)^3 \frac{d^2y}{dx^2} + 2(x^2 - y^2) = 0.$$

4. If $ax^2 + 2hxy + by^2 = 1$, show that

$$\frac{d^2y}{dx^2} = \frac{h^2 - ab}{(hx + by)^3}, \qquad (a, b \text{ and } h \text{ are constants}).$$

5. If $(dy/dx)^2 + y^2 = 1 - 2/y$, show that

$$\frac{d^2y}{dx^2} + y = \frac{1}{y^2}.$$

6. If $y = (x + \sqrt{x^2 + a^2})^m$, prove that

$$(x^2 + a^2)\frac{d^2y}{dx^2} + x\frac{dy}{dx} = m^2y.$$

7. For each of the following equations of motion find the velocity function $v(t)$ and the acceleration function $a(t)$.
   (a) $s = 600 - 20t - 16t^2$
   (b) $s = (4t + 7)^2$
   (c) $s = 2t^3 - 12t^2 + 24t + 7$
   (d) $s = t^3 + 6t^2 - 21t + 15$

8. For each of the functions in problem 7 above
   (a) Find the initial velocity $v(0)$.
   (b) Find the value(s) t^* for which $v(t^*) = 0$.
   (c) Compute the corresponding values of $a(t^*)$.
   (d) Solve the inequalities $v(t) > 0$ and $v(t) < 0$.

9. Let the reaction R of a body as a function of the dosage x of a drug be given by the equation

$$R = 2x^2 \left(\frac{1}{4} - \frac{x}{6} \right).$$

   (a) Compute dR/dx and d^2R/dx^2.
   (b) Find dR/dx at $x = \frac{1}{4}, \frac{1}{3}, \frac{1}{2}, \frac{2}{3}, \frac{3}{4}$.
   (c) Draw a graph of dR/dx vs. x for $0 \leq x \leq 1$.
   (d) Find d^2R/dx^2 at $x = \frac{1}{4}, \frac{1}{3}, \frac{1}{2}, \frac{2}{3}, \frac{3}{4}$.

Sir Isaac Newton

6

More

on

the

Derivative

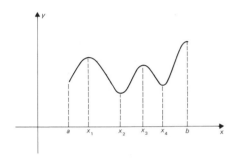

Figure 1

*Increasing and decreasing functions.
A function f is increasing (decreasing) at a
point c, if it is increasing (decreasing) over
an interval containing c.*

1. Increasing and Decreasing Functions

In Chapter 4, we indicated that a continuous function can usually be visualized as a function whose graph can be drawn over an interval without removing the pencil from the paper. Thus, if we know that the given function is continuous, we can sketch its graph by plotting enough points and then connecting the successive points by a curve. The derivative, by telling us about the slope of the tangent lines to the graph, gives us much information about the behavior of the function, and thus helps us in graphing the function. To illustrate other relations between the derivative and the graph, let us consider the function f whose graph is given in Figure 1.

We note from the graph that the function is *increasing* (the graph is "rising") in the intervals (a, x_1), (x_2, x_3), and (x_4, b), and that the function is *decreasing* (the graph is "falling") in the intervals (x_1, x_2) and (x_3, x_4). Moreover, the slope of the tangent line (the derivative) is positive over the intervals where the function is increasing and the slope of the tangent line (the derivative) is negative over the intervals where the function is decreasing. Thus, it seems safe to guess that an increasing function has a positive derivative and a decreasing function has a negative derivative.

Definition 1. A function f is said to be *increasing* over an interval if, for any two points x_1, x_2 of the interval such that $x_1 < x_2$, we have $f(x_1) < f(x_2)$. Similarly, f is *decreasing* over the interval if $x_1 < x_2$ implies $f(x_1) > f(x_2)$.

The following theorem shows how the derivative helps in determining whether a function is increasing or decreasing.

***Theorem 1.* Let *f* be a differentiable function. Then *f* is increasing over any interval on which *f'*(*x*) is positive and *f* is decreasing over any interval on which *f'*(*x*) is negative.**

Proof. Suppose $f'(x) > 0$ for each x in the open interval (a, b). Let x_1, x_2 be any two points in this interval with $x_1 < x_2$. Then by the Mean Value Theorem (see Chapter 5), there exists a point $c \in (a, b)$ such that

(1) $$f(x_2) - f(x_1) = f'(c)(x_2 - x_1).$$

Since $x_1 < x_2$, we have $x_2 - x_1 > 0$ and by hypothesis $f'(c) > 0$. Consequently the right-hand side of (1) is positive. Hence

(2) $$f(x_2) - f(x_1) > 0$$

or

$$f(x_2) > f(x_1).$$

This proves that for any two points x_1, x_2 in (a, b) with $x_1 < x_2$, we have $f(x_1) < f(x_2)$. Therefore, by definition f is increasing over (a, b).

A similar argument proves that if $f'(x) < 0$ for each x in (a, b), then f is decreasing over (a, b).

We point out that the converse of Theorem 1, "If a function f is differentiable and increasing over (a, b), then $f'(x) > 0$ for any $x \in (a, b)$," is not true.

For example, let us consider the function f given by

$$f(x) = x^3 \qquad (-1 < x < 1);$$

then

$$f'(x) = 3x^2.$$

Since $f'(x) > 0$ for all $x \neq 0$, the function is increasing over $(-1, 0) \cup (0, 1)$. In fact, f is increasing over $(-1, 1)$: If $x < 0$, then $x^3 < 0$ and if $x > 0$, then $x^3 > 0$ (see Figure 2). However, $f'(0) = 0$ (not positive).

From the above discussion, we conclude that the points where the derivative is zero or where the derivative is undefined require separate consideration.

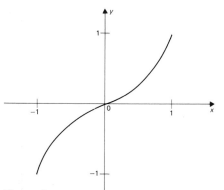

Figure 2
Graph of $f(x) = x^3$

Example 1. Find the intervals in which the function f defined by $f(x) = x^3 - 6x^2 + 9x - 7$ is (a) increasing, (b) decreasing. Sketch the graph of f.

Solution. $f(x) = x^3 - 6x^2 + 9x - 7$; differentiating, we get

$$f'(x) = 3x^2 - 12x + 9 = 3(x^2 - 4x + 3)$$

or

(3) $$f'(x) = 3(x - 1)(x - 3).$$

Now, $f'(x)$ is zero only for the values $x = 1$ and $x = 3$. These two points where the derivative is zero divide the x-axis into three sets:

$$I_1 = \{x | -\infty < x < 1\}$$
$$I_2 = \{x | 1 < x < 3\}$$
$$I_3 = \{x | 3 < x < \infty\}.$$

If $x < 1$, then $x < 3$ and $(x - 1)(x - 3) > 0$. Hence $f'(x) > 0$ and f is increasing. Similarly, if $x > 3$, then $x > 1$ and $(x - 1) \times (x - 3) > 0$. Therefore $f'(x) > 0$ and f is increasing.

If $1 < x < 3$, then $(x - 1) > 0$ and $(x - 3) < 0$, and therefore $(x - 1)(x - 3) < 0$. Hence $f'(x) < 0$ and f is decreasing.

With this information, we can sketch roughly the graph of f as in Figure 3. As we draw the graph from left to right, it rises to the point $(1, -3)$, then falls to the point $(3, -7)$, and rises thereafter.

Example 2. A manufacturer can sell x units of his product per day, if the unit price is $p = 8 - 0.01x$. Determine whether the total revenue is increasing or decreasing when x is 200, 300, 500, and 600. Sketch a graph.

Solution. The total revenue is

$$R(x) = xp(x)$$

or

(4) $$R(x) = 8x - 0.01x^2.$$

Taking the derivative on both sides of equation (4) we obtain

(5) $$\frac{dR}{dx} = 8 - 0.02x.$$

At $x = 200$,

$$\left.\frac{dR}{dx}\right|_{x=200} = 8 - (0.02)(200) = 4.$$

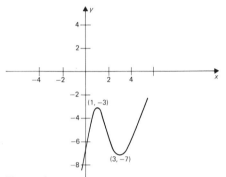

Figure 3

Observe that for $x = 1$, $f(1) = -3$ and for $x = 3$, $f(3) = -7$.

Since this is a positive number, the total revenue is increasing at $x = 200$. Similarly, for $x = 300$, we have

$$\left.\frac{dR}{dx}\right|_{x=300} = 8 - (0.02)(300) = 2.$$

Hence, $R(x)$ is increasing at $x = 300$. At $x = 500$

$$\left.\frac{dR}{dx}\right|_{x=500} = 8 - (0.02)(500) = -2,$$

a negative number, and so the total revenue is decreasing. Finally, at $x = 600$, the total revenue is decreasing, since

$$\left.\frac{dR}{dx}\right|_{x=600} = 8 - (0.02)(600) = -4.$$

In fact we find that

$$\frac{dR}{dx} > 0 \qquad \text{whenever } x < 400$$

and

$$\frac{dR}{dx} < 0 \qquad \text{whenever } x > 400.$$

Thus the total revenue is increasing on any interval where $x < 400$ and is decreasing when $x > 400$. A rough sketch of the graph of R is shown in Figure 4.

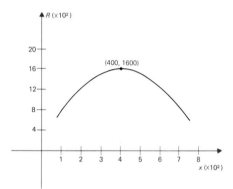

Figure 4

Exercise 1

In problems 1 through 7, find where f is increasing and where it is decreasing, and then sketch the graph of f.

1. $f(x) = x^3 - 3x^2 - 9x + 15$
2. $f(x) = x^3 - 12x$
3. $f(x) = 3x^5 - 5x^3$
4. $f(x) = 4x^3 + 3x^2 - 6x - 1$
5. $f(x) = 2x^4 - x^2$
6. $f(x) = 4x^3 - 15x^2 - 18x + 2$
*7. $f(x) = x^4 - 8x^3 + 22x^2 - 24x + 1$
8. Complete the proof of Theorem 1.
9. Suppose the rate R of an autocatalytic reaction is given by

$$R = kx(4 - x) \qquad 0 \leq x \leq 4$$

where k is a positive constant and x is the amount of substance produced. Find the intervals where the rate R is increasing and where it is decreasing. Sketch the graph.

10. A dress manufacturer has found that the number x of dresses sold per month is related to the price p dollars per dress by the equation

$$p = 2 - \left(\frac{x}{300}\right)^2.$$

Determine whether the marginal revenue is increasing or decreasing when x is 200, 300, 500, and 600. Sketch a graph of the marginal revenue.

2. Extreme Values of a Function

In many practical problems, we are interested in determining the maximum or minimum values of a function. Geometrically, these values correspond, of course, to high and low points on a function's graph. We shall presently see how the derivative of a function is helpful in locating these values. We begin our discussion with

Definition 2. Let f be a differentiable function. A point c for which

$$f'(c) = 0$$

is called a critical value in the domain of f, and the point $(c, f(c))$ is called a critical point on the graph of f.

By our definition, the slope of the tangent line to the graph $\{(x, y) \mid y = f(x)\}$ at a critical point $(c, f(c))$ is zero. Consequently, the tangent line at the critical point is parallel to the x-axis. The function whose graph is given in Figure 5 has three critical points: $(x_1, f(x_1))$, $(x_2, f(x_2))$, and $(x_3, f(x_3))$.

To compute the critical values of a function f, we need to solve for x the equation

$$f'(x) = 0.$$

Example 1. Find the critical values of

(a) $f(x) = 2x^3 - 3x^2 - 36x + 5$

(b) $g(x) = 5 + \dfrac{1}{x}$.

Solution. (a) Differentiating, we get

$$f'(x) = 6x^2 - 6x - 36,$$

Critical points on the graph of a function f

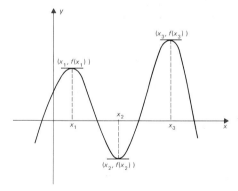

Figure 5

so the critical values must satisfy the equation

$$6x^2 - 6x - 36 = 0$$

or

$$x^2 - x - 6 = 0$$

or

$$(x + 2)(x - 3) = 0.$$

Thus $x = -2$ and $x = 3$ are the critical values of the function f.

(b) Differentiating, we get

$$g'(x) = -\frac{1}{x^2}.$$

Since $-1/x^2$ is never zero, the function g has no critical values.

Relative extremum

The word "relative" as used in the definition means that $f(x_0)$ is an extreme value with respect to "nearby" points.

Note. If x_0 is an interior point in the domain of f, then I is an open interval containing x_0. Whereas, if x_0 is the left endpoint of the domain of f, then $I = \{x | x_0 \leq x < x_0 + h, h > 0\}$. Similarly, if x_0 is the right endpoint, $I = \{x | x - h < x \leq x_0, h > 0\}$.

Note. If a function is constant on an interval I, then at every point of I there is a relative maximum and a relative minimum.

Absolute maximum and minimum

Definition 3. A function f has a *relative maximum value* at x_0 if there is an interval I in the domain of f containing x_0 such that

$$f(x) \leq f(x_0) \text{ for all } x \in I.$$

Similarly, f has a *relative minimum value* at x_1 if there is an interval I in the domain of f containing x_1 such that

$$f(x_1) \leq f(x) \text{ for all } x \in I.$$

The term *relative extremum* covers both situations.

In Figure 5, the function f has relative maxima at the points x_1 and x_3 and a relative minimum at x_2.

Definition 4. A function f is said to have an *absolute maximum* at a if $f(a) \geq f(x)$ for all x in the domain of f. Similarly, f has an *absolute minimum* at b if $f(b) \leq f(x)$ for all x in the domain of f.

Let us discuss some of the above concepts in the following example.

Example 2. Find the extreme values of the functions

(a) $f(x) = x^2$ $(-\infty < x < \infty)$
(b) $f(x) = x^2$ $(-2 \leq x \leq 2)$
(c) $f(x) = x^2$ $(-1 \leq x \leq 3)$.

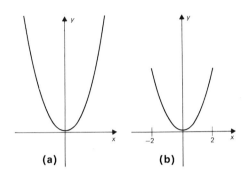

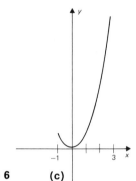

Figure 6   **(c)**

Solution. The graphs of the functions are given in Figure 6.

The relative and in fact the absolute minimum value for (a), (b), and (c) is at $x = 0$, since $f(0) = 0 \le x^2 = f(x)$ for all x in the domain of each of the functions.

Obviously the function $f(x) = x^2$ $(-\infty < x < \infty)$ has no relative or absolute maxima.

The function $f(x) = x^2$ $(-2 \le x \le 2)$ has both relative and absolute maxima at each of the two points $x = -2$ and $x = 2$.

The function $f(x) = x^2$ $(-1 \le x \le 3)$ has relative maxima at $x = -1$ and $x = 3$, and it has an absolute maximum at $x = 3$.

We shall now prove a theorem which gives a connection between the critical values and the extreme values of a function.

Theorem 2. **Let f be differentiable at c, and suppose f has a *relative extremum* at c. Then**

$$f'(c) = 0;$$

in other words, c is a critical value of f.

Proof. We recall that

(6) $$f'(c) = \lim_{h \to 0} \frac{f(c + h) - f(c)}{h}.$$

The limit in (6) is the same if h approaches zero through positive or negative values.

Suppose f has a relative maximum at c. Since f is differentiable at c, it is defined in some open interval I_1 in the domain of f containing c. Also, because f has a relative maximum at c, there must be an open interval I_2 containing c such that

$$f(x) \le f(c) \text{ for all } x \in I_2.$$

Thus, for all h such that $c + h \in I = I_1 \cap I_2$ we have

(7) $$f(c + h) - f(c) \le 0.$$

Now, if $h < 0$, we have

$$\frac{f(c + h) - f(c)}{h} \ge 0.$$

The function $\dfrac{f(c + h) - f(c)}{h}$ is nonnegative; it cannot have a negative limit.

If we let h approach zero through negative values, we have

(8) $$f'(c) = \lim_{h \to 0^-} \frac{f(c + h) - f(c)}{h} \ge 0.$$

Similarly, if $h > 0$, we have (from (7))

$$\frac{f(c + h) - f(c)}{h} \leq 0.$$

So, if we let h approach zero through positive values, we have

(9) $$f'(c) = \lim_{h \to 0^+} \frac{f(c + h) - f(c)}{h} \leq 0.$$

From (8) and (9), we conclude that $f'(c) \leq 0 \leq f'(c)$, whence

$$f'(c) = 0.$$

The proof for the case when f has a relative minimum at c is similar and is left as an exercise.

We advise the reader not to jump to the conclusion that a local extremum always exists when the derivative is zero. The theorem does not say that. The reader should carefully reread the theorem. Let us see why such care is necessary by considering some examples.

Example 3. Find the critical values and the extreme values for the function

(10) $$f(x) = x^3 \qquad (-\infty < x < \infty).$$

Solution. The graph of this function is sketched in Figure 7. Solving the equation

$$f'(x) = 3x^2 = 0$$

to find all the critical values of f, we find that $x = 0$ is the only critical value.

We can easily see that $f(0) = 0$ is *not* a relative extreme value. Since for $h > 0$, $f(-h) = -h^3 < 0$ and $f(h) = h^3 > 0$, $f(0)$ is neither greater nor less than both $f(-h)$ and $f(h)$.

As the graph shows, this function f is an increasing function. It has no extreme values.

The next example shows that a function may have an extremum, either relative or absolute, without the derivative being zero at that point.

Example 4. Find the extreme values of the function

$$f(x) = |x| \qquad (-\infty < x < \infty).$$

Solution. The graph of this function is sketched in Figure 8. Clearly, $f(0) = 0$ is a relative minimum value (in fact, an absolute minimum

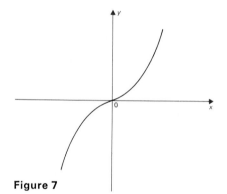

Figure 7

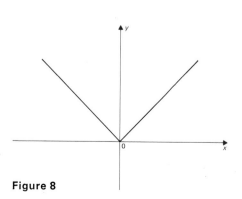

Figure 8

value). However, $f'(0)$ is not equal to zero, because $f'(0)$ does not exist! (See problem 21, Exercise 1, Chapter 5.)

The above examples illustrate that the derivative does not always tell us where the extreme values are. Rather, it tells us where the extreme values aren't. If the derivative f' is positive (or negative) at a point c, then f is increasing (or decreasing) in an interval containing the point c. Thus there cannot be an extreme value of f at c.

Consequently, if we wish to locate the extreme values of a function, only three kinds of points need be examined:

   (i)   points where the derivative is zero,
   (ii) points where the derivative does not exist,
   (iii) endpoints of the function's domain.

Remark. Example 3 above illustrates that a function need not have an extreme value. The following theorem, whose proof is beyond the scope of this book, shows that an important class of functions always has maximum and minimum values.

For a function continuous on a closed interval see problem 30, Exercise 3, Chapter 4.

Theorem 3. **If a function f is continuous on a closed interval, then f has a maximum value and a minimum value on that interval.**

The following theorem is used in many problems to test for relative maxima and minima.

Second derivative test

Theorem 4. **Let a function f have a continuous second derivative at a critical value c.**

   (i) **If $f''(c) > 0$, then f has a relative minimum at c.**
   (ii) **If $f''(c) < 0$, then f has a relative maximum at c.**

The proof of this theorem will be sketched in Section 5.

Example 5. Find the relative maxima and minima for the function

(11) $f(x) = x^3 - 6x^2 + 9x - 7 \ (0 \le x \le 5)$.

Find the absolute extrema.

Solution. We have

$$f'(x) = 3x^2 - 12x + 9$$
$$= 3(x - 1)(x - 3)$$

and the critical values are obtained from the equation

$$f'(x) = 0$$

or

$$3(x - 1)(x - 3) = 0,$$

namely, $x = 1$ and $x = 3$. Now

$$f''(x) = 6x - 12.$$

Thus we have

$$f''(1) = 6(1) - 12 = -6 < 0.$$

Hence $f(1)$ is a relative maximum. Similarly,

$$f''(3) = 6(3) - 12 = 6 > 0.$$

Hence, $f(3)$ is a relative minimum.

To test the endpoints, we note that

$$f'(x) = 3(x - 1)(x - 3) > 0$$

if

$$x < 1 \text{ or } x > 3.$$

Thus by Theorem 1, the function f is increasing on the intervals $(0, 1)$ and $(3, 5)$. Consequently, $f(0)$ is a relative minimum and $f(5)$ is a relative maximum.

To compute the absolute extrema, we note that

$$f(0) = -7$$
$$f(1) = -3$$
$$f(3) = -7$$
$$f(5) = 13.$$

The function has an absolute maximum at $x = 5$ and an absolute minimum at $x = 0$ and $x = 3$.

Example 6. Find the relative maxima and minima for the function

(12) $$f(x) = x + \frac{9}{x}.$$

Solution. We have

$$f'(x) = 1 - \frac{9}{x^2}.$$

Since $f(x)$ is undefined for $x = 0$, the function cannot have a relative maximum or a relative minimum value at 0.

Now from the equation

$$f'(x) = 1 - \frac{9}{x^2} = 0$$

we have

$$x^2 - 9 = 0$$

and the critical values are $x = 3$ and $x = -3$. Since

$$f''(x) = \frac{18}{x^3}$$

we have

$$f''(3) = \frac{18}{(3)^3} = \frac{2}{3} > 0,$$

so $f(3)$ is a relative minimum. Because

$$f''(-3) = \frac{18}{(-3)^3} = -\frac{2}{3} < 0,$$

$x = -3$ is a relative maximum.

Exercise 2

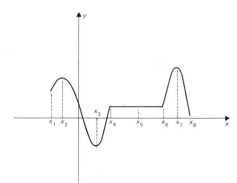

1. Consider the function whose graph is shown at the left. State whether the function has a relative maximum, relative minimum, absolute maximum, or absolute minimum at each of the points x_1, x_2, x_3, x_4, x_5, x_6, x_7, and x_8.

In problems 2 through 14, find the critical values and determine the relative maxima and the relative minima. Find the absolute extrema if any exist.

2. $f(x) = x^2 + 4x - 5 \quad (0 \leq x \leq 2)$
3. $f(x) = 5 - 4x - x^2 \quad (-4 \leq x \leq 4)$
4. $f(x) = 2x^3 - 6x^2 + 9 \quad (-\infty < x < \infty)$
5. $f(x) = 2x^4 + x^2 \quad (-\infty < x < \infty)$
6. $f(x) = x^3 + 3x^2 - 12x + 7 \quad (-\infty < x < \infty)$
7. $f(x) = 2x^4 - x^2 \quad (-1 \leq x \leq 1)$
8. $f(x) = x^3 - 3 \quad (-1 \leq x \leq 1)$
9. $f(x) = x^4 + 3x^2 - 6 \quad (-\infty < x < \infty)$
10. $f(x) = (x - 2)^2(x + 4)^3 \quad (-\infty < x < \infty)$
11. $f(x) = (x^2 - 4)^2 \quad (-4 \leq x \leq 4)$

12. $f(x) = \dfrac{1}{x^2 + 1} \quad (-\infty < x < \infty)$

13. $f(x) = \sqrt[3]{x} + x^3 \quad (-\infty < x < \infty)$

*14. $f(x) = \dfrac{x + 2}{x^2 + 4} \quad (-\infty < x < \infty)$

15. Complete the proof of Theorem 2.

16. Consider the cubic polynomial

$$f(x) = x^3 + ax^2 + bx + c.$$

(a) For what values of a and b will f have critical values at $x = 4$ and $x = -2$?

(b) Determine the nature of the critical points.

17. Consider the function $f(x) = 3x^2 - x + 1$. Show that $f(x) > 0$ for all x. *Hint:* Determine the minimum value of $f(x)$.

18. If $f(x) = -x^2 + x - 4$, show that $f(x) < 0$ for all x.

19. If the selling price x is related to the profit P by the equation

$$P = 5000x - 125x^2$$

(a) For what range of values of x is the profit increasing?

(b) For what range of values of x is the profit decreasing?

(c) Determine the value of x that would yield maximum profit.

3. Applications of Maxima and Minima I

Although they may understand the concepts involved in the theory of maxima and minima for a curve, many students find it difficult to solve word problems. In this and the next section we shall illustrate how the theory developed in Section 2 is applied to practical situations. We shall give examples of how to translate words into abstract symbols.

Example 1. A farmer having 120 ft of fencing wishes to contain a cow in a rectangular plot of land along the bank of a river. What should the dimensions of the rectangle be to provide the cow with maximum grazing ground? (Assume that no fence is needed along the river, which is flowing along a straight edge.)

Solution. We illustrate in steps a method for solving problems such as this one.

Step 1. Draw a figure showing the relevant features.

Step 2. Using the figure, find a relationship between the variables involved and express the quantity to be maximized or minimized (in this case, the area) as a function of one of the variables. Since the fence is 120 ft, we have

(13) $$120 = 2x + y,$$

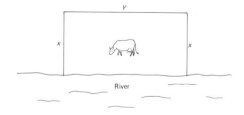

Figure 9

and the area of the rectangle is

(14) $A = xy.$

Solving for y in equation (13) we obtain

(15) $y = 120 - 2x.$

Substitution of (15) into (14) yields

$$A = x(120 - 2x)$$

or

Note. This is a quadratic function and the problem can also be solved by the method of Chapter 2, Section 7.

(16) $A = 120x - 2x^2.$

Thus the area is expressed in terms of a single variable x.

Step 3. Compute the derivative of $A(x)$:

(17) $\dfrac{dA}{dx} = 120 - 4x.$

Step 4. Locate the critical values by solving the equation

$$\dfrac{dA}{dx} = 0.$$

Thus,

$$120 - 4x = 0$$

or

$$x = 30.$$

Step 5. Apply the second derivative test to determine which of the critical points yields the maximum or minimum value of the function.

In this case there exists only one critical value; namely, $x = 30$. From equation (17) we find that

(18) $\dfrac{d^2A}{dx^2} = -4 < 0$

and by Theorem 4, the function A has a maximum value at $x = 30$. (Note again that if more than one critical point is obtained, select the ones that render the problem meaningful.) To solve for y we use equation (15):

$$y = 120 - 2(30) = 60.$$

In conclusion, the farmer encloses maximum area if $x = 30$ and $y = 60$. The area enclosed is

$$A = (30)(60) = 1800 \text{ sq ft.}$$

Example 2. Find the two positive numbers whose sum is 30 having maximum product.

Solution. In this case we do not have the aid of a figure. So we go to Step 2 and find the product as a function of a single variable.

  Let the two numbers be x and y. We have

(19) $x + y = 30$

and the product P is

(20) $P = xy.$

Using equation (19) we express one variable in terms of the other, say

(21) $y = 30 - x.$

Substitution of (21) into (20) yields the function P where

(22) $P(x) = x(30 - x) = 30x - x^2.$

The derivative of P with respect to x is

(23) $\dfrac{dP}{dx} = 30 - 2x.$

Therefore,

$$\frac{dP}{dx} = 0 \text{ when } x = 15.$$

Now, from equation (23) we have

(24) $\dfrac{d^2P}{dx^2} = -2 < 0$

and by Theorem 4, the function P has a maximum value when $x = 15$. From equation (19) we obtain $y = 15$.

Example 3. A soup manufacturing company wishes to pack 25 cu in of mushroom soup in a can in the form of a right circular cylinder. Find the dimensions of the can if the surface area is to be a minimum.

Solution. Again we start by drawing a figure (Figure 10). Clearly, the surface area of the metal used is

(25) $A = 2\pi r^2 + 2\pi rh$

where $2\pi r^2$ is the area of the top and the bottom. Here the area is a function of both r and h. To obtain A as a function of a single variable we use the formula for the volume of the can:

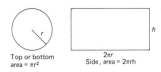

Top or bottom
area = πr^2

$2\pi r$
Side, area = $2\pi rh$

Figure 10

(26) $$V = \pi r^2 h = 25.$$

Solving for h, we have

(27) $$h = \frac{25}{\pi r^2} \quad (r \neq 0).$$

Substitution of (27) into (25) yields the function A where

(28) $$A(r) = 2\pi r^2 + 2\pi r \frac{25}{\pi r^2}$$

or

(29) $$A(r) = 2\pi r^2 + \frac{50}{r}.$$

Therefore we have expressed A as a function of a single variable r. Of course, we could have expressed A as a function of h only. The student should do this and show that the end result is the same.

Next we compute the derivative of A:

(30) $$\frac{dA}{dr} = 4\pi r - \frac{50}{r^2} \quad \text{(why ?)}.$$

Therefore, from the equation $dA/dr = 0$ we have

(31) $$4\pi r - \frac{50}{r^2} = 0$$

or

(32) $$4\pi r^3 - 50 = 0$$

and

$$r^3 = \frac{50}{4\pi} = \frac{25}{2\pi}$$

or

(33) $$r = \sqrt[3]{\frac{25}{2\pi}}.$$

Applying the second derivative test we find that

(34) $$\frac{d^2A}{dx^2} = 4\pi + \frac{100}{r^3}$$

and at $r = \sqrt[3]{25/2\pi}$,

$$\frac{d^2A}{dr^2} = 4\pi + \frac{100}{(\sqrt[3]{25/2\pi})^3}$$

$$= 4\pi + (100)\frac{(2\pi)}{25}$$

$$= 12\pi \quad (> 0).$$

By Theorem 4, therefore, A has a minimum value at $r = \sqrt{25/2\pi}$. The corresponding value of h is obtained from equation (27):

$$h = \frac{25}{\pi(25/2\pi)^{2/3}}$$

$$= \frac{25}{\pi}\left(\frac{2\pi}{25}\right)^{2/3}.$$

We now give an example of an apparently similar problem for which an extremum does not exist.

Example 4. A cylindrical package having 18 cu in. volume is to be wrapped in a rectangular sheet of paper. What should the dimensions of the cylinder be to minimize the surface area of the paper?

Solution. Let the dimensions of the rectangular paper be x by y (see Figure 11). Now the volume of the cylinder is

(35) $$V = \pi r^2 x = 18.$$

Since $r = y/2\pi$, from equation (35) we have

$$\pi\left(\frac{y}{2\pi}\right)^2 x = 18$$

or

(36) $$x = \frac{18(4\pi)}{y^2}.$$

The area A of the paper is

(37) $$A = xy.$$

Substituting (36) into (37) we obtain A in terms of a single variable:

(38) $$A(y) = y\frac{72\pi}{y^2} = \frac{72\pi}{y}.$$

The derivative of A with respect to y is

$$\frac{dA}{dy} = -\frac{72\pi}{y^2}.$$

Since dA/dy is never zero, A has no critical value and hence no minimum value.

y

x

$r = \frac{y}{2\pi}$ x

Figure 11

Exercise 3

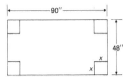

1. A rectangle is to be constructed having a given perimeter P. Find the dimensions of the rectangle of maximum area.

2. A box with square base is to be constructed to hold 64 cu in. Find the dimensions of the box of minimum surface area.

3. A rectangular piece of aluminum sheet is 48″ × 90″. An open rectangular container is to be made by cutting a square from each corner and bending up the sides. Find the dimensions of the container of maximum volume.

4. A sheet of paper for a poster is 18 sq ft in area. The margins at the top and bottom are 9 inches each and the margin on each side is 6 inches. What are the dimensions of the paper if the printed area is maximum?

5. Find the dimensions of the isosceles triangle of maximum area if the perimeter is to be 24 inches.

6. Show that the rectangle of largest area that can be inscribed in a circle is a square.

7. Find the dimensions of the rectangle of largest area that can be inscribed in a semicircle having diameter $2r$.

8. Find the dimensions of the rectangle of largest area that can be inscribed in the ellipse

$$\frac{x^2}{a^2} + \frac{y^2}{b^2} = 1.$$

9. Consider an isosceles triangle with sides 5, 5, and 6. Find the dimensions of the rectangle of largest area that can be inscribed in the triangle so that one side is along the base.

10. A piece of wire 20 inches long is to be cut in two pieces, one to form a circle and the other a square. How should the wire be cut in order that the sum of the two areas enclosed by the wire be minimal?

11. An open cylindrical tank with circular base is to be constructed of sheet metal so as to contain a volume πa^3 of water. Find the height and the radius of the base so that the quantity of sheet metal required may be minimal.

12. A 288 cu ft pool is to have a square top. The sides are to be built of see-through glass and the bottom of mosaic. The cost per unit area of glass is three times the cost of mosaic. Find the dimensions of the pool having minimum cost.

13. Two cars are traveling along two roads which cross each other at right angles at A. Both cars are traveling toward A at 30 ft per sec. Initially their distances from A are 1500 ft and 2100 ft respectively. At what time is the distance between the two cars a minimum? Find this distance.

14. A rectangular box with square bottom and top is to contain 1000 cu ft. The cost of material per square foot for the bottom is 25¢, for the top, 15¢, and for the sides, 20¢. The labor charges for making the box are \$3. Find the dimensions of the box when the cost is minimal.

4. Applications of Maxima and Minima II

The theory of maxima and minima has many applications in a variety of disciplines. In this section, we shall consider examples from the sciences, business, and economics.

Example 1. An experiment shows that the rate of an autocatalytic reaction is proportional to the amount of substance produced times the amount of the original substance. For what value of the substance produced will the rate of the reaction be maximal?

Solution. Let c denote the amount of the original substance at the beginning of the reaction. At a later time, if the amount of the new substance formed is x, then the amount of original substance left is $c - x$. Therefore, if R is the rate of the reaction, then

$$(39) \qquad R = kx(c - x) = ckx - kx^2 \qquad (0 \leq x \leq c)$$

where $k > 0$ is a constant of proportionality. For the rate of the reaction to be a maximum, we must have, for some value of x,

$$(40) \qquad \frac{dR}{dx} = ck - 2kx = 0.$$

Equation (40) is satisfied if $x = c/2$. Thus, when the concentration of the new substance formed is one-half the amount of the original substance at the beginning of the reaction, the rate of the reaction is extremal.

We now show by application of the second derivative test that the rate of the reaction is a maximum when $x = c/2$. From equation (40) we have

$$\frac{d^2R}{dx^2} = -2k.$$

Since $k > 0$, $d^2R/dx^2 < 0$, and by Theorem 4, R attains a maximum at $x = c/2$.

Example 2. We return to the problem of the response of the body to a dose of a drug (see Example 1, Section 4, Chapter 5). Suppose the response R of the body is found to be related to the dosage x of the drug administered according to the equation

(41) $$R = kx^2 - \frac{x^3}{3} \qquad (k > 0).$$

The rate of change of R is

(42) $$\frac{dR}{dx} = 2kx - x^2.$$

Why is the domain $0 < x < 2k$?

Equation (42) tells us that the body is responding as long as $0 < x < 2k$. For what dosage of the drug is the rate of change dR/dx a maximum?

Solution. The function dR/dx has critical points for values of x for which

(43) $$\frac{d}{dx}\left(\frac{dR}{dx}\right) = 0$$

or

(44) $$\frac{d^2R}{dx^2} = 2k - 2x = 0.$$

Equation (44) is satisfied if $x = k$. Application of the second derivative test yields

(45) $$\frac{d^2}{dx^2}\left(\frac{dR}{dx}\right) = \frac{d^3R}{dx^3} = -2.$$

The second derivative test indicates that for $x = k$, the function dR/dx attains a maximum. In other words, when the concentration of the drug in the body is 50% of the amount administered, the greatest changes in the reaction will occur. This type of analysis is used to determine the optimum amount of the drug to be administered to give the greatest changes in the reaction for a small change in dosage.

Example 3. A Florida orange grower finds that an orange tree produces (on the average) 400 oranges per year if no more than 16 trees are planted in a unit area. For each additional tree planted per unit area, he finds that the yield decreases by 20 oranges per tree. How many trees should he plant per unit area to maximize the yield?

Solution. If x is the number of trees planted per unit area, then the number N of oranges per tree depends on x; i.e., $N = N(x)$. Since for every additional tree over 16 each tree produces 20 fewer oranges, we have for $x \geq 16$

(46) $$N = 400 - 20(x - 16)$$

and the total number T of oranges produced per unit area is

$$T = xN = x[400 - 20(x - 16)]$$

or

What assumptions are we making about the function T in order to take its derivative?

(47) $$T = 720x - 20x^2.$$

The first derivative condition for maximum yields

(48) $$\frac{dT}{dx} = 720 - 40x = 0.$$

Solving equation (48) we obtain

$$x = 18.$$

That is, T has a critical point at $x = 18$. Applying the second derivative test we find that

$$\frac{d^2T}{dx^2} = -40,$$

which is a negative number, and so by Theorem 4, if 18 trees are planted per unit area, the yield is maximum.

Example 4. A firm manufactures bikinis and sells its product at $24 per unit. The total cost C (dollars) of producing x bikinis is given by

(49) $$C = 150 + \frac{39}{10} x + \frac{3}{1000} x^2.$$

Write the profit P as a function of x and determine the number of bikinis that the firm should produce and sell to achieve maximum profit.

Solution. The firm's profit is its total revenue minus its total cost; that is,

(50) $$P = R(x) - C(x).$$

The total revenue from the sale of x bikinis at $24 per bikini is

(51) $$R(x) = 24x.$$

Substituting (49) and (51) into (50) we find that

(52) $$P = 24x - \left(150 + \frac{39}{10} x + \frac{3}{1000} x^2\right).$$

To achieve maximum profit, we must have

(53) $$\frac{dP}{dx} = 24 - \frac{39}{10} - \frac{3}{500}x = 0.$$

Solving equation (53) for x we obtain

$$\frac{201}{10} - \frac{3}{500}x = 0.$$

Thus, the critical value of P is $x = 3350$. Applying the second derivative test we find that

$$\frac{d^2P}{dx^2} = -\frac{3}{500},$$

which is a negative number, and so by Theorem 4, a production level of 3350 bikinis yields a maximum profit. The profit in dollars at this level of production is

$$P = 24(3350) - \left[150 + \frac{39}{10}(3350) + \frac{3}{1000}(3350)^2\right]$$

$$= \$33,517.50.$$

Note that if the firm should increase (or decrease) its output by one or more bikini, the firm's profit will decrease.

It is of interest to observe that the marginal revenue is

$$\frac{dR}{dx} = 24$$

and that at a production level of 3350 bikinis, the marginal cost is also

$$\left.\frac{dC}{dx}\right|_{x=3350} = \left.\frac{39}{10} + \frac{3}{500}x\right|_{x=3350}$$

$$= 24.$$

(See problem 9 in the exercises below.)

Exercise 4

1. The rate R of population growth of tuna fish off an island in the tropics is described by the equation

$$R = kN(c - N)$$

where k and c are positive constants and N is the number of tuna. A fishing company is interested in maintaining the population of the tuna fish in equilibrium. What should the annual catch (equilibrium catch) be to keep the population of tuna in equilibrium?

2. The rate R of population growth of a colony of rabbits on an Australian farm is described by

$$R = 200N^2 - \frac{N^3}{6}.$$

What is the equilibrium catch? That is, how many rabbits should be removed from the colony to maintain a fixed number?

3. The concentration of hydrogen ion in a solution is given by

$$X = H + \frac{10^{-5}}{H}.$$

For what value of H is the concentration a minimum?

4. Experiments show that the velocity of the flow of air through the respiratory system during coughing is given by

$$v = \frac{k}{\pi} r^2 (r_0 - r)$$

where k is a constant, r_0 is the radius of the windpipe when there is no pressure, and r is the radius of the windpipe after pressure builds up. Find the value of r for which the velocity of the flow of air is maximum.

5. For each of the following cost functions, find the production level x for which the cost is a minimum.
   (a) $C = 2x^3 - 8x^2 + 10x$
   (b) $C = 800 - \dfrac{x}{c} + \dfrac{1}{1000} x^2$
   (c) $C = \dfrac{x}{2} + \dfrac{5 \cdot 10^5}{x} + 10^3$

6. A firm produces x units of a product. If the profit P in dollars earned in the manufacture and sale of x units is

$$P = 20x - \frac{1}{500} x^2,$$

find the maximum profit.

7. A firm produces mink coats and sells each coat for $1500. The cost of producing x coats per year is

$$C = 9 + 3x + 0.015x^2.$$

   (a) Write an expression for the profit in terms of x.
   (b) Find a number x that yields maximum profit.
   (c) What is the maximum annual profit?

8. A manufacturer finds that he can sell x units per month of a

certain item if the price is $p = 18 - 0.001x$ dollars. Find the production level that maximizes his profits if the cost of producing x items is $C = 4x + 1500$ dollars. What is the price of each item at this production level?

9. A firm's production level is such that the profit is maximum. Find a relationship between the firm's marginal revenue and marginal cost.

10. In a certain area it is found that if 25 apple trees are planted per acre, the average yield per tree is 450 apples, and that for each tree in excess of 25 per acre the average yield is reduced by 15 apples per tree.

   (a) Show that the total number of apples produced per acre is $T = 825x - 15x^2$ $(x \geq 25)$.

   (b) Show that if 28 trees are planted per acre, the yield is maximum. (Remember that you cannot plant a fraction of a tree!)

11. The owner of a day nursery finds that if a monthly fee of $75 per child is charged, 20 children will enroll. To increase the enrollment, she finds that she will have to decrease the fee by $3 for each additional child.

   (a) If x $(x \geq 20)$ is the number of children enrolled, show that the total revenue is $r = 135x - 3x^2$.

   (b) Show that the maximum revenue is obtained by enrolling either 22 or 23 children.

12. A group of students arrange a chartered flight from New York to London, Paris, Heidelberg, Barcelona, and back to New York with a 3-day stay at each European city on the following basis. The charge per person is $499 if 100 students go on the flight. If more than 100 participate, the charge per student is reduced by an amount equal to $4 times the number of students above 100.

   (a) Find the total revenue.

   (b) Find the number of students that will furnish maximum revenue.

   (c) What is the cost per student if the maximum revenue is obtained?

5. The Second Derivative

In Section 2, we stated (without proof) a theorem giving a test for relative maxima and relative minima using the second derivative. In this section we shall sketch a proof of that theorem and study

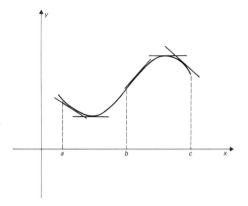

Figure 12

Recall the discussion on concavity of parabolas in Chapter 2, Section 7.

the relationship between the second derivative of a function and its graph.

Consider the graph of the function *f* in Figure 12. As *x* increases from *a* to *b*, the graph of the function bends in a counterclockwise direction, and from *b* to *c* it bends in a clockwise direction. We say that the graph of the function is *concave upward* between *a* and *b* and that it is *concave downward* between *b* and *c*. At the point *x* = *b*, the direction of bending changes. The point (*b*, *f*(*b*)) is called the *point of inflection* of the curve. It can be shown that if the graph of a function *f* is concave upward in an interval *I*, then the slopes of the tangent lines increase as the curve bends in the counterclockwise direction. Similarly, if the graph of a function *f* is concave downward in an interval *J*, then the slopes of the tangent lines decrease as the curve bends in the clockwise direction.

We showed in Section 1 that if the derivative of a function is positive on an interval, then the function is increasing on that interval. Since *f″* is found from *f′* in the same way as *f′* is found from *f*, we note that if *f″* is positive over an interval, then *f′* is increasing on that interval and consequently the graph of the function *f* is concave upward on that interval. Similar remarks apply if *f″* is negative over an interval. We have

Property 1. **If $f''(x) > 0$ on the interval *I*, the graph of *f* is concave upward on *I*. If $f''(x) < 0$ on the interval *I*, the graph of *f* is concave downward on *I*.**

Example 1. Discuss the concavity of the curve

$$y = f(x) = x^4 - 6x^3 + 12x^2 + 5x + 7.$$

Solution. Differentiating, we have

$$f'(x) = 4x^3 - 18x^2 + 24x + 5,$$

and

$$f''(x) = 12x^2 - 36x + 24 = 12(x - 1)(x - 2).$$

For

Explain why.

$$\begin{aligned}
x &< 1, & f''(x) &> 0 \\
1 < x &< 2, & f''(x) &< 0 \\
x &> 2, & f''(x) &> 0.
\end{aligned}$$

Hence the curve is concave upward in the intervals $(-\infty, 1)$ and $(2, \infty)$, and the curve is concave downward in the interval $(1, 2)$.

In the above example, it is easy to see that $f''(1) = 0$. Furthermore, the curve is concave upward on one side of $(1, f(1))$ and concave downward on the other side of $(1, f(1))$. In other words, the concavity changes at $(1, f(1))$. Thus the point $(1, f(1))$ is a point of inflection. Similarly, the point $(2, f(2))$ is also a point of inflection. Formally, we have

Definition 5. **A point $(a, f(a))$ is called a *point of inflection* of the graph of a function f if there exists an interval $(a - c, a + c)$ in the domain of f such that the graph of f is concave upward in $(a - c, a)$ and concave downward in $(a, a + c)$, or vice versa.**

The following theorem enables us to find the possible points of inflection.

Theorem 5. **If $(a, f(a))$ is a point of inflection of the graph of the function f and if f'' is continuous at a, then $f''(a) = 0$.**

The proof follows directly from Property 1.

The above theorem asserts that under the assumption that f'' is continuous, the points of inflection are among those points $(x, f(x))$ for which $f''(x) = 0$. However, if $f''(c) = 0$, it does not mean that $(c, f(c))$ is necessarily a point of inflection, as the following example illustrates.

Example 2. Show that $(0, 0)$ is not a point of inflection for the curve

$$y = f(x) = x^4.$$

Solution. We have

$$f'(x) = 4x^3$$

and

$$f''(x) = 12x^2.$$

Thus $f''(x) = 0$ if $x = 0$. However, $(0, 0)$ is not a point of inflection, since $f''(x) > 0$ for every $x \neq 0$, and therefore the concavity of the graph of f does not change at $x = 0$. Hence $(0,.0)$ is not an inflection point.

We now state and prove a theorem which gives us the sufficient conditions for the extremum of a function.

Theorem 6. Let *f* be a function and *a* be a number such that $f'(a) = 0$. Suppose f'' is continuous at *a*; then:

(i) $f(a)$ is a relative maximum value of *f* if $f''(a) < 0$
(ii) $f(a)$ is a relative minimum value of *f* if $f''(a) > 0$.

Proof. (i) Since f'' is continuous at *a* and $f''(a) < 0$, it follows (see problem 31, Exercise 3, Chapter 4) that there exists an interval $(a - c, a + c)$ such that $f''(x) < 0$ for every *x* in $(a - c, a + c)$. Therefore by Property 1 the graph of *f* is concave downward in the interval $(a - c, a + c)$. Since $f'(a) = 0$, the tangent line at $(a, f(a))$ is horizontal, and the graph of the function *f* is below the tangent line between $x = a - c$ and $x = a + c$. Thus $f(a)$ is a relative maximum value of *f*.

The proof of (ii) is similar.

Example 3. Find the relative extrema and the points of inflection of the graph of the function *f* where

$$f(x) = x^2(x^2 - 6).$$

Solution. Here

$$f'(x) = 4x^3 - 12x$$

and

$$f''(x) = 12(x^2 - 1).$$

Now $f''(x) = 0$ when $x^2 = 1$, that is, when

$$x = 1 \text{ or } x = -1.$$

When $x^2 > 1$, the second derivative is positive and the curve is concave upward. When $x^2 < 1$, the second derivative is negative and the curve is concave downward. Hence $x = 1$ and $x = -1$ give the points of inflection. Thus the points of inflection are $(1, -5)$ and $(-1, -5)$.

Now $f'(x) = 0$ when $4x^3 - 12x = 0$; i.e., when $x = 0$, $x = \sqrt{3}$, and $x = -\sqrt{3}$. Also,

$$f''(0) = 12(0 - 1) = -12 < 0$$
$$f''(\sqrt{3}) = 12(3 - 1) = 24 > 0$$
$$f''(-\sqrt{3}) = 12(3 - 1) = 24 > 0.$$

Thus $f(0)$ is a relative maximum value, while $f(\sqrt{3})$ and $f(-\sqrt{3})$ are relative minimum values for *f*.

Exercise 5

In problems 1 through 10 find the intervals over which the curve is (a) concave upward, (b) concave downward. Also find the points of inflection.

1. $y = x^2 + 2x - 4$
2. $y = 2x^3 + 4x^2 + 2x + 1$
3. $y = x^3 - 5x^2 + 8x - 4$
4. $y = \frac{1}{3}(x^3 + 9x^2)$
5. $y = x^2 - 3x^3 + 3x^4$
6. $y = x + \dfrac{1}{x}$
7. $y = \dfrac{1}{1 + x^2}$
8. $y = \dfrac{x + 1}{x^2 + x + 1}$
9. $(1 + x^2)y = x^2$
10. $(x + 3)^2 y = 2x$
11. In problems 2 through 4 above find the equation of the tangent line to the graph at each point of inflection.
12. Determine the value of k so that the function $f(x) = x^2 + (k/x)$ has an inflection point at $x = 2$.
13. If the cost C dollars of making x units of a product is given by

$$C = 6x - 0.03x^2 + 0.00005x^3,$$

determine whether the curve is concave upward, concave downward, or has an inflection point at $x = 100$, $x = 200$, $x = 300$, and $x = 400$.
14. (a) Find the points of inflection of the graph of the function $f(x) = (x + 1)/(x^2 + 1)$.
    (b) Show that the points lie on a line.
15. Prove Theorem 5.
16. Prove part (ii) of Theorem 6.
17. Consider the fourth degree polynomial $f(x) = x^4 + ax^3 + bx^2 + cx + d$ and suppose $3a^2 = 4b$. Show that f has no inflection points.

Augustin Louis Cauchy (1789–1857) was born in Paris at the outset of the Revolution. His father was a lawyer and later became Secretary of the Senate under Napoleon. The first eleven years of Cauchy's life were spent in the village of Arcueil where the family had gone to escape the Reign of Terror.

At eighteen he entered civil engineering school, and two years later Napoleon commissioned him to participate in the construction of the harbor at Cherbourg. In 1815 he was appointed professor at the École Polytechnique and at the Sorbonne. With the revolution of 1830 he left France in voluntary exile. When he returned to Paris eight years later he resumed his posts without giving allegiance to the government.

Cauchy introduced rigorous methods in mathematical analysis. He created the theory of functions of a complex variable and produced over 500 papers on all branches of mathematics.

6. The Differential

Let $y = f(x)$ define a differentiable function of x. In the previous chapter we introduced the notation dy/dx to represent the derivative of f with respect to x without clarifying the meaning of the symbols dy and dx. During the nineteenth century much controversy shrouded these symbols. Historically, the symbols date back to Leibniz, who attempted to define *dy* and *dx* as infinitesimals (very small quantities) "greater than zero but less than any positive real number." Leibniz was aware that there were no such real numbers. Nevertheless, he was very successful in solving many problems using these symbols. Some years later, Cauchy offered an alternate definition of the symbols. In this section we shall define these symbols and show how they are used in mathematics and the sciences.

Let us return to the definition of the derivative of a function. Suppose *f* given by $y = f(x)$ is a differentiable function of *x*. Let Δx be the *increment*, or change, in the variable *x*; then the *increment* Δy in the functional value $f(x)$ is given by

Here we are using the customary Δx (read "delta x") in place of h as used in Chapter 5.

$$(54) \qquad \Delta y = f(x + \Delta x) - f(x).$$

For a particular function *f*, Δy depends on both the numbers *x* and Δx.

We may write the derivative *f'* as

$$(55) \qquad f'(x) = \lim_{\Delta x \to 0} \frac{f(x + \Delta x) - f(x)}{x}$$

$$= \lim_{\Delta x \to 0} \frac{\Delta y}{\Delta x}.$$

In many practical problems (see Example 3 below), we are interested in computing Δy. But the computation of Δy may be quite difficult or tedious. We therefore use a close approximation to Δy, called the *differential*.

Definition 6. **Let $y = f(x)$ define a differentiable function of x. Then**

(i) **The *differential dx* of the independent variable x is defined by $dx = \Delta x$, where Δx is any real number.**

(ii) **The *differential dy* of the dependent variable y is defined by $dy = f'(x)\, dx$.**

A few remarks are in order:

Remark 1. The definition $dx = \Delta x$ is completely arbitrary and is therefore an independent variable whose range is the set of real numbers.

Remark 2. As we have indicated earlier, *dy* is used as an approximation for Δy (see examples below), and in many cases it will be a relatively simpler matter to calculate *dy* rather than Δ*y*.

Remark 3. An important result of this definition is the fact that the ratio of *dy* to *dx* is *f'*(*x*).

Example 1. Let *y* = *f*(*x*) = *x²* − 3*x* + 2. Compare Δ*y* and *dy* for

$$\text{(a)} \;\; x = 3, \Delta x = -\tfrac{1}{2}$$
$$\text{(b)} \;\; x = 3, \Delta x = \tfrac{1}{8}.$$

Solution. Since

$$f(x) = x^2 - 3x + 2,$$
$$f'(x) = 2x - 3.$$

By definition:

(56)
$$dy = f'(x)\,dx$$
$$= (2x - 3)\,dx.$$

Also

$$\Delta y = f(x + \Delta x) = f(x)$$
$$= [(x + \Delta x)^2 - 3(x + \Delta x) + 2] - [x^2 - 3x + 2]$$

or

(57)
$$\Delta y = 2x\,\Delta x + (\Delta x)^2 - 3\,\Delta x.$$

(a) Substituting *x* = 3 and *dx* = Δ*x* = −½ into equations (56) and (57) respectively, we obtain

$$dy = (2 \cdot 3 - 3)(-\tfrac{1}{2}) = -\tfrac{3}{2}$$
$$\Delta y = 2 \cdot 3(-\tfrac{1}{2}) + (-\tfrac{1}{2})^2 - 3(-\tfrac{1}{2})$$
$$= -3 + \tfrac{1}{4} + \tfrac{3}{2}$$
$$= -\tfrac{3}{2} + \tfrac{1}{4}.$$

Thus

(58)
$$|\Delta y - dy| = \tfrac{1}{4}.$$

(b) Setting *x* = 3 and Δ*x* = ⅛, we get

$$dy = (2 \cdot 3 - 3)\tfrac{1}{8} = \tfrac{3}{8}$$

and

$$\Delta y = 2 \cdot 3 \cdot \tfrac{1}{8} + (\tfrac{1}{8})^2 - 3 \cdot \tfrac{1}{8}$$
$$= \tfrac{3}{8} + \tfrac{1}{64},$$

Note that in this example |Δ*y* − *dy*| = (Δ*x*)².

or

(59)
$$|\Delta y - dy| = \tfrac{1}{64}.$$

We observe that even in such simple cases, it is easier to calculate *dy* than Δ*y*. We also note in this example that *dy* approximates Δ*y* better when Δ*x* is changed from ½ to ⅛ while *x* = 3 is kept fixed.

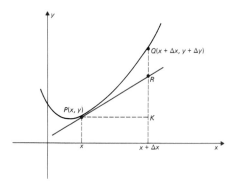

Figure 13

Let us now consider the geometrical meaning of the differential. Let $P(x, y)$ and $Q(x + \Delta x, y + \Delta y)$ be two points on the graph of $y = f(x)$ (see Figure 13).

Then the slope of the tangent line at P is given by $f'(x) = dy/dx$. But the slope of the tangent line is also given by the ratio RK/PK, and since $PK = \Delta x = dx$, it follows that $RK = f'(x)\, dx = dy$.

Now $\Delta y = KQ$. Thus the difference between Δy and dy is represented geometrically by the segment RQ.

Applications. In many problems we come across a quantity x which is measured by a faulty instrument. Due to this error in the increment of x, there will be a corresponding error in the computed value of y, where $y = f(x)$. The error in y, i.e., Δy, is approximated by dy. If the error in y is dy, then dy/y is called the *relative error* in y and $100 dy/y$ is called the *percentage error* in y.

Example 2. Calculate approximately the fourth root of 626.

Solution. Since we know the fourth root of 625 is 5, we shall approximate the amount by which $x^{1/4}$ increases when x is increased by 1. Let

$$y = x^{1/4} \quad \text{and} \quad y + \Delta y = \sqrt[4]{x + \Delta x}.$$

Then

$$dy = \tfrac{1}{4} x^{-3/4}\, dx$$

or

$$dy = \frac{dx}{4x^{3/4}}.$$

Take $x = 625$ and $dx = \Delta x = 1$; then

$$y = (625)^{1/4} = 5$$

and

$$dy = \frac{1}{4(625)^{3/4}} = \frac{1}{4(125)} = \frac{1}{500} = 0.002.$$

Now

$$(626)^{1/4} = y + \Delta y$$
$$\approx y + dy$$
$$\approx 5 + 0.002$$
$$\approx 5.002.$$

Example 3. A student in a physics laboratory measures the side

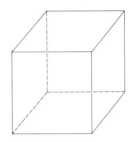

Figure 14

of a cube to be 2 ft with a possible error of 0.002 ft. What is the greatest possible error in his computed volume?

Solution. The possible error of 0.002 ft in the side x of the cube signifies that the value 2 ft is subject to a correction which may amount to as much as 0.002 ft in either direction. That is, the error in x may be as much as $dx = \Delta x = \pm 0.002$. Then since Δx is small, the computed value of the volume

$$V = f(x) = x^3$$

is subject to an approximate error given by

$$dV = f'(x)\ dx = 3x^2\ dx.$$

Here $x = 2$ and $\Delta x = \pm 0.002$.
   Therefore

$$dV = \pm 3(2)^2(0.002) = \pm 0.024.$$

Thus the computed volume 8 cu ft may be too large or too small by 0.024 cu ft, approximately.
   The relative error in this case is

$$\frac{dV}{V} = \frac{0.024}{8} = 0.003.$$

The relative error in x is

$$\frac{dx}{x} = \frac{0.002}{2} = 0.001.$$

   The percentage error in the side of the cube and its volume will be 0.1 and 0.3 respectively.

Exercise 6

In problems 1 through 5, find dy, Δy, and $|\Delta y - dy|$ when $x = 4$ and $\Delta x = 0.5$.

1. $y = x^3$ **2.** $y = 2x^3 - 7x$

3. $y = x^2 + 5x + 3$ **4.** $y = \sqrt{x}$

5. $y = \dfrac{2x}{3 + x}$

6. Compute $\sqrt[5]{33}$ approximately.

7. Compute approximately the square root of
   (a) 123   (b) 0.0143   (c) 255   (d) 10.34.

8. Compute $(1.98)^5$ approximately.

9. By writing $25 = 27 - 2$, calculate approximately $\sqrt[3]{25}$.

10. The diameter of a circle is 4 ft. By approximately how much does the area decrease if the radius decreases by $\frac{1}{30}$ inches?

11. The diameter of a sphere is 6 inches with a possible error of 0.05 in. What is the greatest possible error in the computed area and the relative possible error?

12. A cubic box is to hold 1000 cu ft. What is the allowable error in the edge of the cube if the error in the volume is not to exceed 2 cu ft?

13. If the area of a circle increases at a constant rate, show that the rate of increase of the perimeter varies inversely as the radius.

14. The total cost of producing x units is C dollars where

$$C = \tfrac{1}{100}x^2 + 5x + 200.$$

(a) Find the differential dC.

(b) When the level of production changes from 100 to 101, what is the value of dC?

(c) Determine the error involved when dC is used as an approximation for ΔC.

(d) Determine the relative and percentage errors.

15. The total cost C (dollars) of producing x shirts is

$$C = \frac{x^3}{1500} - \frac{2}{5}x^2 + 45x$$

and each shirt is sold at $10.

(a) Find the total profit P.

(b) Find the differential dP.

(c) When the production level changes from $x = 350$ to $x = 355$, what will be an approximate change in P?

Georg Friedrich Bernhard Riemann

7

The

Integral

Elementary calculus is often divided into two major branches: differential calculus and integral calculus. The former deals with the concept of the derivative. The latter deals with another basic concept—that of the *integral*. We have seen that the derivative, from a geometrical point of view, is related to the slope of a tangent line to a curve. We have used this notion to study problems in the sciences and in economic theory. From a geometrical point of view, the integral is related to the area of a region. In this chapter we shall study some elementary aspects of the integral and develop techniques for calculating areas of regions bounded by curves. Also, we shall show how the integral is related to certain problems arising in the sciences and economic theory. First we shall introduce the notion of antiderivatives and then a definition of the definite integral will be given in Section 2. We shall relate the problem of tangents to the problem of areas.

The problem of finding the area of a region bounded by curves is an old one. Archimedes (287–212 B.C.) successfully solved one case (see Figure 1); however, his method could not be applied generally. The problem was not solved until Newton and Leibniz created the calculus.

1. The Antiderivative

Consider the function f defined by $f(x) = 2x$. Then the function F defined by $F(x) = x^2$ has the property that

$$F'(x) = f(x)$$

for all x.

Similarly, if $f(x) = 3x^2$, then $F(x) = x^3$ has the property that

$$F'(x) = f(x).$$

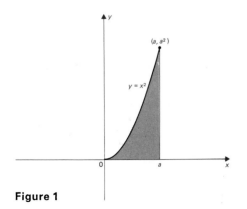

Figure 1

199

Definition 1. Suppose *f* is a continuous function defined on an open interval *I*. Then a function *F* is called an *antiderivative* of *f* on *I* if

$$F'(x) = f(x)$$

for all $x \in I$.

Thus, in each example above the function *F* is an antiderivative of *f*.

We observe that for $f(x) = 2x$ the function $G(x) = x^2 + C$ where *C* is a constant is also an antiderivative of *f*. We find that

$$G'(x) = (x^2 + C)' = 2x = f(x).$$

In general we have

Theorem 1. Suppose

(i)   *f* is a continuous function on an open interval *I*.
(ii) *F* is an antiderivative of *f*.

Then the function *G* is also an antiderivative of *f* if

$$G(x) = F(x) + C$$

where *C* is a constant.

Proof. If $G(x) = F(x) + C$, then

$$G'(x) = (F(x) + C)' = F'(x) + 0 = F'(x) = f(x)$$

and so by Definition 1, *G* is an antiderivative of *f*.

Notation. If *F* is an antiderivative of *f*, then every antiderivative of *f* is of the form $F + C$ where *C* is a constant. We write

$$\int f = F + C$$

and call

$$\int \qquad \text{the integral sign}$$

$$f \qquad \text{the integrand}$$

$$\int f \qquad \text{the indefinite integral of } f.$$

The constant *C* is called the constant of integration. Another notation we will use for the indefinite integral of f is

$$\int f(x)\ dx.$$

Example 1. Find the antiderivative of f where $f(x) = x^2$.

Solution. We have

$$\int f(x)\ dx = \int x^2\ dx = \tfrac{1}{3}x^3 + C.$$

It is an easy task to check one's answer. Remember that the derivative of the indefinite integral should be equal to the integrand. Thus in Example 1 we have

$$\frac{d}{dx}\left(\frac{x^3}{3} + C\right) = \frac{d}{dx}\left(\frac{x^3}{3}\right) + \frac{d}{dx}(C) = x^2 + 0 = x^2.$$

We observe that

$$\int 4x^3\ dx = x^4 + C$$

$$\int x^4\ dx = \frac{x^5}{5} + C$$

$$\int x^5\ dx = \frac{x^6}{6} + C$$

$$\int x^{-1/2}\ dx = \frac{x^{1/2}}{1/2} + C.$$

In general we have

Note. $\dfrac{d}{dx}\left[\dfrac{x^{r+1}}{r+1} + C\right] = x^r$, *for $r \neq -1$.*

(1) $$\int x^r\ dx = \frac{x^{r+1}}{r+1} + C \qquad (r \neq -1).$$

The case $r = -1$ will be discussed in Chapter 8.

Some important properties of the indefinite integral are stated in

Theorem 2. If f and g have antiderivatives and k is a constant, then

(i) $$\int kf(x)\ dx = k\int f(x)\ dx$$

(ii) $$\int [f(x) + g(x)]\ dx = \int f(x)\ dx + \int g(x)\ dx.$$

Proof of (i). We have

$$\frac{d}{dx}\left[k\int f(x)\ dx\right] = k\frac{d}{dx}\left[\int f(x)\ dx\right] = kf(x).$$

Therefore, by Definition 1, $k \int f$ is the antiderivative of kf.

The proof of (ii) is left for the student as an exercise (see problem 22 below).

Example 2. Calculate each of the following indefinite integrals:

(a) $\displaystyle\int (2 + x)\, dx$ (b) $\displaystyle\int \sqrt{x}\, dx$

(c) $\displaystyle\int \frac{x + 3}{x^3}\, dx,$ $\quad x \neq 0$ (d) $\displaystyle\int \frac{x^2 - 4}{x - 2}\, dx,$ $\quad x \neq 2.$

Solution.

Application of Theorem 2

(a) $\displaystyle\int (2 + x)\, dx = \int 2\, dx + \int x\, dx$

Application of equation (1) yields

$$\int dx = \int x^0\, dx = \frac{x^{0+1}}{0 + 1} = x.$$

$$= 2\int dx + \int x\, dx$$

$$= 2x + C_1 + \frac{x^2}{2} + C_2$$

$C = C_1 + C_2$

$$= 2x + \frac{x^2}{2} + C.$$

(b) $\displaystyle\int \sqrt{x}\, dx = \int x^{1/2}\, dx$

$$= \frac{x^{1/2+1}}{\frac{1}{2} + 1} + C$$

$$= \tfrac{2}{3}x^{3/2} + C$$

(c) $\displaystyle\int \frac{x + 3}{x^3}\, dx = \int \left(\frac{1}{x^2} + \frac{3}{x^3}\right) dx,$ $\quad x \neq 0$

$$= \int (x^{-2} + 3x^{-3})\, dx$$

$$= \int x^{-2}\, dx + 3\int x^{-3}\, dx$$

The two constants of integration are combined.

$$= \frac{x^{-1}}{-1} + 3\left(\frac{x^{-2}}{-2}\right) + C$$

$$= -\frac{2x + 3}{2x^2} + C$$

(d) $\displaystyle\int \frac{x^2 - 4}{x - 2}\, dx = \int \frac{(x - 2)(x + 2)}{x - 2}\, dx,$ $\quad x \neq 2$

$$= \int (x + 2)\, dx$$

$$= \int x \, dx + 2 \int dx$$

$$= \frac{x^2}{2} + 2x + C.$$

Exercise 1

In problems 1 through 20 calculate the given indefinite integrals.

1. $\int \frac{1}{2} \, dx$ **2.** $\int \frac{1}{2}x \, dx$

3. $\int 3x^3 \, dx$ **4.** $\int (x + x^3) \, dx$

5. $\int (x^2 + 2x + 1) \, dx$ **6.** $\int x(x - 2) \, dx$

7. $\int \frac{1}{x^2} \, dx, \quad x \neq 0$ **8.** $\int \frac{3}{x^3} \, dx, \quad x \neq 0$

9. $\int (x - 1)^2 \, dx$ **10.** $\int \frac{x + 4}{x^4} \, dx, \quad x \neq 0$

11. $\int 2\sqrt{x} \, dx$ **12.** $\int \frac{1}{\sqrt{x}} \, dx, \quad x > 0$

13. $\int x^{-1/3} \, dx, \quad x \neq 0$ **14.** $\int (3x^{1/3} + 2x^{-1/3}) \, dx, \quad x \neq 0$

15. $\int \frac{x^2 - 1}{x + 1} \, dx, \quad x \neq -1$ **16.** $\int x^{1/3}(2 + x^{1/3}) \, dx$

17. $\int \frac{x^3 - 2}{\sqrt{x}} \, dx, \quad x > 0$ **18.** $\int \frac{x^3 - 27}{x - 3} \, dx, \quad x \neq 3$

19. $\int \frac{x - 4}{\sqrt{x} + 2} \, dx$ **20.** $\int \frac{x + 1}{x^{1/3} + 1} \, dx, \quad x \neq -1$

21. Give examples of functions f and g having antiderivatives such that

(a) $\int f(x)g(x) \, dx \neq \left(\int f(x) \, dx \right) \left(\int g(x) \, dx \right)$

(b) $\int xf(x) \, dx \neq x \int f(x) \, dx.$

22. Prove Theorem 2, part (ii).

2. The Definite Integral

What is meant by the area of a region in the plane? If the region is bounded by straight line segments, it is easy to calculate its area by dividing it into rectangles and triangles. Thus in Figure 2 the area $A(R)$ of the region R is given by

$$A(R) = A(R_1) + A(R_2) + A(R_3) + A(R_4)$$

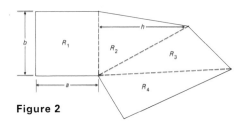

Figure 2

where the area of the rectangular region R_1 with length a and height b is

$$A(R_1) = ab,$$

the area of the triangular region R_2 with base b and height h is

$$A(R_2) = \tfrac{1}{2}bh,$$

and so on.

The problem of finding areas is more complicated if the region has curvilinear boundaries (see Figure 3). In this section we shall find that the area of a region in the plane with curvilinear boundaries can be represented as the limit of a sequence. Our exposition will be intuitive in nature and will give the general idea of the precise definition which follows.

Consider the region R bounded by the graph of $f(x) = x^2$ and the lines $\{(x, y)\,|\, x = 2\}$ and $\{(x, y)\,|\, y = 0\}$ (see Figure 4). We see that R is contained in a rectangular region whose area is 8 square units. Then assuming that R has an area,

$$0 < A(R) < 8.$$

This gives an approximation of the area of the region R.

Suppose we divide the interval $[0, 2]$ into two parts and consider the point $(1, 1)$ on the graph of f. We draw two sets of rectangles as shown in Figure 5. Obviously,

$$A(R_1) < A(R) < A(R_2) + A(R_3)$$

where

$$A(R_1) = (1)(1) = 1$$
$$A(R_2) = (1)(1) = 1$$

and

$$A(R_3) = (1)(4) = 4.$$

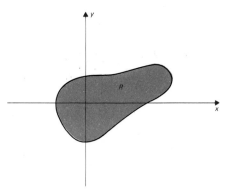

Figure 3

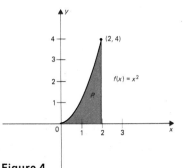

Figure 4

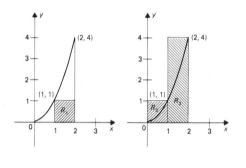

(a)   **(b)**

Figure 5

Thus,

$$1 < A(R) < 1 + 4$$

or

$$1 < A(R) < 5.$$

Next we divide the interval $[0, 2]$ into four parts and consider the points $(\frac{1}{2}, \frac{1}{4})$, $(1, 1)$, and $(\frac{3}{2}, \frac{9}{4})$ on the graph of f. We draw two sets of rectangles as shown in Figure 6. Here we find that

$$A(R_1) + A(R_2) + A(R_3) < A(R)$$
$$< A(R_4) + A(R_5) + A(R_6) + A(R_7)$$

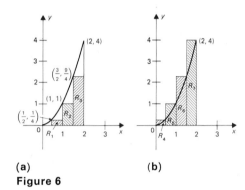

(a)   **(b)**

Figure 6

where

$$A(R_1) = \left(\tfrac{1}{2}\right)\left(\tfrac{1}{4}\right) = \tfrac{1}{8}$$
$$A(R_2) = \left(\tfrac{1}{2}\right)(1) = \tfrac{1}{2}$$
$$A(R_3) = \left(\tfrac{1}{2}\right)\left(\tfrac{9}{4}\right) = \tfrac{9}{8}$$
$$A(R_4) = \left(\tfrac{1}{2}\right)\left(\tfrac{1}{4}\right) = \tfrac{1}{8}$$
$$A(R_5) = \left(\tfrac{1}{2}\right)(1) = \tfrac{1}{2}$$
$$A(R_6) = \left(\tfrac{1}{2}\right)\left(\tfrac{9}{4}\right) = \tfrac{9}{8}$$
$$A(R_7) = \left(\tfrac{1}{2}\right)(4) = 2.$$

Hence

$$\tfrac{1}{8} + \tfrac{1}{2} + \tfrac{9}{8} < A(R) < \tfrac{1}{8} + \tfrac{1}{2} + \tfrac{9}{8} + 2$$

or

$$\tfrac{14}{8} < A(R) < \tfrac{30}{8}.$$

It appears that by increasing the number of subintervals of $[0, 2]$, the areas of the corresponding rectangles yield a better approximation for $A(R)$.

We divide the interval $[0, 2]$ into n subintervals by choosing numbers

$$0 = x_0 < x_1 < x_2 < \ldots < x_{n-1} < x_n = 2.$$

This is called a *partition* of the interval $[0, 2]$. To simplify our problem we choose a partition of n equal subintervals. Then each subinterval is of length $2/n$ (Why?).

Now the set of rectangles in Figure 7(a) all have length $2/n$ and height $f(x_{k-1})$, $k = 1, 2, \ldots, n$. Thus, starting at the left endpoint we have

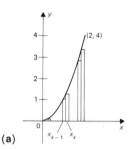

Figure 7   (a)

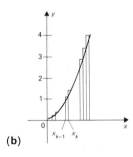

(b)

area of
first rectangle $= \dfrac{2}{n} f(x_0) = \dfrac{2}{n} f(0) = \dfrac{2}{n} \cdot 0 = 0$

area of
second rectangle $= \dfrac{2}{n} f(x_1) = \dfrac{2}{n} f\left(\dfrac{2}{n}\right) = \dfrac{2}{n}\left(\dfrac{2}{n}\right)^2$

area of
third rectangle $= \dfrac{2}{n} f(x_2) = \dfrac{2}{n} f\left(\dfrac{4}{n}\right) = \dfrac{2}{n}\left(\dfrac{4}{n}\right)^2$

$$\vdots$$

area of
kth rectangle $= \dfrac{2}{n} f(x_{k-1}) = \dfrac{2}{n} f\left(\dfrac{2k-2}{n}\right) = \dfrac{2}{n}\left[\dfrac{2(k-1)}{n}\right]^2.$

Therefore the sum of the areas of the rectangles in Figure 7(a) is given by

$$s_n = 0 + \frac{2}{n}\left(\frac{2}{n}\right)^2 + \frac{2}{n}\left(\frac{4}{n}\right)^2 + \ldots + \frac{2}{n}\left[\frac{2(k-1)}{n}\right]^2$$

$$+ \ldots + \frac{2}{n}\left[\frac{2(n-1)}{n}\right]^2$$

$$= \sum_{k=1}^{n} \frac{2}{n}\left[\frac{2(k-1)}{n}\right]^2$$

$$= \frac{8}{n^3} \sum_{k=1}^{n} (k-1)^2.$$

The rectangles in Figure 7(b) all have length $2/n$ and height $f(x_k)$, $k = 1, 2, \ldots, n$. And so,

area of kth rectangle $= \dfrac{2}{n} f(x_k) = \dfrac{2}{n} f\left(\dfrac{2k}{n}\right) = \dfrac{2}{n}\left(\dfrac{2k}{n}\right)^2.$

Therefore, the sum of the areas of the rectangles in Figure 7(b) is given by

$$S_n = \frac{2}{n}\left(\frac{2}{n}\right)^2 + \frac{2}{n}\left(\frac{4}{n}\right)^2 + \ldots + \frac{2}{n}\left(\frac{2n}{n}\right)^2$$

$$= \frac{2}{n}\sum_{k=1}^{n}\left(\frac{2k}{n}\right)^2$$

$$= \frac{8}{n^3}\sum_{k=1}^{n}k^2.$$

Using the formula

Use mathematical induction to verify this result. The following formulas will also be needed in this section.

$$(2) \quad \sum_{k=1}^{n}k^2 = 1^2 + 2^2 + 3^2 + \ldots + n^2 = \frac{n(n+1)(2n+1)}{6},$$

we can write

$(2a) \displaystyle\sum_{k=1}^{n}k = 1 + 2 + 3 + \ldots + n = \frac{n(n+1)}{2}$

$$S_n = \frac{8}{n^3}\left[\frac{n(n+1)(2n+1)}{6}\right] = \frac{8}{3} + \frac{4}{n} + \frac{4}{3n^2}$$

$(2b) \displaystyle\sum_{k=1}^{n}c = c + c + c + \ldots + c = cn$

and

$$s_n = \frac{8}{n^3}\left[\frac{(n-1)n\{2(n-1)+1\}}{6}\right] = \frac{8}{3} - \frac{4}{n} + \frac{4}{3n^2}.$$

Obviously,

$$s_n < A(R) < S_n$$

or

$$\frac{8}{3} - \frac{4}{n} + \frac{4}{3n^2} < A(R) < \frac{8}{3} + \frac{4}{n} + \frac{4}{3n^2}.$$

As *n* increases indefinitely, it is clear that

$$\lim_{n\to\infty} s_n = \lim_{n\to\infty} S_n$$

$$= \lim_{n\to\infty}\left[\frac{8}{3} + \frac{4}{n} + \frac{4}{3n^2}\right]$$

$$= \tfrac{8}{3}.$$

Here we say that

$$A(R) = \tfrac{8}{3}.$$

We now generalize this procedure for computing the area of a region having curvilinear boundaries. Consider a *continuous non-negative* function *f* defined on a closed interval [*a*, *b*]. Let

$$a = x_0 < x_1 < x_2 < \ldots < x_{n-1} < x_n = b$$

be a partition of $[a, b]$ and choose numbers $c_1, c_2, \ldots, c_n$ such that

$$x_{k-1} \leq c_k \leq x_k, \quad k = 1, 2, \ldots, n.$$

Next we form rectangles (see Figure 8) with

$$\text{length} = x_k - x_{k-1}, \quad k = 1, 2, \ldots, n$$

and

$$\text{height} = f(c_k), \quad k = 1, 2, \ldots, n.$$

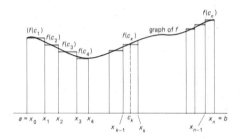

Figure 8

The sum of the areas of the rectangles is given by

$$f(c_1)(x_1 - x_0) + f(c_2)(x_2 - x_1) + \ldots + f(c_k)(x_k - x_{k-1})$$
$$+ \ldots + f(c_n)(x_n - x_{n-1}) = \sum_{k=1}^{n} f(c_k)(x_k - x_{k-1}).$$

Note that each term represents the area of one of the rectangles. The sum is called a *Riemann sum*. Let

$$(3) \qquad S_n = \sum_{k=1}^{n} f(c_k) \, dx_k$$

where $dx_k = x_k - x_{k-1}$, for $k = 1, 2, \ldots, n$.

Question. What happens to S_n as n increases?

It can be shown that S_n tends to a number A as $n \to \infty$ and each $dx_k \to 0$. Furthermore, the number A is independent of the partition used and the choice of the numbers c_k. The number A represents the area of this region. Note the restriction on the partition: As $n \to \infty$ we require that each $dx_k \to 0$ for $k = 1, 2, \ldots, n$.

The limit A is called the *definite integral* of f over $[a, b]$ and is denoted by the symbol

$$\int_a^b f(x) \, dx.$$

That is,

$$(4) \qquad \int_a^b f(x) \, dx = \lim_{\substack{n \to \infty \\ dx_k \to 0}} \sum_{k=1}^n f(c_k) \, dx_k.$$

The function f is said to be *integrable* on $[a, b]$. The numbers a and b are called the *lower* and *upper limits of integration* respectively.

Example 1. Calculate $\int_0^1 f(x) \, dx$ where $f(x) = x + 1$.

Solution. (See Figure 9.)

Step 1. Divide the interval $[0, 1]$ into n equal subintervals. Each subinterval is of length

$$dx_k = \frac{1 - 0}{n} = \frac{1}{n}.$$

Step 2. Select c_k. In this case let c_k be the right hand endpoint of the kth subinterval. That is,

$$c_1 = \frac{1}{n}, \, c_2 = \frac{2}{n}, \, c_3 = \frac{3}{n}, \, \dots, \, c_k = \frac{k}{n}, \, \dots, \, c_n = 1.$$

Step 3. Calculate $\sum_{k=1}^n f(c_k) \, dx_k$. Here we have

$$\sum_{k=1}^n f(c_k) \, dx_k = \sum_{k=1}^n f\left(\frac{k}{n}\right) dx_k$$

$$= \sum_{k=1}^n \left(\frac{k}{n} + 1\right)\frac{1}{n}$$

$$= \sum_{k=1}^n \left(\frac{k}{n^2} + \frac{1}{n}\right)$$

$$= \frac{1}{n^2} \sum_{k=1}^n k + \frac{1}{n} \sum_{k=1}^n 1$$

$$= \frac{1}{n^2}\left[\frac{n(n+1)}{2}\right] + \frac{1}{n} \cdot n$$

$$= \frac{1}{2} + \frac{1}{2n} + 1 = \frac{3}{2} + \frac{1}{2n}.$$

Step 4. Calculate $\int_0^1 (x + 1) \, dx$. We find that

$$\int_0^1 (x + 1) \, dx = \lim_{\substack{n \to \infty \\ dx_k \to 0}} \sum_{k=1}^n f(c_k) \, dx_k$$

Remark. The similarity between $\int_a^b f(x) \, dx$ and the antiderivative symbol is intentional. A remarkable relationship exists between the definite integral and the antiderivative, and this relationship is stated precisely in the Fundamental Theorem of Calculus (see Section 3).

We used equal subintervals to simplify the problem.

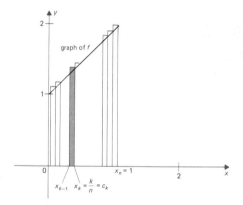

Figure 9

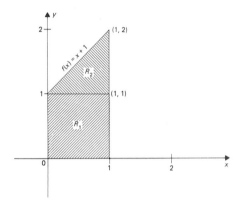

Figure 10

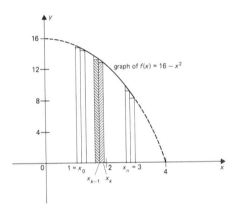

Figure 11

$$= \lim_{\substack{n \to \infty \\ dx_k \to 0}} \left(\frac{3}{2} + \frac{1}{2n} \right) = \frac{3}{2}.$$

Note that the area of the region bounded by the graph of f and the lines $\{(x, y) \mid x = 1\}$, $\{(x, y) \mid x = 0\}$ and $\{(x, y) \mid y = 0\}$ is given by (see Figure 10)

$$A(R) = A(R_1) + A(R_2)$$

where

$$A(R_1) = (1)(1) = 1$$
$$A(R_2) = \tfrac{1}{2}(1)(1) = \tfrac{1}{2}.$$

Hence

$$A(R) = 1 + \tfrac{1}{2} = \tfrac{3}{2} = \int_0^1 (x + 1) \, dx.$$

Example 2. Calculate $\int_1^3 (16 - x^2) \, dx$.

Solution. The region is indicated in Figure 11. It is bounded by the graph of $f(x) = 16 - x^2$ and the lines $\{(x, y) \mid x = 1\}$, $\{(x, y) \mid x = 3\}$, and $\{(x, y) \mid y = 0\}$.

Step 1. Divide the interval $[1, 3]$ into n equal subintervals. Hence each subinterval is of length

$$dx_k = \frac{3 - 1}{n} = \frac{2}{n}.$$

Step 2. Select c_k to be the right hand endpoint of the kth subinterval. Then

$$c_1 = 1 + \frac{2}{n}, \qquad\qquad c_2 = 1 + \frac{4}{n}, \ldots,$$

$$c_k = 1 + \frac{2k}{n}, \ldots, \qquad\qquad c_n = 1 + \frac{2n}{n} = 3.$$

Step 3. Calculate $\sum_{k=1}^{n} f(c_k) \, dx_k$. We have

$$\sum_{k=1}^{n} f(c_k) \, dx_k = \sum_{k=1}^{n} f\left(1 + \frac{2k}{n} \right) dx_k$$

$$= \sum_{k=1}^{n} \left[16 - \left(1 + \frac{2k}{n} \right)^2 \right] \frac{2}{n}$$

$$= \sum_{k=1}^{n} \left(\frac{30}{n} - \frac{8k}{n^2} - \frac{8k^2}{n^3} \right)$$

$$= \frac{30}{n} \sum_{k=1}^{n} 1 - \frac{8}{n^2} \sum_{k=1}^{n} k - \frac{8}{n^3} \sum_{k=1}^{n} k^2$$

$$= \frac{30}{n} \cdot n - \frac{8}{n^2} \left[\frac{n(n+1)}{2} \right]$$

$$- \frac{8}{n^3} \left[\frac{n(n+1)(2n+1)}{6} \right]$$

$$= \frac{70}{3} - \frac{8}{n} - \frac{4}{3n^2}.$$

Step 4. Calculate $\int_1^3 (16 - x^2)\, dx$. Here we have

$$\int_1^3 (16 - x^2)\, dx = \lim_{\substack{n \to \infty \\ dx_k \to 0}} \sum_{k=1}^{n} f(c_k)\, dx_k$$

$$= \lim_{\substack{n \to \infty \\ dx_k \to 0}} \left(\frac{70}{3} - \frac{8}{n} - \frac{4}{3n^2} \right)$$

$$= \frac{70}{3}.$$

Therefore, the area of the region in Figure 11 is $\frac{70}{3}$.

Exercise 2

In problems 1 through 6 divide the interval $[a, b]$ into eight equal parts and evaluate s_8 and S_8 such that

$$s_8 < \int_a^b f(x)\, dx < S_8.$$

1. $\int_1^2 (1 + x)\, dx$ 2. $\int_1^3 \frac{x}{2}\, dx$

3. $\int_0^4 x^2\, dx$ 4. $\int_{-1}^1 x^2\, dx$

5. $\int_1^2 \frac{1}{x}\, dx$ 6. $\int_1^4 \sqrt{x + 5}\, dx$

7. Use mathematical induction to establish formulas (2a) and (2b).
8. Repeat Example 1 by selecting c_k to be the left hand endpoint of the kth subinterval.
9. Repeat Example 1 by dividing the interval $[0, 1]$ into $2n$ equal subintervals.
10. Repeat Example 2 by dividing the interval $[1, 3]$ into $2n$ equal subintervals.

In problems 11 through 20 evaluate the indicated definite integral. Sketch the region involved in each case and show the kth rectangle in the approximating sum.

11. $\displaystyle\int_0^2 x\,dx$ **12.** $\displaystyle\int_1^2 (2 + x)\,dx$

13. $\displaystyle\int_0^1 x^2\,dx$ **14.** $\displaystyle\int_1^3 (1 + x^2)\,dx$

15. $\displaystyle\int_2^5 (x - 1)\,dx$ **16.** $\displaystyle\int_{-1}^3 (4 - x)\,dx$

17. $\displaystyle\int_{-1}^2 (1 + 2x^2)\,dx$ **18.** $\displaystyle\int_{-1}^1 (2 - x)^2\,dx$

19. $\displaystyle\int_1^4 (2 - x)^2\,dx$ **20.** $\displaystyle\int_0^1 x^3\,dx$

21. Show that $\displaystyle\int_a^b x\,dx = \frac{b^2 - a^2}{2}$.

*22. Show that $\displaystyle\int_0^1 x^2\,dx = \frac{b^3 - a^3}{3}$.

23. If $f(x) = x^2$ and $a < c < b$, show that

    (a) $\displaystyle\int_a^b f(x)\,dx = \int_a^c f(x)\,dx + \int_c^b f(x)\,dx$

    (b) $\displaystyle\int_a^b mf(x)\,dx = m\int_a^b f(x)\,dx$ (m is a constant).

24. Consider the region bounded by the graph of $f(x) = x^3$ and the lines $\{(x, y)\,|\,x = -1\}$, $\{(x, y)\,|\,x = 1\}$, and $\{(x, y)\,|\,y = 0\}$. Divide the interval $[-1, 1]$ into 4 equal subintervals and construct the approximating four rectangles.
    (a) Find the area of the four rectangles.
    (b) Using formula (4), evaluate $\int_{-1}^1 x^3\,dx$.
    (c) Does $\int_{-1}^1 x^3\,dx$ represent the area of the region under consideration? Explain.

3. The Fundamental Theorem of Calculus

Obviously, the evaluation of a definite integral can be tedious. Fortunately, there is a technique available to help. It is given in the Fundamental Theorem of Calculus, which relates the definite integral to the antiderivative of a given function. This provides a simple method for evaluating the definite integral. We merely state this theorem; for a proof the student should consult more advanced calculus texts.

The Fundamental Theorem of Calculus

Theorem 3. **If *f* is a function continuous on [*a*, *b*] and *F* is an antiderivative of *f*, then**

(5)
$$\int_a^b f = F(b) - F(a).$$

*This basic relationship between the area under the curve y = f(x) and the antiderivative of f was noted first by Newton's teacher, **Isaac Barrow** (1630–1677). Barrow was a mathematician and a theologian. He held a mathematics chair at Cambridge and in 1669 he relinquished his post to the young Newton.*

Another notation commonly used is

$$\int_a^b f(x)\ dx = F(x)\Big|_a^b = F(b) - F(a).$$

The symbol $\int_a^b f(x)\ dx$ reminds us that the definite integral is the limit of a sequence of sums. We note that $F(b) - F(a)$ does not involve the variable *x*, and so we may write

$$\int_a^b f(x)\ dx = \int_a^b f(r)\ dr = \int_a^b f(t)\ dt$$

where the letters *x*, *r*, and *t* are called *dummy variables*.

Now if *G* is an antiderivative of *f*, then by Theorem 1

$$G = F + C$$

and so

$$\int_a^b f = G(b) - G(a)$$
$$= (F + C)(b) - (F + C)(a)$$
$$= F(b) + C - F(a) - C$$
$$= F(b) - F(a).$$

Therefore, for a given function *f* we can compute the definite integral of *f* over [*a*, *b*] by finding *any* antiderivative of *f*, evaluating this antiderivative at *b* and at *a*, and then subtracting the second value from the first.

Example 1. Evaluate $\int_{-1}^2 2x^3\ dx$.

Solution. Here $f(x) = 2x^3$ and an antiderivative of *f* is

$$F(x) = \frac{x^4}{2}.$$

Evaluating *F* at $x = 2$ and at $x = -1$ and subtracting we obtain

$$F(2) - F(-1) = \frac{x^4}{2}\Big|_{-1}^2 = \frac{(2)^4}{2} - \frac{(-1)^4}{2} = \frac{15}{2}.$$

Therefore,

$$\int_{-1}^{2} 2x^3 \, dx = \frac{15}{2}.$$

Example 2. Evaluate $\int_{1}^{4} \left(x^2 - \sqrt{x} + \frac{1}{x^2} \right) dx.$

Solution. An antiderivative of f where $f(x) = x^2 - \sqrt{x} + \frac{1}{x^2}$ is

$$F(x) = \frac{x^3}{3} - \frac{2}{3} x^{3/2} - \frac{1}{x}.$$

Therefore,

$$\int_{1}^{4} \left(x^2 - \sqrt{x} + \frac{1}{x^2} \right) dx = \left(\frac{x^3}{3} - \frac{2}{3} x^{3/2} - \frac{1}{x} \right)\Big|_{1}^{4}$$

$$= \left[\frac{(4)^3}{3} - \frac{2}{3} (4)^{3/2} - \frac{1}{4} \right]$$

This is F(4) − F(1).

$$- \left[\frac{(1)^3}{3} - \frac{2}{3} (1)^{3/2} - \frac{1}{1} \right]$$

$$= \frac{205}{12}.$$

The following properties of the definite integral can be deduced readily by application of Theorem 3.

If F is an antiderivative of f, then

$$\int_{a}^{a} f(x) \, dx = F(a) - F(a) = 0.$$

(6) $$\int_{a}^{a} f(x) \, dx = 0$$

(7) $$\int_{a}^{b} f(x) \, dx = -\int_{b}^{a} f(x) \, dx$$

The student is encouraged to establish properties (7) and (8).

(8) $$\int_{a}^{b} f(x) \, dx = \int_{a}^{c} f(x) \, dx + \int_{c}^{b} f(x) \, dx, \quad a \leq c \leq b.$$

In the previous section we indicated that if f is nonnegative on $[a, b]$, then the definite integral of f from a to b defines the area of the region bounded by the graph of f, the x-axis, and the lines $\{(x, y) | x = a\}$ and $\{(x, y) | x = b\}$. In Theorem 3 we did not restrict f to be positive or negative on $[a, b]$. The question then arises as to what $\int_{a}^{b} f$ does represent if $f(x) \geq 0$.

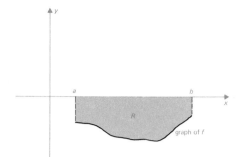

Figure 12

Theorem 4. If f is a continuous function on $[a, b]$ and $f(x) \leq 0$ on $[a, b]$, then

$$\int_a^b f = -A(R)$$

where $A(R)$ is the area of the region R bounded by the graph of f, the x-axis, and the lines $\{(x, y) \mid x = a\}$ and $\{(x, y) \mid x = b\}$ (see Figure 12).

That is,

$$(9) \qquad A(R) = -\int_a^b f.$$

Thus in general, if f is a continuous function on $[a, b]$ whose graph is as shown in Figure 13, then

$$\int_a^b f = A(R_1) - A(R_2) + A(R_3).$$

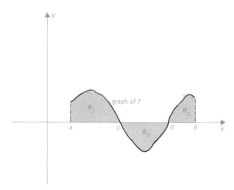

Figure 13

In this case we find that

$$A(R_1) = \int_a^c f$$

$$A(R_2) = -\int_c^d f$$

By Theorem 4

and

$$A(R_3) = \int_d^b f.$$

Thus

$$A(R_1) - A(R_2) + A(R_3) = \int_a^c f + \int_c^d f + \int_d^b f$$

$$= \int_a^d f + \int_d^b f$$

$$= \int_a^b f.$$

Application of property (8)

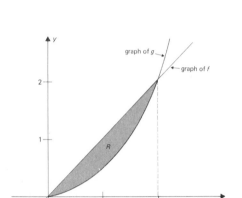

Figure 14

Example 3. Find the area of the region bounded by the graphs of functions f and g where $f(x) = x$ and $g(x) = \frac{1}{2}x^2$.

Solution. Step 1. Sketch a graph of the region R as shown in Figure 14.

Step 2. Find the points of intersection of the graphs of f and g. That is, find

$$\{(x, y) \mid y = x \text{ and } y = \tfrac{1}{2}x^2\}.$$

Here we have to solve the equation

$$x = \tfrac{1}{2}x^2$$

or

$$2x - x^2 = 0.$$

Thus $x = 0$ and $x = 2$, and the points of intersection are $(0, 0)$ and $(2, 2)$. Note that the points of intersection will define the region R.

Step 3. We observe that the area $A(R)$ of the region R is given by

$$A(R) = A(R_1) - A(R_2)$$

where $A(R_1)$ is the area of the region R_1 bounded by the graph of f, the x-axis, and the line $\{(x, y) \mid x = 2\}$; and $A(R_2)$ is the area of the region R_2 bounded by the graph of g, the x-axis, and the line $\{(x, y) \mid x = 2\}$. Now,

$$A(R_1) = \int_0^2 f(x) \, dx = \int_0^2 x \, dx = \frac{x^2}{2}\Big|_0^2 = \frac{(2)^2}{2} - 0 = 2$$

and

$$A(R_2) = \int_0^2 g(x) \, dx = \int_0^2 \frac{1}{2}x^2 \, dx = \frac{1}{2}\frac{x^3}{3}\Big|_0^2 = \frac{(2)^3}{6} - 0 = \frac{4}{3}.$$

Therefore, the area of the region in question is

$$A(R) = \int_0^2 x \, dx - \int_0^2 \tfrac{1}{2}x^2 \, dx = 2 - \tfrac{4}{3} = \tfrac{2}{3}.$$

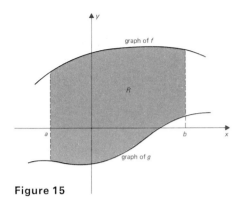

Figure 15

In the previous section we indicated that

$$A(R) = \int_a^b f(x)\ dx$$

is the area of the region R bounded by the graph of f, the x-axis, and the lines $\{(x, y)\mid x = a\}$ and $\{(x, y)\mid x = b\}$ provided f is non-negative on $[a, b]$. We now generalize this definition.

If f and g are continuous functions on $[a, b]$ and $f(x) \geq g(x)$ for each point on $[a, b]$, then the area of the region R bounded by the graphs of f and g and the lines $\{(x, y)\mid x = a\}$ and $\{(x, y)\mid x = b\}$ is defined by

(10) $$A(R) = \int_a^b [f(x) - g(x)]\ dx.$$

Such a region is illustrated in Figure 15.

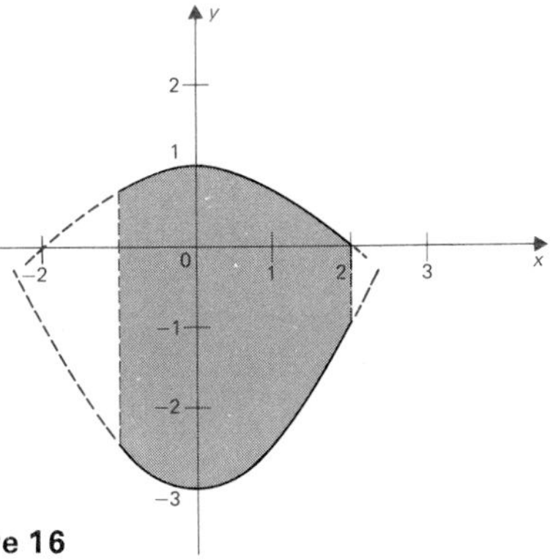

Example 4. Compute the area of the region for which $x \in [-1, 2]$ and which is bounded by the graphs of the function f and g where

$$f(x) = -\tfrac{1}{4}x^2 + 1 \qquad \text{and} \qquad g(x) = \tfrac{1}{2}x^2 - 3.$$

Solution. The region R is illustrated in Figure 16. Since $f(x) \geq g(x)$ for each x on $[-1, 2]$, we have

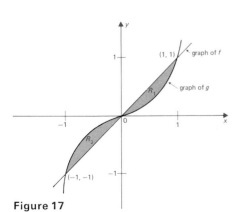

Figure 16

$$
\begin{aligned}
A(R) &= \int_{-1}^2 [f(x) - g(x)]\ dx \\
&= \int_{-1}^2 [(-\tfrac{1}{4}x^2 + 1) - (\tfrac{1}{2}x^2 - 3)]\ dx \\
&= \int_{-1}^2 (-\tfrac{3}{4}x^2 + 4)\ dx \\
&= \left(-\frac{3}{4} \cdot \frac{x^3}{3} + 4x\right)\Bigg|_{-1}^2 \\
&= \left[-\frac{(2)^3}{4} + 4(2)\right] - \left[\frac{(-1)^3}{4} + 4(-1)\right] \\
&= \frac{41}{4}.
\end{aligned}
$$

Example 5. Compute the area of the region R bounded by the graphs of the functions f and g where

$$f(x) = x \qquad \text{and} \qquad g(x) = x^3.$$

Solution. We follow the steps indicated in Example 3. First we sketch a graph of the region R. This is illustrated in Figure 17. We

Figure 17

find the points of intersection of the graphs of f and g by solving the equation

$$x = x^3,$$

from which we obtain $x = 0$, $x = 1$, and $x = -1$. Thus the points of intersection are $(0, 0)$, $(1, 1)$, and $(-1, -1)$.

Here we see that there are two regions to be considered.

In R_1, $x \in [0, 1]$ and $f(x) \geq g(x)$ for each x.

In R_2, $x \in [-1, 0]$ and $g(x) \geq f(x)$ for each x.

Hence we treat the problem in two parts. First we find $A(R_1)$:

$$A(R_1) = \int_0^1 [f(x) - g(x)]\, dx = \int_0^1 (x - x^3)\, dx = \left(\frac{x^2}{2} - \frac{x^4}{4}\right)\Big|_0^1$$

$$= \left[\frac{(1)^2}{2} - \frac{(1)^4}{4}\right] - \left[\frac{0}{2} - \frac{0}{4}\right] = \frac{1}{4}.$$

Next,

$$A(R_2) = \int_{-1}^0 [g(x) - f(x)]\, dx = \int_{-1}^0 (x^3 - x)\, dx = \left(\frac{x^4}{4} - \frac{x^2}{2}\right)\Big|_{-1}^0$$

$$= \left[\frac{0}{4} - \frac{0}{2}\right] - \left[\frac{(-1)^4}{4} - \frac{(-1)^2}{2}\right] = \frac{1}{4}.$$

The area $A(R) = A(R_1) + A(R_2)$. Therefore

$$A(R) = \tfrac{1}{4} + \tfrac{1}{4} = \tfrac{1}{2}.$$

Example 6. Suppose $F(x) = \int_1^x t^3\, dt$. Find $F'(x)$.

Solution.

$$F(x) = \frac{t^4}{4}\Big|_1^x = \frac{x^4}{4} - \frac{1}{4}.$$

Thus

$$F'(x) = \frac{d}{dx}\left(\frac{x^4}{4} - \frac{1}{4}\right) = x^3.$$

Exercise 3

In problems 1 through 12, evaluate the given definite integrals.

1. $\displaystyle\int_1^3 (x^2 - 3)\, dx$

2. $\displaystyle\int_{-1}^2 (2x^2 + 4x)\, dx$

3. $\displaystyle\int_{-1}^2 (r + 1)^2\, dr$

4. $\displaystyle\int_1^2 (x + 2)(x - 2)\, dx$

5. $\int_4^9 3\sqrt{t}\ dt$

6. $\int_1^3 \dfrac{2}{s^2}\ ds$

7. $\int_{-2}^4 x(x+1)\ dx$

8. $\int_2^5 \dfrac{1}{\sqrt{u}}\ du$

9. $\int_{-1}^1 (2x^2 + x^{1/3} - 6)\ dx$

10. $\int_0^2 \dfrac{1 + 2v^{1/3}}{v^{2/3}}\ dv$

11. $\int_1^3 \left(\dfrac{1}{u}\right)^4 du$

12. $\int_0^{16} 2x^{1/4}\ dx$

13. Establish formulas (7) and (8).

In problems 14 through 22 find the area of the region in the plane bounded by the given curves. Sketch a graph of the region.

14. $\{(x, y) \mid y = 2x + 1\}$, $\{(x, y) \mid x = 1\}$, $\{(x, y) \mid x = 3\}$, and the x-axis

15. $\{(x, y) \mid y = 9 - x^2\}$ and the x-axis

16. $\{(x, y) \mid y = x^2 + 1\}$, $\{(x, y) \mid x = 2\}$, $\{(x, y) \mid x = -3\}$, and the x-axis

17. $\{(x, y) \mid y = x^2\}$ and $\{(x, y) \mid y = 2x\}$

18. $\{(x, y) \mid y = x^2\}$ and $\{(x, y) \mid y = x + 2\}$

19. $\{(x, y) \mid y = \sqrt{x}\}$ and $\{(x, y) \mid y = x^2\}$

20. $\{(x, y) \mid y = x^2\}$ and $\{(x, y) \mid y = 2 - x^2\}$

21. $\{(x, y) \mid y = 4x - x^2\}$ and $\{(x, y) \mid y = 3\}$

22. $\{(x, y) \mid y = 4x - x^2\}$ and $\{(x, y) \mid y = -5\}$

In problems 23 through 30 determine $F'(x)$.

23. $F(x) = \int_1^x t\ dt$

24. $F(x) = \int_{-2}^x u^2\ du$

25. $F(x) = \int_0^x (r - 1)^2\ dr$

26. $F(x) = \int_1^x 2\sqrt{t}\ dt$

27. $F(x) = \int_x^2 u^2\ du$

28. $F(x) = \int_x^{3x} r^2\ dr$

29. $F(x) = \int_x^{x+2} r^2\ dr$

30. $F(x) = \int_x^{x+2} r^2\ dr + \int_{x+2}^x r^2\ dr$

In problems 31 through 34 determine the values for x which satisfy the given equation.

31. $\int_0^x 3t^2\ dt = 8$

32. $\int_0^x (4t - 5)\ dt = 2$

33. $\int_{-x}^x \dfrac{u^6}{2}\ du = \dfrac{128}{7}$

34. $\int_x^{2x^2} dt = 6$

4. Applications of the Integral

We now show how the process of integration is used in solving problems arising in the sciences. Many problems in the sciences are described as rates of change of a function. The simplest such problems are of the form

$$(11) \qquad\qquad y'(t) = f(t).$$

Such an equation is called a *differential equation*. In general, a differential equation is an equation involving a variable t (or x or s, etc.), an unknown function $y(t)$, and certain derivatives of y with respect to t. The problem is usually to determine a function $y(t)$ that satisfies the given equation. Such a function is called a *solution* of the differential equation. In many applications, equations such as (11) are usually associated with additional conditions that the function $y(t)$ is to satisfy. For instance, it may be required that at $t = t_0$

$$(12) \qquad\qquad y(t_0) = k.$$

This is called an *initial condition*. Equation (11) together with (12) form an *initial-value problem*. The problem involves finding a function $y(t)$ that satisfies the differential equation (11) as well as the initial condition (12).

The following examples illustrate the procedure for solving such problems.

Example 1. The Goldegg Chicken Farm produces $E(t)$ dozen eggs in t days. Its *production rate* is $E'(t)$ and this can be estimated by

$$E'(t) \approx \frac{E(t + h) - E(t)}{h}.$$

If the production rate (in dozen/day) is given by

$$(13) \qquad\qquad E'(t) = 150 + \tfrac{2}{5}t,$$

find

   (a) the number of eggs produced in t days assuming $E(0) = 0$
   (b) the number of eggs produced in one year (365 days)
   (c) the average daily production.

Solution. (a) We shall use two methods for solving part (a).
   First Method. The production E in t days is

$$E(t) = \int E'(t)\, dt$$

$$= \int (150 + \tfrac{2}{5}t)\, dt$$

(14)
$$= 150t + \frac{t^2}{5} + C.$$

The student can show readily that $E(t)$ satisfies equation (13). We now satisfy the initial condition. From (14) we have

$$E(0) = (150)(0) + \frac{0^2}{5} + C = 0$$

which implies that

$$C = 0.$$

Therefore, the solution to the initial-value problem is

(15)
$$E(t) = 150t + \frac{t^2}{5}.$$

Second Method. Integrating both sides of equation (13) from 0 to t we have

Note. (a) The lower limit of integration is the initial value for t.
(b) The dummy variable s is used to avoid confusion with the upper limit of integration.

(16)
$$\int_0^t E'(s)\, ds = \int_0^t (150 + \tfrac{2}{5}s)\, ds.$$

This yields

$$E(s)\Big|_0^t = \left(150s + \frac{s^2}{5}\right)\Big|_0^t$$

or

$$E(t) - E(0) = \left[150t + \frac{t^2}{5}\right] - \left[(150)(0) + \frac{0^2}{5}\right]$$

or

Observe that E(0) = 0.

$$E(t) = 150t + \frac{t^2}{5}$$

as before.

(b) To obtain the production in one year we have

(17) $\quad E(365) = (150)(365) + \dfrac{(365)^2}{5} = 81{,}395$ dozen eggs.

(c) The average daily production is

(18)
$$\frac{E(365)}{365} = \frac{81{,}395}{365} = 223 \text{ dozen/day.}$$

Example 2. Consider an (idealized) experiment in which a colony of live bacteria is introduced to a limited food supply. (See Chapter 5, Section 2, Example 4.) Suppose the rate of change in the number N of live bacteria with respect to time t is given by

(19) $$N'(t) = 6000t^2 - 75t^4.$$

Find the size $N(t)$ of the population of bacteria at time t if initially 1000 bacteria were introduced to the food supply.

Solution. We have an initial-value problem involving the differential equation (19) and initial condition

(20) $$N(0) = 1000.$$

Integrating both sides of equation (19) we have

$$\int_0^t N'(s)\ ds = \int_0^t (6000s^2 - 75s^4)\ ds$$

$$N(s)\Big|_0^t = (2000s^3 - 15s^5)\Big|_0^t$$

$$N(t) - N(0) = [2000t^3 - 15t^5] - [(2000)(0)^3 - (15)(0)^5],$$

that is,

(21) $$N(t) = 2000t^3 - 15t^5 + 1000.$$

The student is encouraged to show that the function defined by (21) satisfies equations (19) and (20).

A rough sketch of the graph of N is shown in Figure 18, indicating the period of adaptation (p.a.), period of reproduction (p.r.), and period of dying (p.d.).

The student should solve this initial-value problem using the first method described in Example 1 above.

We now consider the motion of an object defined by

(22) $$s = f(t)$$

where s is the distance as a function of time t. We recall that the *velocity function*, $v(t)$, is defined by

(23) $$v(t) = \frac{ds}{dt}$$

which is the rate of change of the distance with respect to time; and the *acceleration function*, $a(t)$, is defined by

(24) $$a(t) = \frac{dv}{dt} = \frac{d^2s}{dt^2}$$

which is the rate of change of the velocity with respect to time.

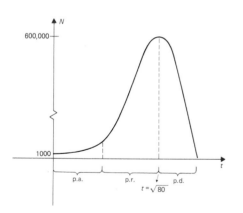

N(0) = 1000

Figure 18

We assume that air resistance or other forces do not act on the object.

Example 3. An object on the ground is projected vertically with initial velocity of 96 ft/sec. If the acceleration $a(t) = -32$ ft/sec², find

(a) the velocity function,
(b) the distance at time t,
(c) the height the object will attain,
(d) the height of the object in 5 seconds.

Solution. (a) Here we are dealing with the initial-value problem

(25) $$v'(t) = -32$$

(26) $$v(0) = 96.$$

Integrating on both sides of equation (25) we obtain

(27) $$\int_0^t v'(r)\, dr = -32 \int_0^t dr$$

or

$$v(r)\Big|_0^t = -32r\Big|_0^t,$$

that is,

$$v(t) - v(0) = -32[t - 0].$$

Since $v(0) = 96$, we have

(28) $$v(t) = -32t + 96.$$

(b) We observe that at $t = 0$ the object is at ground level: The initial condition is $s(0) = 0$. Since

$$s'(t) = v(t) = -32t + 96,$$

we have

$$\int_0^t s'(r)\, dr = \int_0^t (-32r + 96)\, dr$$

$$s(r)\Big|_0^t = (-16r^2 + 96r)\Big|_0^t$$

$$s(t) - s(0) = [-16t^2 + 96t] - [(-16)(0)^2 + (96)(0)];$$

and since $s(0) = 0$, we have

(29) $$s(t) = 16t^2 + 96t.$$

(c) The object will continue to move as long as $v(t) \neq 0$, and it will be at instantaneous rest when $v(t) = 0$. Therefore, from equation (28) we have

$$v(t) = -32t + 96 = 0,$$

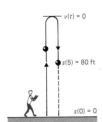

Figure 19

$v(t) = 0$

$s(5) = 80$ ft

$s(0) = 0$

which is satisfied when $t = \frac{96}{32} = 3$. That is, 3 seconds after the object is projected upward it comes to instantaneous rest and then begins its journey downward. To find the distance traveled by the object in 3 seconds we use equation (29) and obtain

$$s(3) = (-16)(3)^2 + (96)(3) = 144,$$

that is, the object will rise to a height of 144 ft above the ground.

(d) From equation (29) we have

$$s(5) = (-16)(5)^2 + (96)(5) = 80;$$

in 5 seconds the object will be 80 ft above ground level. We observe that $s(6) = 0$. Hence, the object will be back on the ground in 6 seconds, and for $t > 6$, $v(t) \equiv 0$ and $s(t) \equiv 0$. (Why?)

Example 4. A water supply tank of volume 1500 cu ft develops a leak. Suppose the rate at which the water is flowing out of the tank is given by

The negative sign indicates that the volume is decreasing.

(30) $$V'(t) = -\tfrac{1}{30}t$$

where t is measured in minutes.

(a) Find the volume of the water in the tank at time t
(b) If the tank develops the leak at 9:05 P.M., at what time will the tank be empty?

Solution. (a) We assume that time is measured starting at 9:05 P.M. Hence, at $t = 0$, $V = 1500$ and we have the initial-value problem

$$V'(t) = -\tfrac{1}{30}t$$
$$V(0) = 1500.$$

Then

$$\int_0^t V'(s) \, ds = -\tfrac{1}{30} \int_0^t s \, ds,$$

which yields

$$V(s) \Big|_0^t = -\frac{1}{30} \frac{s^2}{2} \Big|_0^t$$

$$V(t) - V(0) = -\tfrac{1}{60}t^2$$

(31) $$V(t) = -\tfrac{1}{60}t^2 + 1500.$$

(b) When the tank is empty, $V = 0$. From equation (31) we have

$$-\tfrac{1}{60}t^2 + 1500 = 0$$

which implies

$$t = 300.$$

This means that 300 minutes after the leak develops the tank will be empty. This occurs at 2:05 A.M.

It is important to note that the volume function defined by (31) has domain $\{t \mid 0 \leq t \leq 300 \text{ minutes}\}$, since we cannot have negative volume. For $t > 300$, we define $V(t) = 0$.

Exercise 4

In problems 1 through 10 find a solution of the initial-value problem. Use two methods.

1. $y'(t) = 1 - 3t,$ $\quad y(0) = 4$

2. $\dfrac{dy}{dx} = x^2 + 4x^3,$ $\quad y(1) = -2$

3. $y'(t) = t^3 + t^{-2},$ $\quad y(-2) = 1$

4. $y'(t) = \sqrt{t} + 4,$ $\quad y(4) = 3$

5. $\dfrac{dy}{dx} = \dfrac{3x^2 + x}{x^5},$ $\quad y(1) = -1$

6. $\dfrac{dy}{dx} = 2 - x^{3/2},$ $\quad y(9) = 2$

7. $y' = x^{1/3}(1 + 3x^{1/3}),$ $\quad y(-8) = -\frac{1}{4}$

8. $\sqrt{t}\, y'(t) = 2,$ $\quad y(1) = 3$

9. $\dfrac{\sqrt{x}}{x^2 - 1} \dfrac{dy}{dx} + 1 = 0,$ $\quad y(4) = -\frac{1}{2}$

10. $\dfrac{dy}{dx} = 3x^{1/3}(1 - x^{1/3}),$ $\quad y(8) = -1$

11. Suppose the marginal cost in producing x units of an item is given by

$$C'(x) = 4x - 350.$$

  (a) Find the total cost if there is a fixed cost of \$100 (i.e., $C(0) = 100$).
  (b) Find the profit if each item produced sells at \$10.
  (c) How many units should be sold to produce maximum profit?

12. Suppose the marginal cost of producing x gallons of Slick oil is $10\sqrt{x}$ dollars. Find the cost function $C(x)$, if there is an overhead cost of \$350. (Note that the overhead cost is independent of x.)

13. Suppose the marginal revenue of producing and selling x units of a product is $20 + \dfrac{x}{50}$. If the revenue for the first 500 units is \$12,500, find the revenue function and the price function.

14. A snowball in the form of a perfect sphere has radius 72 inches. Suppose it melts so that the radius decreases at the rate of 4 in/min. (Assume that the snowball retains its spherical shape until it is melted completely.)

    (a) Find the time t when the diameter is 72 inches.

    (b) What is the diameter of the snowball after 12 minutes?

    (c) At what time will the snowball be melted completely?

15. Three astronauts are marooned in a spacecraft. There are 350 units of oxygen available, and the rate of consumption in units per minute is

$$A'(t) = -0.03 \sqrt{t}.$$

    (a) Determine the amount A of oxygen in the spacecraft as a function of time.

    (b) Find the time at which the supply of oxygen is exhausted.

16. A driver sees a dog crossing the highway ahead, and he applies the brakes when his car is 100 feet from the dog. If the car's velocity after application of the brakes is $(70 - 26t)$ ft/sec, will he stop in time?

17. Find the velocity and distance functions under the following conditions.

    (a) $a(t) = t + 1,$ $v(0) = 2,$ $s(0) = 1$

    (b) $a(t) = t^2 - 5,$ $v(0) = -1,$ $s(0) = 4$

In the problems below, assume that $a(t) = 32$ ft/sec².

18. An object is projected vertically from the surface of the earth with initial velocity $v(0) = 288$ ft/sec.

    (a) Find the velocity and distance functions.

    (b) How high will the object rise?

19. An object projected from the surface of the earth rises 288 ft. before starting to fall. Find the time it took the object to rise to this height.

64 ft

20. An object is thrown upward from the roof of a building 64 ft above the ground. If the initial velocity is 48 ft/sec, determine when the object will hit the ground.

    (2) Rocket fired from silo

5. Methods of Integration

In this section we shall introduce two methods of calculating anti-derivatives. These will aid the student in evaluating definite or indefinite integrals involving certain complicated functions.

A. THE METHOD OF SUBSTITUTION OF VARIABLES

See Chapter 5, Section 6.

Suppose we wish to find $\int (2x^3 + 5)^{137}6x^2 \, dx$. We could expand $(2x^3 + 5)^{137}$ and multiply the result by $6x^2$. Then the integral of the expression obtained can be evaluated. Undoubtedly the student will agree that this is tedious work to say the least. To avoid such a cumbersome task we use the chain rule.

Consider a function F expressible as a composite function

$$F(x) = G(u(x)).$$

By the chain rule

$$F'(x) = G'(u(x))u'(x)$$

and thus

$$\int F'(x) \, dx = \int G'(u(x))u'(x) \, dx.$$

The notation dx or du has an interesting history. Originally, Leibniz defined them as "infinitesimals" or "very small quantities greater than zero but less than any positive number." This led to much confusion since there are no such numbers.

If dx denotes a positive real number, Cauchy defined

$$du = u'(x) \, dx,$$

from which we obtain

$$(32) \qquad \int F'(x) \, dx = \int G'(u) \, du$$
$$= G(u(x)) + C = F(x) + C.$$

We illustrate this procedure in the following examples.

Example 1. Evaluate $\int (2x^3 + 5)^{137}6x^2 \, dx$.

Solution. Let $u = 2x^3 + 5$. Then

$$du = u'(x) \, dx = 6x^2 \, dx.$$

Using these substitutions we obtain

$$\int (2x^3 + 5)^{137}6x^2 \, dx = \int u^{137} \, du$$

Application of formula (1)

$$= \frac{u^{138}}{138} + C$$
$$= \frac{(2x^3 + 5)^{138}}{138} + C.$$

Example 2. Evaluate $\int \sqrt{2x + 1} \, dx$.

Solution. Let $u = 2x + 1$. Then

$$du = u'(x) \, dx = 2dx.$$

Multiply and divide by 2. We have used the fact that

$$c \int f(x) \, dx = \int cf(x) \, dx \quad (c = \text{constant}).$$

Using these substitutions we obtain

$$\int \sqrt{2x + 1} \, dx = \tfrac{1}{2} \int \sqrt{2x + 1} \, 2dx$$

$$= \tfrac{1}{2} \int u^{1/2} \, du$$

Using formula (1)

$$= \frac{1}{2} \frac{u^{3/2}}{3/2} + C$$

$$= \tfrac{1}{3} u^{3/2} + C$$

$$= \tfrac{1}{3} (2x + 1)^{3/2} + C.$$

Example 3. Evaluate $\int (2x + 3)(x^2 + 3x + 5)^{-3/2} \, dx$.

Solution. Let $u = x^2 + 3x + 5$. Then

$$du = u'(x) \, dx = (2x + 3) \, dx.$$

Thus we have

$$\int (2x + 3)(x^2 + 3x + 5)^{-3/2} \, dx = \int u^{-3/2} \, du$$

$$= -2u^{-1/2} + C$$

$$= -2(x^2 + 3x + 5)^{-1/2} + C.$$

In the examples above it is important to observe how the function u was selected. In each case we chose u in such a way that changed the form of the integrand to u^n. That is, $\int f(x) \, dx$ became $\int u^n \, du, \; n \neq -1$.

We now consider the evaluation of definite integrals. Obviously, the method involves first finding an antiderivative of the integrand and then evaluating this antiderivative at the limits of integration. There are two possible ways of doing this and we shall illustrate both methods.

Example 4. Evaluate $\int_1^2 (1 - x^{-1})^3 x^{-2} \, dx$.

Solution 1. We first evaluate the indefinite integral

$$\int (1 - x^{-1})^3 x^{-2} \, dx.$$

We let

$$u = 1 - x^{-1}$$
$$du = u'(x) \, dx = x^{-2} \, dx.$$

Substitution yields

$$\int (1 - x^{-1})^3 x^{-2} \, dx = \int u^3 \, du$$

$$= \tfrac{1}{4} u^4 + C$$

$$= \tfrac{1}{4} (1 - x^{-1})^4 + C.$$

We use the antiderivative of the integrand $\frac{1}{4}(1 - x^{-1})^4$:

$$\int_1^2 (1 - x^{-1})^3 x^{-2} \, dx = \frac{1}{4}(1 - x^{-1})^4 \Big|_1^2$$

$$= [\tfrac{1}{4}(1 - \tfrac{1}{2})^4] - [\tfrac{1}{4}(1 - 1)^4]$$

$$= \tfrac{1}{64}.$$

Solution 2. Let

$$u = 1 - x^{-1}$$
$$du = x^{-2} \, dx.$$

We note that the limits of the definite integral for values of x are $x = 1$ and $x = 2$. We find the corresponding values of u:

$$\text{when } x = 1, \quad u = 1 - 1 = 0$$
$$\text{when } x = 2, \quad u = 1 - \tfrac{1}{2} = \tfrac{1}{2}.$$

Therefore, by substitution we have

$$\int_1^2 (1 - x^{-1})^3 x^{-2} \, dx = \int_0^{1/2} u^3 \, du$$

$$= \tfrac{1}{4} u^4 \Big|_0^{1/2}$$

$$= [\tfrac{1}{4}(\tfrac{1}{2})^4] - [\tfrac{1}{4}(0)^4]$$

$$= \tfrac{1}{64}.$$

In Solution 2 we are using the formula

$$\int_a^b g(u(x))u'(x) \, dx = \int_{u(a)}^{u(b)} g(u) \, du.$$

Under certain conditions on $u(x)$ and $g(x)$, this formula is valid. The conditions will be satisfied for the functions considered in this text.

Reviewing the examples above, it is clear that the method of substitution of variables is not always applicable. For instance, in Example 1 the factor $6x^2$ plays an important role. To evaluate

$$\int (2x^3 + 5)^{137} \, dx$$

we will have to use the binomial theorem to expand the integrand and then find the antiderivative. Note that if we let

$$u = 2x^3 + 5,$$

then

$$du = 6x^2 \, dx$$

and

$$\int (2x^3 + 5)^{137} \, dx = \int u^{137} \frac{1}{6x^2} \, du.$$

Be careful!

$$\int (2x^3 + 5)^{137} \, dx \neq \frac{1}{6x^2} \int (2x^3 + 5)^{137} 6x^2 \, dx$$

In general,

$$\int f(x) \, dx = \frac{1}{g(x)} \int f(x)g(x) \, dx$$

if and only if g is a constant function.

The factor $1/6x^2$ makes it impossible to evaluate the integral on the right-hand side. The method of substitution of variables does not work in this case.

Another method of integration is

B. INTEGRATION BY PARTS

Suppose f and g are differentiable functions of x. Then by the rule of the derivative of a product we have

(33) $$(f \cdot g)' = f \cdot g' + f' \cdot g.$$

Integrating both sides of equation (33) with respect to x we obtain

(34) $$\int \frac{d}{dx} [f(x)g(x)] \, dx = \int [f(x) \cdot g'(x) + f'(x)g(x)] \, dx$$

or

(35) $$f(x)g(x) = \int f(x)g'(x) \, dx + \int g(x)f'(x) \, dx.$$

From this we have

(36) $$\int f(x)g'(x) \, dx = f(x)g(x) - \int g(x)f'(x) \, dx.$$

Formula (36) is called *integration by parts*.

For the definite integral we have

(37) $$\int_a^b f(x)g'(x) \, dx = f(x)g(x) \Big|_a^b - \int_a^b g(x)f'(x) \, dx.$$

At first glance, it might appear that we are complicating the problem. However, integration by parts facilitates the evaluation of antiderivatives of a large class of functions.

Example 5. Evaluate $\int x\sqrt{x-1} \, dx$.

Solution. There are two choices we can make for f and g'. First, we let

$$f(x) = \sqrt{x-1} \quad \text{and} \quad g'(x) = x;$$

then

$$f'(x) = \tfrac{1}{2}(x-1)^{3/2} \quad \text{and} \quad g(x) = \tfrac{1}{2}x^2,$$

and integration by parts yields

$$\int x\sqrt{x-1} \, dx = \tfrac{1}{2}x^2\sqrt{x-1} - \int \tfrac{1}{4}x^2(x-1)^{-1/2} \, dx.$$

We note that the problem is getting more complicated. Therefore, this choice of f and g' is discarded.

A second choice is

$$f(x) = x \quad \text{and} \quad g'(x) = \sqrt{x-1};$$

To evaluate $\int \frac{2}{3}(x-1)^{3/2} \, dx$ we use the method of substitution of variables with $u = x - 1$ and $du = dx$.

then

$$f'(x) = 1 \quad \text{and} \quad g(x) = \tfrac{2}{3}(x-1)^{3/2}.$$

Substitution into formula (36) yields

$$\int x\sqrt{x-1}\ dx = \tfrac{2}{3}x(x-1)^{3/2} - \int \tfrac{2}{3}(x-1)^{3/2}\ dx$$

$$= \tfrac{2}{3}x(x-1)^{3/2} - \tfrac{2}{3} \cdot \tfrac{2}{5}(x-1)^{5/2} + C$$

$$= \frac{2x}{3}(x-1)^{3/2} - \frac{4}{15}(x-1)^{5/2} + C.$$

Example 6. Evaluate $\int_0^1 x^3\sqrt{1+x^2}\ dx$.

Solution. Let

$$f(x) = x^2 \quad \text{and} \quad g'(x) = x\sqrt{1+x^2}.$$

We choose g' to be the most complicated factor in the integrand that can be integrated readily.

Then

$$f'(x) = 2x \quad \text{and} \quad g(x) = \tfrac{1}{3}(1+x^2)^{3/2}.$$

To find g we use the method of substitution of variables with $u = 1 + x^2$ and $du = 2x\ dx$.

Therefore,

$$\int_0^1 x^3\sqrt{1+x^2}\ dx = \int_0^1 x^2(x\sqrt{1+x^2})\ dx$$

Integration by substitution of variables

$$= \tfrac{1}{3}x^2(1+x^2)^{3/2}\Big|_0^1 - \int_0^1 \tfrac{2}{3}x(1+x^2)^{3/2}\ dx$$

$$= \tfrac{1}{3}x^2(1+x^2)^{3/2}\Big|_0^1 - \tfrac{1}{3} \cdot \tfrac{2}{5}(1+x^2)^{5/2}\Big|_0^1$$

$$= \left[\frac{2\sqrt{2}}{3} - 0\right] - \frac{2}{15}[4\sqrt{2} - 1]$$

$$= \tfrac{2}{15}(\sqrt{2} + 1).$$

Example 7. Evaluate $\int x^2(x-1)^{43}\ dx$.

Solution. Let

$$f(x) = x^2 \quad \text{and} \quad g'(x) = (x-1)^{43}.$$

Then

$$f'(x) = 2x \quad \text{and} \quad g(x) = \frac{(x-1)^{44}}{44},$$

and formula (36) yields

$$\int x^2(x-1)^{43}\ dx = \frac{x^2}{44}(x-1)^{44} - \int \frac{2x}{44}(x-1)^{44}\ dx.$$

Remark. *Sometimes it is necessary to apply the integration by parts formula more than once.*

We now apply integration by parts to the integral

$$\int x(x - 1)^{44} \, dx.$$

Let

$$u(x) = x \quad \text{and} \quad v'(x) = (x - 1)^{44},$$

then

$$u'(x) = 1 \quad \text{and} \quad v(x) = \frac{(x - 1)^{45}}{45}.$$

Hence,

$$\int x(x - 1)^{44} \, dx = \frac{x}{45} (x - 1)^{45} - \int \frac{(x - 1)^{45}}{45} \, dx$$

$$= \frac{x}{45} (x - 1)^{45} - \frac{(x - 1)^{46}}{(45)(46)}.$$

Therefore,

$$\int x^2 (x - 1)^{43} \, dx$$

$$= \frac{x^2}{44} (x - 1)^{44} - \frac{1}{22} \int x(x - 1)^{44} \, dx$$

$$= \frac{x^2}{44} (x - 1)^{44} - \frac{1}{22} \left[\frac{x}{45} (x - 1)^{45} - \frac{(x - 1)^{46}}{2070} \right] + C$$

$$= \frac{x^2 (x - 1)^{44}}{44} - \frac{x(x - 1)^{45}}{990} + \frac{(x - 1)^{46}}{45540} + C.$$

In evaluating an integral the student might try the techniques of integration in the following order: (1) guessing, (2) integration by substitution, (3) integration by parts.

There are tables of integrals published for engineers and others who need various integrals in their daily work.

Exercise 5

In problems 1 through 14 use the method of substitution of variables to evaluate the given integral.

1. $\int (x + 4)^2 \, dx$

2. $\int (x - 5)^6 \, dx$

3. $\int 3\sqrt{3x + 4} \, dx$

4. $\int \frac{1}{\sqrt{3x + 4}} \, dx$

5. $\int (4 - x)^3 \, dx$

6. $\int \left(\frac{x - 3}{2} \right)^2 \, dx$

7. $\int (4x - 1)^{-2} \, dx$ **8.** $\int (2x + 3)(x^2 + 3x - 1)^{-3} \, dx$

9. $\int x^4 (4x^5 + 6)^{100} \, dx$ **10.** $\int (x + 1)(x^2 + 2x - 5)^3 \, dx$

11. $\int x\sqrt{x^2 - 4} \, dx$ **12.** $\int x\sqrt{4 - x^2} \, dx$

13. $\int x^{-2}\sqrt{1 - x^{-1}} \, dx$ **14.** $\int x^{2/3}(2 + x^{5/3})^{1/2} \, dx$

In problems 15 through 20 evaluate the indicated definite integrals.

15. $\int_0^1 (3x - 1) \, dx$ **16.** $\int_{-2}^1 \sqrt{1 - x} \, dx$

17. $\int_3^7 (x - 1)^{-1/2} \, dx$ **18.** $\int_{-1}^0 (1 - 3x)^{3/2} \, dx$

19. $\int_0^1 x(x^2 + 1)^{-2} \, dx$ **20.** $\int_0^1 x(1 - x^2)^{-1/2} \, dx$

In problems 21 through 30, use the formula for integration by parts to evaluate each of the given integrals.

21. $\int x(2x - 1) \, dx$ **22.** $\int x\sqrt{1 + x} \, dx$

23. $\int x(x + 5)^6 \, dx$ **24.** $\int x(2x + 1)^{-1/2} \, dx$

25. $\int_0^2 x(1 + x)^{3/2} \, dx$ **26.** $\int x^3(1 - x^2)^{1/2} \, dx$

27. $\int x^3(1 + x^2)^{3/2} \, dx$ **28.** $\int_0^{1/2} \frac{2x^3}{\sqrt{1 - x^2}} \, dx$

29. $\int_0^3 x\sqrt{25 - 3x} \, dx$ **30.** $\int_0^8 x^2(1 + x)^{-1/2} \, dx$

Augustin Louis Cauchy

8

Exponential

and

Logarithmic

Functions

John Napier (1550–1617) was born at Mer-chiston, near Edinburgh, Scotland. His early edu-cation was received at St. Andrews and he completed his studies in Europe. In 1572 he married and had two children. His wife Elizabeth died seven years later. He remarried and fathered ten more children.

Theology and mathematics occupied Napier. He took an active part in church politics and was the author of a theological treatise. In 1839 he published De Arte Logistica. *His work with equations involved imaginary roots which he referred to as "a great algebraic secret."*

Napier devised several mechanical aids for calculations including the Rabologia (familiarly known as "Napier's bones"). He spent many years preparing the Table of Logarithms which was published in 1614. This invention was welcomed by the mathematical world. Napier also wrote the Constructio *and* Descripto *which explained logarithms.*

1. Definitions and Basic Properties

The invention of logarithms by John Napier was a milestone in the advancement of mathematics. Whereas most mathematical concepts evolve from previous ideas, the logarithm is an exception. Napier invented logarithms to simplify the computation of cumbersome algebraic operations, though his original work was unwieldy. However, together with Henry Briggs (1561–1631), he constructed tables for $\log_{10} x$. These tables were completed and published by Briggs after Napier's death. The invention of logarithms paved the way for the famous astronomer Johann Kepler (1571–1630) to complete his calculations that led to the third law of planetary motion. When Kepler received Napier's tables he refused to use them until he derived the formula used for their construction. Kepler invented other formulas and constructed his own tables, which he published in 1624. Napier's invention led to the logarithmic and exponential functions which today play an important role in the sciences.

In this section we shall define the exponential and logarithmic functions and indicate the relationship between them.

A. THE EXPONENTIAL FUNCTION

Descartes introduced the notation

$$b^n = \underbrace{b \cdot b \cdot b \cdots b}_{n \text{ factors}}$$

where b is a real number and n is a positive integer. For example,

$$5^2 = 5 \cdot 5, \qquad 2^3 = 2 \cdot 2 \cdot 2.$$

It is obvious that if m and n are positive integers and a and b are real numbers, then

(1)
$$b^m \cdot b^n = \underbrace{b \cdot b \cdots b}_{m \text{ factors}} \cdot \underbrace{b \cdot b \cdots b}_{n \text{ factors}} = b^{m+n}$$

Note. $(2^3)^2 = (2 \cdot 2 \cdot 2)(2 \cdot 2 \cdot 2) = 2^6$

(2)
$$(b^m)^n = b^{mn}$$

(3)
$$(a \cdot b)^m = a^m \cdot b^m.$$

If we define

(4)
$$b^0 = 1 \qquad (b \neq 0),$$

then (1), (2), and (3) also hold when the integer 0 is allowed as an exponent. A natural extension is to the negative integers as exponents. We observe that

$$b^n \cdot b^{-n} = b^{n-n} = b^0 = 1,$$

hence we define

The expressions 0^0 and 0^{-n} are not defined.

(5)
$$b^{-n} = \frac{1}{b^n}$$

where n is a positive integer and $b \neq 0$.

In our attempt to preserve the above rules of exponents we define $b^{1/n}$ as the unique positive number r such that $r^n = b$ where n is a positive integer and $b > 0$. This extends the meaning of exponentials to the case where the exponent is a rational number. Thus if m and n are integers and $n > 0$, then $b^{m/n}$ is the nth root of b^m and we write

Note. $\underbrace{b^{1/n} \cdot b^{1/n} \cdots b^{1/n}}_{n \text{ factors}} = (b^{1/n})^n$

$$= b^{(1/n) \cdot n} = b^1 = b.$$

(6)
$$b^{m/n} = \sqrt[n]{b^m} = (\sqrt[n]{b})^m.$$

Hence b^x is defined for any rational number x.

What of the case b^x where x is an irrational number? What do the expressions $5^{\sqrt{2}}$ and 3^π mean? Preserving the definition of exponent appears impossible: What does it mean to say that 5 is multiplied by itself $\sqrt{2}$ times?

It turns out that if x is an irrational number we can define b^x as the limit of a sequence of numbers. It can be shown that if x_i is a sequence of rational numbers such that $x_i \to x$, then the sequence of numbers b^{x_i} has a limit, and this limit is the same for any other sequence b^{y_i} where $y_i \to x$. This common limit is defined to be b^x.

We have thus defined b^x for all real numbers x, and we shall assume that the rules of exponents (1)–(5) are valid. This can be proved, but we shall not do so in this text.

Suppose b is a positive real number. For each number x there is a unique number b^x. Hence we have a function f defined by

The exponential function

(7) $$f(x) = b^x, \text{ for all } x \in R.$$

The function defined by (7) is called an *exponential function*. The exponential function has the following basic properties:

(8) $$f(x_1 + x_2) = f(x_1)f(x_2)$$

(9) $$f(x_1 - x_2) = f(x_1)/f(x_2)$$

Property (10) states that f is increasing for all x.

(10) $$f(x_1) < f(x_2) \text{ if and only if } x_1 < x_2$$

(11) $$f(x) \to +\infty \text{ as } x \to +\infty$$

(12) $$f(x) \to 1 \text{ as } x \to 0.$$

Graphs of $f(x) = 2^x$ and $f(x) = b^x$ are shown in Figures 1 and 2 respectively. In Figure 3 we illustrate the graph of the function $f(x) = b^{-x}$. Note that the graph of $f(x) = b^{-x}$ is the reflection about the y-axis of the graph of $f(x) = b^x$.

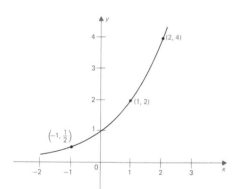

Figure 1

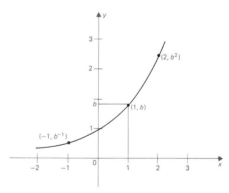

Figure 2
Graph of $f(x) = b^x$ ($b > 1$)

B. THE LOGARITHMIC FUNCTION

Let $b > 0$ ($b \neq 1$). Then the function $f(x) = b^x$ has domain $\{x \mid x \in R\}$ and range $\{y \mid y > 0\}$. Obviously, f is one-to-one, therefore it has an inverse f^{-1} with domain $\{x \mid x > 0\}$ and range $\{y \mid y \in R\}$. We denote $f^{-1}(x)$ by

(13) $$f^{-1}(x) = \log_b x$$

read *logarithm to the base b of x*. The function f^{-1} is called the

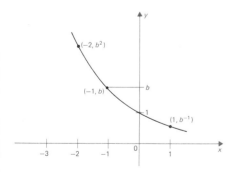

Figure 3
Graph of $f(x) = b^{-x}$ ($b > 1$)

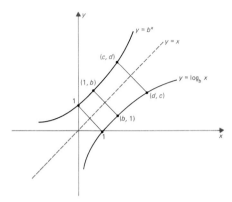

Figure 4

What happens if we allow b = 1 ?

Check the definition of one-to-one in Chapter 2.

logarithmic function. The graphs of f and f^{-1} are illustrated in Figure 4. Observe that the graph of f^{-1} is the reflection of the graph of f through the line $y = x$.

Now if f^{-1} is the inverse of the function f, we have

$$x = f(f^{-1}(x)).$$

Therefore for the exponential function we have

(14) $$x = b^{f^{-1}(x)}.$$

From (13) and (14) we obtain

(15) $$x = b^{\log_b x}.$$

Formula (15) implies the following:

If b is a positive real number and $b \neq 1$, then the equation $y = \log_b x$ is equivalent to the equation $x = b^y$.

With the aid of equation (15) we can establish the following theorem easily.

Theorem 1. If a, b, and c are positive numbers and $b > 1$, then

(16) $$\log_b (ac) = \log_b a + \log_b c$$

(17) $$\log_b \frac{a}{c} = \log_b a - \log_b c$$

(18) $$\log_b 1 = 0$$

(19) $$\log_b b = 1$$

(20) $$\log_b a^r = r \log_b a \qquad \text{if } r \text{ is a real number.}$$

We prove (16), (18), and (20) and leave (17) and (19) for the student as an exercise.

Proof of (16). Let $u = \log_b a$ and $v = \log_b c$. We have

$$a = b^u \qquad \text{and} \qquad c = b^v$$

and

(21) $$a \cdot c = b^u \cdot b^v = b^{u+v}.$$

Equation (21) is equivalent to

$$\log_b (a \cdot c) = u + v$$
$$= \log_b a + \log_b c.$$

Proof of (18). Let $u = \log_b 1$. Then $b^u = 1$, and from (4) we have $u = 0$. Therefore $\log_b 1 = 0$.

Proof of (20). Let $u = \log_b a$. Then

$$(22) \qquad\qquad b^u = a$$

and

$$(23) \qquad\qquad (b^u)^r = a^r \quad \text{or} \quad b^{ur} = a^r.$$

From equations (15) and (23) we have

$$(24) \qquad\qquad a^r = b^{\log_b a^r}$$

and

$$(25) \qquad\qquad b^{ur} = b^{\log_b a^r}.$$

Since the exponential function is one-to-one,

$$ur = \log_b a^r$$

and

$$r \log_b a = \log_b a^r.$$

We now consider some examples involving the exponential and logarithmic functions.

> $b^x = b^y \Rightarrow x = y, \ (b > 1)$
>
> Substitution of $u = \log_b a$

Example 1. Simplify the following expressions.

(a) $(27^{1/2})^{2/3}$ (b) $(3^{1/2} - 3^{-1/2})^2$
(c) $\log_2 4^2$ (d) $\log_3 (9^{1/2} \cdot 3^{1/4})$

Solution.

(a) $(27^{1/2})^{2/3} = 27^{(1/2) \cdot (2/3)} = 27^{1/3} = 3$

(b) $(3^{1/2} - 3^{-1/2})^2 = (3^{1/2})^2 - 2(3^{1/2})(3^{-1/2}) + (3^{-1/2})^2$
$= 3^{(1/2) \cdot 2} - 2 \cdot 3^{(1/2)-(1/2)} + 3^{-(1/2) \cdot 2}$
$= 3^1 - 2 \cdot 3^0 + 3^{-1}$
$= 3 - 2 + \tfrac{1}{3}$
$= \tfrac{4}{3}$

> *Using the binomial theorem*
> Note that $(a + b)^2 \neq a^2 + b^2$ if $a \cdot b \neq 0$.
> $3^1 = 3, \quad 3^0 = 1, \quad \text{and} \quad 3^{-1} = \dfrac{1}{3^1} = \dfrac{1}{3}.$

(c) $\log_2 4^2 = \log_2 (2^2)^2$
$= \log_2 2^4$
$= 4 \log_2 2$
$= 4$

(d) $\log_3 (9^{1/2} \cdot 3^{1/4}) = \log_3 9^{1/2} + \log_3 3^{1/4}$
$= \log_3 (3^2)^{1/2} + \log_3 3^{1/4}$
$= \log_3 3 + \tfrac{1}{4} \log_3 3$
$= 1 + \tfrac{1}{4} = \tfrac{5}{4}$

Example 2. Solve for x in each of the following equations.

(a) $3^x = \tfrac{1}{27}$
(b) $\log_x 8 = 3.$

Solution. (a) $3^x = 27^{-1} = (3^3)^{-1} = 3^{-3}$. Therefore $x = -3$.

(b) Since $\log_x 8 = 3$ is equivalent to

$$8 = x^3,$$

we write

$$2^3 = x^3.$$

Hence $x = 2$.

Exercise 1

In problems 1 through 16 simplify the given expression as much as possible.

1. $\dfrac{a^{5x} \cdot a^{-4x}}{a^x}$

2. $(4^{1/3})^{3/2}$

3. $(5^{-2/3} \cdot 5^{4/3})^3$

4. $(2^{1/2} + 2^{-1/2})^2$

5. $b^y \cdot b^{-3y} \cdot c^y$

6. $(7^{3/2} + 6^{5/7} \cdot 3^{2/3})^0$

7. $\log_c c^5$

8. $\log_3 9^2$

9. $\log_3 \dfrac{9^{2/3}}{3^{1/2}}$

10. $\log_4 (16^{1/5} \cdot 64^{1/4})$

11. $\log_2 \dfrac{2^3}{(64)^{1/2}}$

12. $3^{\log_3 3}$

13. $4^{\log_4 2}$

14. $27^{\log_8 2}$

15. $b^{\log_b 10 - \log_b 5}$

16. $\log_b (\log_b b^b)$

In problems 17 through 26 solve for x.

17. $10^x = 1000$

18. $2^x = \frac{1}{16}$

19. $4^{-x} = \frac{1}{64}$

20. $(\frac{1}{2})^x = 32$

21. $b^x - b^{-x} + 1 = 0$

22. $4 = b^{\log_b x}$

23. $\log_2 x = 4$

24. $\log_x 625 = 4$

25. $\log_2 \frac{1}{8} = x$

26. $\log_{10} 10^{-\log_{10} x^2} = -4$

27. Show that $\log_b \dfrac{a}{c} = \log_b a - \log_b c$.

28. Show that $\log_b b = 1$.

29. Show that $x^x = b^{x \log_b x}$.

30. Show that $\log_b x = \dfrac{\log_a x}{\log_a b}$.

Hint: Let $N = \log_b x$, then $x = b^N \Rightarrow \log_a x = \log_a b^N$.

2. Differentiation of the Logarithmic and Exponential Functions

Consider the expression

$$(1 + p)^{1/p}$$

We find its value for various values of p:

$$p = 1 \qquad (1 + 1)^1 = 2.0000$$
$$p = \tfrac{1}{2} \qquad (1 + \tfrac{1}{2})^2 = 2.2500$$
$$p = \tfrac{1}{4} \qquad (1 + \tfrac{1}{4})^4 = 2.4414$$
$$p = \tfrac{1}{100} \qquad (1 + \tfrac{1}{100})^{100} = 2.7048$$
$$p = \tfrac{1}{1000} \qquad (1 + \tfrac{1}{10000})^{1000} = 2.7181$$

It can be shown that $\lim\limits_{p \to 0} (1 + p)^{1/p}$ exists and is an irrational number. It is denoted by e. That is,

See Theorem 2 in Chapter 3.

(26) $$\lim_{p \to 0} (1 + p)^{1/p} = e \approx 2.71828 \ldots .$$

To obtain a formula for the derivative of $\log_b x$, equation (26) will be needed. We now prove

Theorem 2. If $f(x) = \log_b x$, then $f'(x) = \dfrac{1}{x} \log_b e \ (x > 0)$.

Proof. We write the difference quotient:

$$\frac{f(x + h) - f(x)}{h} = \frac{1}{h} [\log_b (x + h) - \log_b x]$$

Using property (17)

$$= \frac{1}{h} \left[\log_b \left(\frac{x + h}{x} \right) \right]$$

$$= \frac{1}{h} \log_b \left(1 + \frac{h}{x} \right)$$

$$= \frac{1}{x} \frac{x}{h} \log_b \left(1 + \frac{h}{x} \right)$$

$$= \frac{1}{x} \log_b \left(1 + \frac{h}{x} \right)^{x/h} .$$

Let $p = h/x$, then

$$\frac{f(x + h) - f(x)}{h} = \frac{1}{x} \log_b (1 + p)^{1/p} .$$

As $h \to 0$, $p \to 0$ and

$$\lim_{h \to 0} \frac{f(x + h) - f(x)}{h} = \lim_{p \to 0} \left[\frac{1}{x} \log_b (1 + p)^{1/p} \right] .$$

Hence

$$f'(x) = \frac{1}{x} \lim_{p \to 0} \log_b (1 + p)^{1/p}$$

This is permissible because of the continuity of the logarithmic function.

$$= \frac{1}{x} \log_b \lim_{p \to 0} (1 + p)^{1/p}$$

Using equation (26)

$$= \frac{1}{x} \log_b e.$$

If u is a differentiable function of x, we can use the chain rule to extend the result of Theorem 2 and obtain

For u(x) > 0   (27)

$$\frac{d}{dx} [\log_b u] = \frac{1}{u} (\log_b e) \frac{du}{dx}.$$

Example 1. If $f(x) = \log_b (2x^2 - 3x)$, find $f'(x)$.

Solution. Let $u = 2x^2 - 3x$, then $\dfrac{du}{dx} = 4x - 3$.

Using formula (27) we obtain

$$f'(x) = \frac{1}{2x^2 - 3x} (\log_b e)(4x - 3)$$

For 2x² − 3x > 0

$$= \frac{4x - 3}{2x^2 - 3x} \log_b e.$$

We see that the expression in (27) can be simplified if we selected $b = e$:

logₑ e = 1   (28)

$$\frac{d}{dx} [\log_e u] = \frac{1}{u} (\log_e e) \frac{du}{dx} = \frac{1}{u} \frac{du}{dx}.$$

The expression $\log_e x$ is called the *natural logarithm* (or *Napierian logarithm*) of x and the following notation is commonly used:

$$\log_e x = \ln x$$

read *ell-en of x*.

Example 2. Find $f'(x)$.
  (a) $f(x) = \ln \sqrt{x^2 + 1}$
  (b) $f(x) = \ln \dfrac{(x^2 + 1)^{20}(3x - 1)^3}{(2x^2 + 3)^4}$

We could simplify the problem by writing

$$\ln \sqrt{x^2 + 1} = \ln (x^2 + 1)^{1/2}$$
$$= \frac{1}{2} \ln (x^2 + 1).$$

Solution. (a) Let $u = \sqrt{x^2 + 1}$, then $\dfrac{du}{dx} = \dfrac{x}{\sqrt{x^2 + 1}}$.

Therefore

$$f'(x) = \frac{1}{\sqrt{x^2 + 1}} \cdot \frac{x}{\sqrt{x^2 + 1}} = \frac{x}{x^2 + 1}.$$

(b) First we simplify the function by using the rules of logarithms. Thus

$$\ln \frac{(x^2 + 1)^{20}(3x - 1)^3}{(2x^2 + 3)^4}$$

Note. $\ln \dfrac{AB}{C} = \ln A + \ln B - \ln C$

$$= 20 \ln (x^2 + 1) + 3 \ln (3x - 1) - 4 \ln (2x^2 + 3).$$

Therefore,

$$f'(x) = \frac{d}{dx} [20 \ln (x^2 + 1) + 3 \ln (3x - 1) - 4 \ln (2x^2 + 3)]$$

$$= 20 \frac{d}{dx} [\ln (x^2 + 1)] + 3 \frac{d}{dx} [\ln (3x - 1)]$$

$$- 4 \frac{d}{dx} [\ln (2x^2 + 3)]$$

Note. $\dfrac{d}{dx} [\ln (x^2 + 1)] = \dfrac{1}{x^2 + 1} \cdot \dfrac{d}{dx} (x^2 + 1)$

$= \dfrac{1}{x^2 + 1} \cdot 2x$

$$= 20 \cdot \frac{2x}{x^2 + 1} + 3 \cdot \frac{3}{3x - 1} - 4 \frac{4x}{2x^2 + 3}$$

$$= \frac{40x}{x^2 + 1} + \frac{9}{3x - 1} - \frac{16x}{2x^2 + 3}.$$

Now we develop a formula for the derivative of the function $f(x) = b^x$.

Theorem 3. If $f(x) = e^x$, then $f'(x) = e^x$.

Proof. Let $y = e^x$. Then $\ln y = x$ and by differentiating we find that

We are using implicit differentiation.

$$\frac{d}{dx} [\ln y] = \frac{d}{dx} [x]$$

$$\frac{1}{y} \frac{dy}{dx} = 1$$

or

$$\frac{dy}{dx} = y.$$

That is,

$$f'(x) = e^x.$$

It is easy to show that if u is a differentiable function of x, then by the chain rule

(29) $$f(x) = e^u \Longrightarrow f'(x) = e^u \frac{du}{dx}.$$

Example 3. Find $f'(x)$.

(a) $f(x) = e^{x^2+4}$ (b) $f'(x) = 3e^{1/x}$

Solution. (a) Let $u = x^2 + 4$, then $\dfrac{du}{dx} = 2x$.

Using (29) we find
$$f'(x) = 2xe^{x^2+4}.$$

(b) Let $u = \dfrac{1}{x}$, then $\dfrac{du}{dx} = -\dfrac{1}{x^2}$.

Therefore
$$f'(x) = -\frac{3}{x^2} e^{1/x}.$$

Theorem 4. If b is a positive number and $f(x) = b^x$, then $f'(x) = b^x \ln b$.

Recall that $b = e^{\ln b}$.

Proof. Let $y = b^x$. Then
$$y = e^{(\ln b)x}.$$

Set $u = (\ln b)x$ and $\dfrac{du}{dx} = \ln b$. Then
$$y = e^u,$$

and by differentiating we find
$$\frac{dy}{dx} = e^u \frac{du}{dx}$$
$$= e^{(\ln b)x} \ln b$$
$$= b^x \ln b.$$

That is,
$$f'(x) = b^x \ln b.$$

In general, if u is a differentiable function of x, then

(30) $f(x) = b^u \Rightarrow f'(x) = b^u (\ln b) \dfrac{du}{dx}$.

Example 4. Find $f'(x)$.
(a) $f(x) = 3^x$
(b) $f(x) = 5^{3x^2-2x}$
(c) $f(x) = 3^{x-1} \cdot 2^x$

Solution.

(a) $f'(x) = 3^x \ln 3$

(b) $f'(x) = 5^{3x^2-2x} (\ln 5)(6x - 2)$

(c) $f'(x) = (3^{x-1}) \dfrac{d}{dx}(2^x) + 2^x \dfrac{d}{dx}(3^{x-1})$

$= 3^{x-1} \cdot 2^x \ln 2 + 2^x \cdot 3^{x-1} \ln 3$

$= 3^{x-1} \cdot 2^x[\ln 2 + \ln 3]$

$= 3^{x-1} \cdot 2^x \ln 6$

We can make use of logarithms to differentiate complicated functions, as in the following example.

Example 5. If $y = \dfrac{(x^2 + 3)^{10}(x - 1)^{1/2}}{(x^2 + 3x + 5)^{80}}$, find $\dfrac{dy}{dx}$.

Solution. We have

$$\ln y = \ln \frac{(x^2 + 3)^{10}(x - 1)^{1/2}}{(x^2 + 3x + 5)^{80}}$$

$$= 10 \ln (x^2 + 3) + \tfrac{1}{2} \ln (x - 1) - 80 \ln (x^2 + 3x + 5).$$

Differentiation yields

$$\frac{1}{y} \frac{dy}{dx} = 10 \frac{1}{x^2 + 3} \frac{d}{dx}(x^2 + 3) + \frac{1}{2} \frac{1}{x - 1} \frac{d}{dx}(x - 1)$$

$$- 80 \frac{1}{x^2 + 3x + 5} \frac{d}{dx}(x^2 + 3x + 5)$$

$$= \frac{20x}{x^2 + 3} + \frac{1}{2(x - 1)} - \frac{80(2x + 3)}{x^2 + 3x + 5}.$$

Substituting for y and solving for $\dfrac{dy}{dx}$ we obtain

$$\frac{dy}{dx} = \frac{(x^2 + 3)^{10}(x - 1)^{1/2}}{(x^2 + 3x + 5)^{80}} \left[\frac{20x}{x^2 + 3} + \frac{1}{2(x - 1)} - \frac{80(2x + 3)}{x^2 + 3x + 5} \right].$$

Exercise 2

In problems 1 through 24 find $f'(x)$.

1. $f(x) = e^{4x}$ 2. $f(x) = 4e^{4x^2+5}$

3. $f(x) = e^{2/x^2}$ 4. $f(x) = e^{3x^2-4x+1}$

5. $f(x) = x^3 e^{2x}$ 6. $f(x) = (2x^2 + 3)e^x$

7. $f(x) = e^{-2x^2}$ 8. $f(x) = (e^{-2x})^2$

9. $f(x) = \ln (x^2 + 4)$ 10. $f(x) = \ln (e^x + 1)$

11. $f(x) = e^{x \ln x}$ **12.** $f(x) = \dfrac{e^x - 1}{e^x + 1}$

13. $f(x) = \ln (e^{2x^2+4} + 1)$ **14.** $f(x) = (e^{x^2} + 4)e^{5x^2-1}$

15. $f(x) = 5^x$ **16.** $f(x) = 5^{-x}$

17. $f(x) = 2^{x-1}5^x$ **18.** $f(x) = \dfrac{2^{x-1}}{5^x}$

19. $f(x) = e^{2x} + (2x)^e$ **20.** $f(x) = \ln (x + \sqrt{x^2 + 1})^2$

21. $f(x) = \log_{10} x$ **22.** $f(x) = \log_{10} (\log_{10} x)$

23. $f(x) = \log_3 (x^2 + e^x)$ **24.** $f(x) = \log_2 [\log_5 (x^3 + x^{1/2})]$

In problems 25 through 27, use the rules of logarithms to simplify the expression and find $f'(x)$.

25. $f(x) = \ln \dfrac{(x + 1)^{16}(2x^2 + x)^8}{\sqrt{x^2 + 4}}$

26. $f(x) = \ln \dfrac{(x^2 + 5)^7(x^3 + 1)^{1/3}}{(x^2 + 1)^{5/4}}$

27. $f(x) = \ln \dfrac{(e^{2x} + 6)^7\sqrt{x + 4}}{(e^{-x} + e^x)^5}$

In problems 28 through 30 use the procedure of Example 5 to find $\dfrac{dy}{dx}$.

28. $y = \dfrac{(3x^2 + 1)^5(x^2 + 1)^{3/2}}{(x^2 + 4)^{10}}$

29. $y = \dfrac{(x^2 + x + 1)^{60}(x^2 - 5x)^4}{\sqrt{x^2 - 4x + 10}}$

30. $y = \dfrac{(10x^2 + 5x + 1)^{5/2}(x^2 - 3x + 1)^{9/7}e^{-x^3+5+7}}{(x^3 + 2x^2 + 5)^8}$

3. Integration Involving e^u and $\ln u$

Since $\dfrac{d}{dx}(e^x) = e^x$, we have

(31) $$\int e^x \, dx = e^x + C.$$

If u is a differentiable function of x, then

$$\frac{d}{dx}(e^u) = e^u \frac{du}{dx}.$$

From this we obtain

(32) $$\int e^u \frac{du}{dx} \, dx = e^u + C.$$

Example 1. Evaluate the integral $\int x e^{2x^2+1} \, dx$.

Solution. Let $u = 2x^2 + 1$, then $du/dx = 4x$. Hence

$$\int x e^{2x^2+1} \, dx = \tfrac{1}{4} \int 4x e^{2x^2+1} \, dx$$

$$= \tfrac{1}{4} \int e^u \frac{du}{dx} \, dx$$

$$= \tfrac{1}{4} e^u + C$$

$$= \tfrac{1}{4} e^{2x^2+1} + C.$$

Example 2. Evaluate the integral $\int_0^2 x e^x \, dx$.

Solution. We use integration by parts. Set

$$f(x) = x \qquad g'(x) = e^x$$
$$f'(x) = 1 \qquad g(x) = e^x.$$

Using formula (37) in Chapter 7 we have

In this case we are unable to write the integral in the form given in (32).

$$\int_a^b f(x) g'(x) \, dx = f(x) g(x) \Big|_a^b - \int_a^b g(x) f'(x) \, dx$$

$$\int_0^2 x e^x \, dx = x e^x \Big|_0^2 - \int_0^2 e^x \, dx$$

$$= x e^x \Big|_0^2 - e^x \Big|_0^2$$

$$= [2e^2 - 0] - [e^2 - e^0]$$

$$= 2e^2 - e^2 + 1$$

$$= e^2 + 1.$$

We have seen that $d/dx \, (\ln x) = 1/x$ for $x > 0$. Therefore

(33) $$\int \frac{1}{x} \, dx = \ln x + C$$

provided $x > 0$.

If u is a differentiable function of x and $u(x) > 0$, then

$$\frac{d}{dx} (\ln u) = \frac{1}{u} \frac{du}{dx}.$$

This yields

(34) $$\int \frac{1}{u} \frac{du}{dx} \, dx = \ln u + C$$

provided $u(x) > 0$.

Example 3. Evaluate $\int \dfrac{2x}{x^2 + 4} \, dx$.

Solution. Let $u = x^2 + 4$. Then $du/dx = 2x$. Since $u(x) > 0$ for all x, we have

$$\int \frac{2x}{x^2 + 4} \, dx = \int \frac{1}{u} \frac{du}{dx} \, dx$$

$$= \ln u + C$$

$$= \ln (x^2 + 4) + C.$$

It can be shown that $d/dx \, (\ln |x|) = 1/x$, $x \neq 0$, so that we can generalize formula (34) and obtain

(35) $$\int \frac{1}{u} \frac{du}{dx} \, dx = \ln |u| + C, \qquad u \neq 0.$$

Example 4. Evaluate $\int \dfrac{2x^2 + 2x + 1}{x + 1} \, dx$, $x \neq -1$.

Solution. We first divide the integrand to obtain

$$\int \frac{2x^2 + 2x + 1}{x + 1} \, dx = \int \left(2x + \frac{1}{x + 1} \right) dx$$

$$= \int 2x \, dx + \int \left(\frac{1}{x + 1} \right) dx$$

$$= x^2 + \ln |x + 1| + C.$$

We now consider the equation

(36) $$\frac{dy}{dt} = ky$$

when k is a constant. This is a differential equation. Such equations arise in many areas of the sciences and economic theory; they define problems such as the growth of human cells, population growth, and the decay of radioactive substances mathematically. The Italian mathematician Vitto Volterra was one of the first to use such equations to study growth processes.

Equation (36) can be written

$$\frac{1}{y} \frac{dy}{dt} = k, \qquad y \neq 0$$

and can be solved readily by direct integration with respect to t. Thus

(37) $$\int \frac{1}{y} \frac{dy}{dt} \, dt = k \int dt.$$

$$\frac{2x^2 + 2x + 1}{x + 1} = 2x + \frac{1}{x + 1}$$

Vito Volterra (1860–1940), the only child of Abramo Volterra and Angelica Almagia, was born at Ancona, Italy. When he was two years old his father died and his maternal uncle, Alfonso Almagia, took care of the family. Vito spent most of his childhood in Florence.

At an early age, Vito showed strong inclinations toward mathematics. He began his studies at the Faculty of Natural Sciences at Florence and in 1878 moved to the University of Pisa. In 1880 he entered the Scuola Normale Superiore and in 1882 he received the doctorate degree. He wrote his first original paper while still a student. In 1883 he was promoted to full professorship in the University of Pisa.

Volterra's contributions to mathematics are outstanding. He generalized Riemann integration and contributed immensely to potential theory and differential equations. He is best known for his investigations of integral equations. He related mathematics to many areas of study, and after World War I he applied mathematics to biology and studied the problem of competition between species.

In 1905 he was given the honor of Senator of the Kingdom of Italy. He designed the mounting for guns fired from airships. For this he was decorated with the War Cross. When Mussolini took power, Volterra refused to sign an oath of allegiance required of all teachers and left the University. Volterra received countless honors from almost every National Academy and Mathematical Society in the world

We assume that the population p is a continuous and differentiable function of t. Also we assume that no death is occurring during this time interval. In real growth processes, over a very short period of time the size of the population might not change ; the rate of change would be zero during certain time intervals and nonzero during other intervals. Experimental results indicate, however, that over a long period of time our assumption yields fairly accurate results.

This yields

$$\ln |y| = kt + C_1$$

or

$$|y(t)| = e^{kt+C_1}.$$

Since e^{kt+C_1} is positive, we obtain

(38)
$$y(t) = e^{kt+C_1}.$$

Noting that $e^{kt+C_1} = e^{C_1}e^{kt}$, we replace e^{C_1} by a constant C and write equation (38) in the form

(39)
$$y(t) = Ce^{kt}.$$

Example 5. In a laboratory it is observed that the population of bacteria grows from 1000 to 3000 in 10 hours. If the rate of change of the population p is assumed to be proportional to the number present at any time t, find the size of the population after 5 hours.

Solution. Since the rate of change of p is proportional to p we have

(40)
$$\frac{dp}{dt} = kp$$

where k is a constant of proportionality. We choose the reference time $t = 0$, when the population is 1000; that is,

(41)
$$p(0) = 1000.$$

We are dealing with an initial-value problem defined by equations (40) and (41).

Using equation (39) we obtain solutions of equation (40). Thus

(42)
$$p(t) = Ce^{kt}.$$

From the initial condition (41) we obtain

$$p(0) = Ce^{k \cdot 0} = Ce^0 = C = 1000.$$

Therefore (42) becomes

(43)
$$p(t) = 1000e^{kt}.$$

Also we are given that at $t = 10$, $p = 3000$, which helps us determine the constant of proportionality k:

$$p(10) = 1000e^{10k} = 3000$$

or

$$e^{10k} = 3$$

or

$$k = \tfrac{1}{10} \ln 3,$$

and thus

(44)
$$p(t) = 1000e^{(1/10 \ln 3)t}.$$

From equation (44) we can determine the size of the population of bacteria at any time $t \geq 0$ for which equation (40) is valid. Therefore the size of the population at $t = 5$ hours is given by

$$\begin{aligned} p(5) &= 1000e^{(1/10 \ln 3)5} \\ &= 1000e^{1/2 \ln 3} \\ &\approx 1735. \end{aligned}$$

We shall study more applications in section 5.

Exercise 3

In problems 1 through 15 calculate the indicated integrals.

1. $\displaystyle \int e^{3x}\, dx$

2. $\displaystyle \int_0^2 e^{2x+1}\, dx$

3. $\displaystyle \int_1^2 xe^{x^2}\, dx$

4. $\displaystyle \int xe^{2x}\, dx$

5. $\displaystyle \int \frac{2}{x^2} e^{1/x}\, dx$

6. $\displaystyle \int x^2 e^x\, dx$

7. $\displaystyle \int_1^4 \frac{e^{\sqrt{x}}}{\sqrt{x}}\, dx$

8. $\displaystyle \int (e^x + e^{-x})^2\, dx$

9. $\displaystyle \int \frac{1}{x-2}\, dx$

10. $\displaystyle \int \frac{x}{2x^2 + 3}\, dx$

11. $\displaystyle \int \frac{2x+1}{x^2 + x - 6}\, dx$

12. $\displaystyle \int_1^2 \frac{(\ln x)^2}{x}\, dx$

13. $\displaystyle \int_2^3 \frac{x^2 - x - 1}{x - 1}\, dx$

14. $\displaystyle \int_0^1 \frac{6x^2 + 3x - 2}{2x + 1}\, dx$

15. $\displaystyle \int \frac{x^3 + 5x^2 + 10x - 8}{x + 2}\, dx$

16. The rate of increase in the population p of a colony of ants is directly proportional to the number present.
    (a) Set up the differential equation.
    (b) Find solutions of the equation.
    (c) If the population is 10^3 at $t = 0$, and 24 weeks later the population is 10^4, find the size of the population 10 weeks measured from $t = 0$.

17. The city of Bloomville has a population of 150,000. Twenty years ago the population was 50,000. Assume that the population p satisfies the equation $dp/dt = kp$. Find the size of the population 5 years in the future.

18. In yeast culture the number of bacteria doubles in one hour. If

the rate of increase of the population of bacteria is proportional to the number present, find the number of bacteria that will be present in 5 hours.

4. Integration by Partial Fractions

In this section we present a method for finding the antiderivative of a function of the form

$$(45) \qquad \int \frac{P(x)}{Q(x)} \, dx$$

Remark. In case P has degree greater than or equal to the degree of Q, one simply divides P by Q. Thus there is no loss of generality in assuming that the degree of P is less than the degree of Q.

where P and Q are polynomial functions and P is of degree less than the degree of Q. The method is called *integration by partial fractions*, and the procedure depends on Q being factorable. We shall consider the cases in which $Q(x)$ can be expressed as a product of linear and/or quadratic functions.

Case 1: The function Q can be written as a product of linear functions

$$Q(x) = a(x - a_1)(x - a_2) \ldots (x - a_n)$$

in which $a_i \neq a_j$ for $i \neq j$. In this case we can write

$$(46) \qquad \frac{P(x)}{Q(x)} = \frac{A_1}{x - a_1} + \frac{A_2}{x - a_2} + \ldots + \frac{A_n}{x - a_n}$$

where A_i ($i = 1, 2, \ldots, n$) are constants that can be determined.

Example 1. Express the following function as a sum of partial fractions as in (46) and find its antiderivative:

$$\frac{10x - 4}{x^3 - 4x}, \qquad x \neq 0, \pm 2.$$

Solution. We write

$$\frac{10x - 4}{x^3 - 4x} = \frac{10x - 4}{x(x - 2)(x + 2)}$$

$$(47) \qquad = \frac{A}{x} + \frac{B}{x - 2} + \frac{C}{x + 2}.$$

Multiplying equation (47) by $x(x - 2)(x + 2)$ we obtain

$$(48) \quad 10x - 4 = A(x - 2)(x + 2) + Bx(x + 2) + Cx(x - 2)$$

or

$$(49) \qquad 10x - 4 = (A + B + C)x^2 + 2(B - C)x - 4A.$$

The polynomial on the left of equation (49) is identical to the polynomial on the right if and only if every coefficient on the left is equal to the corresponding coefficient on the right. Thus we have

$$A + B + C = 0$$
$$2(B - C) = 10$$
$$-4A = -4.$$

These equations can be solved for A, B, and C. An alternate method for finding A, B, and C is as follows. We evaluate equation (48) at the roots of $Q(x)$:

For $x = 0$, $\qquad -4 = -4A$ $\qquad$ or $\qquad A = 1.$
For $x = 2$, $\qquad 16 = 8B$ $\qquad$ or $\qquad B = 2.$
For $x = -2$, $\qquad -24 = 8C$ $\qquad$ or $\qquad C = -3.$

Hence (47) may be written as

(50) $$\frac{10x - 4}{x^3 - 4x} = \frac{1}{x} + \frac{2}{x - 2} - \frac{3}{x + 2}.$$

Therefore integration with respect to x on both sides of equation (50) yields

$$\int \frac{10x - 4}{x^3 - 4x}\, dx = \int \left(\frac{1}{x} + \frac{2}{x - 2} - \frac{3}{x + 2} \right) dx$$

$$= \int \frac{1}{x}\, dx + 2 \int \frac{1}{x - 2}\, dx - 3 \int \frac{1}{x + 2}\, dx$$

Using formula (35)
$$= \ln |x| + 2 \ln |x - 2| - 3 \ln |x + 2| + C$$

Using formulas (16), (17), and (20)
$$= \ln \left| \frac{x(x - 2)^2}{(x + 2)^3} \right| + C.$$

Case 2: The function Q can be written as a product of linear functions

$$Q(x) = a(x - a_1)(x - a_2) \ldots (x - a_n)$$

in which two or more a_i's are equal. The following are examples involving case 2:

(i) $$\frac{P(x)}{(x - a)^3} = \frac{A}{x - a} + \frac{B}{(x - a)^2} + \frac{C}{(x - a)^3}$$

(ii) $$\frac{P(x)}{(x - b)(x - a)^4}$$

$$= \frac{A}{x - b} + \frac{B}{x - a} + \frac{C}{(x - a)^2} + \frac{D}{(x - a)^3} + \frac{E}{(x - a)^4}$$

(iii) $\dfrac{P(x)}{(x-a)^2(x-b)^3}$

$$= \frac{A}{x-a} + \frac{B}{(x-a)^2} + \frac{C}{x-b} + \frac{D}{(x-b)^2} + \frac{E}{(x-b)^3}$$

Example 2. Express the following function as a sum of partial fractions and find its antiderivative:

$$\frac{x+4}{x^3+x^2-5x+3}, \qquad x \neq 1, -3$$

Solution. In this case we have

We find the roots of Q using Theorem 4 (the "rational-root theorem") in Appendix 3.

$$\frac{x+4}{x^3+x^2-5x+3} = \frac{x+4}{(x+3)(x-1)^2}$$

(51)
$$= \frac{A}{x+3} + \frac{B}{x-1} + \frac{C}{(x-1)^2}.$$

Multiplying equation (51) by $(x+3)(x-1)^2$ we obtain

(52) $\quad x+4 = A(x-1)^2 + B(x+3)(x-1) + C(x+3).$

Setting $x = 1$ we have

$$5 = 4C \quad \text{or} \quad C = \tfrac{5}{4},$$

setting $x = -3$ we have

$$1 = 16A \quad \text{or} \quad A = \tfrac{1}{16}.$$

We still need to evaluate B. Since A and C have been determined, choose any number for x and solve for B. To simplify the evaluation, we set $x = 0$ in equation (52) and use the determined values for A and C to obtain

$$4 = \tfrac{1}{16} - 3B + \tfrac{15}{4} \quad \text{or} \quad B = -\tfrac{1}{16}.$$

Hence equation (51) may be written in the form

(53) $\dfrac{x+4}{x^3+x^2-5x+3} = \dfrac{1}{16(x+3)} - \dfrac{1}{16(x-1)} + \dfrac{5}{4(x-1)^2}$

Therefore,

$$\int \frac{x+4}{x^3+x^2-5x+3}\,dx$$

$$= \int \left[\frac{1}{16(x+3)} - \frac{1}{16(x-1)} + \frac{5}{4(x-1)^2} \right] dx$$

$$= \frac{1}{16}\left[\int \frac{1}{x+3}\, dx - \int \frac{1}{x-1}\, dx + 20 \int \frac{1}{(x-1)^2}\, dx \right]$$

$$= \frac{1}{16}\left[\ln|x+3| - \ln|x-1| - 20\frac{1}{x-1} \right] + C$$

$$= \frac{1}{16}\ln\left|\frac{x+3}{x-1}\right| - \frac{5}{4(x-1)} + C.$$

The quadratic function is assumed to have no real roots.

Case 3: The function Q can be written as a product of linear and quadratic functions

$$Q(x) = (x-a)\cdot(x^2 + bx + c).$$

In this case we have

Note carefully the form of the numerator in the term involving $\frac{1}{x^2+bx+c}$.

$$\frac{P(x)}{(x-a)(x^2+bx+c)} = \frac{A}{x-a} + \frac{Bx+C}{x^2+bx+c}.$$

Example 3. Express the following function as a sum of partial fractions and find its antiderivative:

$$\frac{3x^2 - 8x + 1}{(x-4)(x^2+1)}, \qquad x \neq 4$$

Solution. We have

(54) $$\frac{3x^2 - 8x + 1}{(x-4)(x^2+1)} = \frac{A}{x-4} + \frac{Bx+C}{x^2+1}.$$

Multiplying equation (54) by $(x-4)(x^2+1)$ we obtain

$$3x^2 - 8x + 1 = A(x^2+1) + (Bx+C)(x-4)$$

or

(55) $$3x^2 - 8x + 1 = A(x^2+1) + Bx(x-4) + C(x-4).$$

Setting $x = 4$ we obtain

$$17 = 17A \quad \text{or} \quad A = 1,$$

setting $x = 0$ we obtain

$$1 = 1 - 4C \quad \text{or} \quad C = 0,$$

setting $x = 1$ we obtain

$$-4 = 2 - 3B \quad \text{or} \quad B = 2.$$

Hence

$$\frac{3x^2 - 8x + 1}{(x-4)(x^2+1)} = \frac{1}{x-4} + \frac{2x}{x^2+1}.$$

Integration yields

$$\int \frac{3x^2 - 8x + 1}{(x - 4)(x^2 + 1)} \, dx = \int \left[\frac{1}{x - 4} + \frac{2x}{x^2 + 1} \right] dx$$

$$= \int \frac{1}{x - 4} \, dx + \int \frac{2x}{x^2 + 1} \, dx$$

$$= \ln |x - 4| + \ln (x^2 + 1) + C$$

$$= \ln |x - 4|(x^2 + 1) + C.$$

The second integral is of the form $\int \frac{1}{u} \frac{du}{dx} \, dx.$

Do not confuse this C (constant of integration) with the C in equation (54).

A chemical reaction in which the product aids in its own formation is called an autocatalytic reaction.

Example 4. Suppose in an autocatalytic reaction 1 unit of a substance is present initially. Let q denote the amount of the product. If the rate of formation of q is given by

Note that (1 − q) represents the amount of the substance and q represents the amount of the product.

$$(56) \qquad \frac{dq}{dt} = kq(1 - q)$$

where k is a constant,

    (a) find the amount of the product at time t

    (b) find $q(t)$ given that at $t = 0$ the amount of the product is $\frac{1}{2}$ units.

Solution. (a) We write equation (56) in the form

We observe that 0 < q < 1. Why?

$$(57) \qquad \frac{1}{q(1 - q)} \frac{dq}{dt} = k.$$

Integrating both sides of equation (57) with respect to t we obtain

$$(58) \qquad \int \frac{1}{q(1 - q)} \frac{dq}{dt} \, dt = \int k \, dt.$$

We note that

$$\frac{1}{q(1 - q)} = \frac{A}{q} + \frac{B}{1 - q}.$$

Solving for A and B we find

$$1 = A(1 - q) + Bq.$$

Setting $q = 0$ we have $A = 1$, setting $q = 1$ we have $B = 1$. Hence, equation (58) may be written as

$$\int \left(\frac{1}{q} + \frac{1}{1 - q} \right) dq = \int k \, dt$$

$$\ln q - \ln (1 - q) = kt + C$$

Using property (17)

$$(59) \qquad \ln \frac{q}{1 - q} = kt + C.$$

We find the exponential of each side of equation (59):

$$\frac{q}{1-q} = e^{kt+C}$$

or

$$\frac{q}{1-q} = Me^{kt}.$$

Solving for q we obtain

(60)
$$q(t) = \frac{Me^{kt}}{1 + Me^{kt}}.$$

(b) At $t = 0$, $q = \frac{1}{2}$. Hence from equation (60) we have

$$\frac{1}{2} = \frac{Me^0}{1 + Me^0}.$$

Solving for M we have $M = 1$. Therefore

(61)
$$q(t) = \frac{e^{kt}}{1 + e^{kt}}.$$

Exercise 4

In problems 1 through 14, find the antiderivative.

1. $\displaystyle\int \frac{5x - 3}{x^2 - 2x - 3}\, dx$

2. $\displaystyle\int \frac{7x - 13}{(x + 1)(x - 3)}\, dx$

3. $\displaystyle\int \frac{6x^2 - 11x - 14}{(x - 3)(x - 2)(x + 4)}\, dx$

4. $\displaystyle\int \frac{2x - 3}{x^2 - 8x + 16}\, dx$

5. $\displaystyle\int \frac{x^2 + 2x + 5}{x(x + 1)(x - 1)}\, dx$

6. $\displaystyle\int \frac{x + 3}{(x - 1)^2(x + 2)}\, dx$

7. $\displaystyle\int \frac{4 - 2x}{(x^2 - 1)(x - 1)}\, dx$

8. $\displaystyle\int \frac{x^4 + 1}{x^3 - x}\, dx$

9. $\displaystyle\int \frac{x + 4}{(x + 1)^2(x - 2)}\, dx$

10. $\int \dfrac{2x - 3}{(x + 2)(x - 1)^3} \, dx$

11. $\int \dfrac{x^3 + 5}{(x^2 - 4)^2} \, dx$

12. $\int \dfrac{2x^2}{x^4 - 5x^2 + 4} \, dx$

13. $\int \dfrac{3x^2 - 1}{(x - 1)^2(x + 3)} \, dx$

14. $\int \dfrac{x^3 + x^2 - 2x - 5}{(x + 1)^2(x - 2)^2} \, dx$

15. In Example 4, if $q = \frac{1}{4}$ at $t = 1$, find $q(t)$.

16. Two chemicals react to form a new substance x. The rate of formation of x is given by

$$\frac{dx}{dt} = k(1 - x)(2 - x)$$

where k is a constant.
(a) Find $x(t)$ given that $x(0) = 0$.
(b) Find k if $x(1) = \frac{1}{2}$.

17. Let P_0 be the population of a city struck by an epidemic. Suppose the rate of spreading of the disease is given by

$$\frac{dP}{dt} = P(P_0 - P)$$

where P is the population that has contacted the disease.
(a) Determine $P(t)$ if $P(0) = 0$.
(b) If t is measured in days, how long will it take before half the population is affected by the disease.

5. Some Applications

We consider some examples from the sciences and economic theory involving the exponential and logarithmic functions.

Let A dollars be invested earning 5 percent per annum. If P is the principal at the end of one year, then

$$P = A\left(1 + \frac{5}{100}\right) \quad \text{if the interest is compounded annually}$$

$$P = A\left(1 + \frac{5}{(100)(2)}\right)^2 \quad \text{if the interest is compounded semiannually}$$

$$P = A\left(1 + \frac{5}{(100)(12)}\right)^{12} \quad \text{if the interest is compounded monthly.}$$

In general, if the annual interest rate is r percent compounded n times per annum, then the principal P at the end of one year is

(62)
$$P = A\left(1 + \frac{r}{100n}\right)^n,$$

and at the end of m years

(63)
$$P = A\left[\left(1 + \frac{r}{100n}\right)^n\right]^m.$$

What happens to P when the interest is compounded more and more frequently? If the number n increases without limit, we obtain from equation (63)

$$P = \lim_{n \to \infty} A\left[\left(1 + \frac{r}{100n}\right)^n\right]^m$$

$$= \lim_{n \to \infty} A\left[\left(1 + \frac{r}{100n}\right)^{100n/r}\right]^{mr/100}$$

or

See Theorem 2 in Chapter 3.

(64)
$$P = Ae^{(r/100)m}.$$

Finally, if $m = t$ years we have

(65)
$$P = Ae^{(r/100)t}.$$

Formula (65) tells us that the largest return from a deposit of A dollars at r percent interest for t years cannot exceed $Ae^{(r/100)t}$ irrespective of the number of times the interest is compounded.

A bank that uses formula (65) is said to compound the interest *continuously*.

Example 1. How long will it take a $1000 deposit to double if it is compounded continuously at 5 percent interest per annum?

Solution. Using formula (65) we have

$$A = \$1000, \quad r = 5,$$

and we need to find t when $P = \$2000$. Thus

$$2000 = 1000e^{0.05t}$$

or

$$2 = e^{0.05t}.$$

We take the natural logarithm to obtain

$$\ln 2 = 0.05t$$

or

$$t = \frac{\ln 2}{0.05} = \frac{0.6931}{0.05} \approx 13.86 \text{ years.}$$

Remark. To double a given deposit requires t years which is dependent on the interest paid and not on the amount deposited.

It is interesting to note that $1000 at 5% per annum compounded semiannually doubles itself in about 14.03 years. Therefore continuous compounding of interest is not as advantageous as some banks might have you believe!

Suppose an object A with temperature T_A is immersed in a medium B with constant temperature T_0. It is found experimentally that the rate of change of temperature of A is proportional to the difference in temperature between A and B. Mathematically this is expressed as

(66) $$\frac{dT_A}{dt} = k(T_0 - T_A)$$

where $k > 0$ is a constant of proportionality. We illustrate this in

Example 2. An object at a temperature of $180°$ is allowed to cool in a medium at a constant temperature of $60°$. If in one minute the temperature of the object is $120°$, in how many minutes will its temperature be $90°$?

Solution. Here $T_0 = 60°$. If T represents the temperature of the object at any time t, then

(67) $$T(0) = 180°$$

and

(68) $$T(1) = 120°.$$

Using the differential equation (66) we have

(69) $$\frac{dT}{dt} = k(60 - T).$$

Equations (67) and (69) form an initial-value problem. We solve to obtain

$$\int_0^t \frac{1}{60 - T(s)} \frac{dT}{ds} ds = k \int_0^t ds$$

$$-\ln |60 - T(s)|\Big|_0^t = ks\Big|_0^t$$

$$-\ln |60 - T(t)| + \ln |60 - 180| = kt$$

$$\ln \frac{120}{T(t) - 60} = kt$$

Note. (1) Since $T \geq 60$, $|60 - T| = T - 60$.

(2) $\dfrac{120}{T - 60} = e^{kt}$

and thus

(70) $$T(t) = 60(1 + 2e^{-kt}).$$

In (70) the number k is still unknown. To evaluate k we use (68) and (70). Thus

$$120 = 60(1 + 2e^{-k})$$

$$\tfrac{1}{2} = e^{-k},$$

and

$$\ln \tfrac{1}{2} = -k.$$

Equation (70) becomes

(71) $$T(t) = 60[1 + 2e^{(\ln \frac{1}{2})t}].$$

To find t when $T = 90°$ we use equation (71) and obtain

$$90 = 60[1 + 2e^{(\ln \frac{1}{2})t}]$$

$$\tfrac{1}{4} = e^{(\ln \frac{1}{2})t},$$

and

$$\ln \tfrac{1}{4} = (\ln \tfrac{1}{2})t.$$

Hence

$$t = \frac{\ln \frac{1}{4}}{\ln \frac{1}{2}} = \frac{-\ln 4}{-\ln 2} = \frac{2 \ln 2}{\ln 2} = 2.$$

That is, in 2 minutes the temperature of the body decreases to 90°.

Substances which decrease in quantity continuously at a rate proportional to the amount present are said to *decay* exponentially. In mathematical symbols we write

(72) $$\frac{dQ}{dt} = -kQ$$

where $k > 0$ and Q is the amount of substance at time t.

Example 3. Suppose a radioactive substance decays according to the law in (72). If 20 lbs are stored now, 18 lbs would be left at the end of 1 century (100 years). Find the amount present at time t.

Solution. We have the initial-value problem

(73)
$$\frac{dQ}{dt} = -kQ$$

(74)
$$Q(0) = 20.$$

Integration yields

$$\int_0^t \frac{1}{Q} \frac{dQ}{ds} \, ds = - \int_0^t k \, ds$$

$$\ln Q(s) \Big|_0^t = -ks \Big|_0^t$$

$$\ln Q(t) - \ln Q(0) = -kt$$

$$\ln \frac{Q(t)}{20} = -kt,$$

and

(75)
$$Q(t) = 20e^{-kt}.$$

Since at $t = 1$, $Q = 18$, we have

$$18 = 20e^{-k}$$
$$0.9 = e^{-k},$$

and

$$-k = \ln 0.9.$$

Therefore, the desired function is defined by

$$Q(t) = 20e^{(\ln 0.9)t}.$$

Carbon 14 is a radioactive substance. The date at which an organism lived may be estimated by analyzing the remains of carbon 14 in the organism, since all living organisms contain carbon 12 and carbon 14. From the moment the organism dies, carbon 14 continues to decay and is no longer replaced, while carbon 12 remains unchanged. The change in the amount of carbon 14 relative to carbon 12 makes it possible to calculate the (approximate) time at which the organism lived.

Dr. Willard F. Libby (1908–   ) developed a method for determining the age of ancient objects, using 5600 years as the half-life of carbon 14. (This figure differs among chemists.) Libby was awarded the 1960 Nobel Prize for Chemistry for his method of carbon dating.

The half-life of a substance is the time required for one-half the amount of the substance to decay.

Example 4. An archaeological expedition working in the Gobi desert came across some ancient objects. It was determined that about 30% of the original amount of C^{14} was still present in the objects. If C^{14} decays continuously, find the approximate age of the object (assume the half-life of C^{14} to be 5600 years).

Solution. Let $A(t)$ be the amount of C^{14} at time t. If $A(t)$ satisfies the differential equation (72), then

(76) $$A(t) = Ce^{-kt}$$

where C is a constant. At $t = 0$ the amount present is A_0. Hence

$$A(0) = Ce^0 \Rightarrow C = A_0$$

and

(77) $$A(t) = A_0e^{-kt}.$$

After 5600 years the amount of C^{14} present is $A_0/2$. Thus

$$A(5600) = A_0e^{-5600k} = \frac{A_0}{2}$$

$$\tfrac{1}{2} = e^{-5600k},$$

or

$In \frac{1}{2} = -In\ 2$

$$In\ \tfrac{1}{2} = In\ e^{-5600k}$$

$$k = \tfrac{1}{5600}\ In\ 2.$$

Thus

$$A(t) = A_0e^{(-In\ 2/5600)t}.$$

In this case 30% of the C^{14} is present. Therefore

$$\tfrac{30}{100}A_0 = A_0e^{(-In\ 2/5600)t}$$

$$\tfrac{3}{10} = e^{(-In\ 2/5600)t},$$

and

$$In\ \frac{3}{10} = -\frac{In\ 2}{5600}t.$$

Hence

$$t = -5600\ \frac{In\ \frac{3}{10}}{In\ 2}$$

$$= -5600\ \frac{In\ 3 - In\ 10}{In\ 2}$$

$$= -5600\ \frac{1.0986 - 2.3026}{.6931},$$

or

$$t \approx 9700\ \text{years}.$$

This means that the objects found are about 9700 years old.

Exercise 5

1. A deposit of A dollars is invested at 4 percent per annum compounded continuously. In how many years will A double itself?

In how many years will A double itself at 4 percent per annum compounded semiannually?

2. A deposit of $1000 at an interest compounded continuously doubles itself in 10 years. What is the interest rate?

3. A deposit of $500 earns 5 percent interest per annum compounded continuously. What is the principal after 20 years?

4. A deposit of A dollars is invested at 5 percent interest compounded continuously. If after 30 years the principal is $50,000, find the amount A deposited.

5. A sum of $2000 is deposited in a bank and it draws interest at the annual rate of 6 percent, compounded continuously. Find the amount the depositor will receive at the end of 8 years and 4 months. Assume $\sqrt{e} = 1.649$.

In problems 6 through 9 the rate of change of temperature is proportional to the difference of temperatures between two mediums.

6. At a room temperature of 25° the temperature of a liquid is observed to be 75°. Fifteen minutes later, the temperature of the liquid is observed to be 65°. Find the temperature of the liquid 90 minutes after the first observation.

7. Carol prepares an apple pie for her friends Bob, Ted, and Alice. It is removed from the oven at 420° and left to cool at a room temperature of 70°. In $\frac{1}{2}$ hour the pie is at 210°. Eager to have a piece of the pie, Bob calculates the time when the pie will be at 100°. How long will they have to wait?

8. A thermometer reading 75° is placed outdoors. Four minutes later the reading is 30°. If the outside temperature is fixed at 20°, find
   (a) the thermometer reading 8 minutes after it was placed outdoors,
   (b) the time it takes for the reading to drop to 21°.

9. An object is placed in a medium with constant temperature. Initially the temperature of the object is 40° higher than the medium. In 8 minutes the temperature of the object is 20° higher than the medium.
   (a) Determine k.
   (b) In how many minutes will the difference in temperature be 10°?

10. The population of a country increases continuously ($dP/dt = kP, k > 0$) at 4 percent per year. If the present population is 200 million, what will it be in 10 years? 20 years? When will it have doubled? tripled?

11. The birth rate of a population of ants is 3.5 percent per week.

If there are no deaths among the ants, what is the population in 20 weeks when it is known that at $t = 0$, $p = 10^3$?

12. According to the 1940 census the population of Nuroville was 64,000 and in 1960 it was 80,000. If the rate of increase of population is proportional to the population, find in what year the population will be 160,000. (Use 2 ln 2 = ln 4 = 1.38628, ln 5 = 1.60944.)

13. A population P of mosquitoes increases by 3000 in one day. If P satisfies the initial-value problem

$$\frac{dP}{dt} = kP, \qquad P(0) = 1200,$$

(a) find k,

(b) find the size of the population after 12 hours.

14. Suppose the half-life of a substance that decays continuously is 1200 years. What is the amount of the substance left in 750 years? 500 years?

15. Suppose the rate of change in the amount of oil in this country is proportional to the amount present. If the constant of proportionality is 5 percent per year and the amount present in 1800 was 10^{12} barrels, find the amount that was present in 1900, 1970.

16. A chemical plant discharges fumes into the atmosphere. The level h is 5 feet above the stacks at 6 A.M. and 20 feet at noon. If the rate of change in the level is proportional to h what is the level at 5 P.M.?

17. Find the point on the curve $y = 2 \ln (x - 3)$ at which the tangent to the curve has slope $\frac{1}{2}$.

18. A particle moves according to the law

$$s = ae^{kt} + be^{-kt}$$

where a, b, and k are constants. Show that its acceleration is proportional to its distance from the origin.

Joseph Louis Lagrange

9

Functions

of

Several

Variables

1. Functions of Two Variables

In this chapter we shall study functions of two variables. The theory for the function of three or more variables can be developed along similar lines. Functions of several (two or more) variables arise in many disciplines. For example, the area A of a rectangle is given by (see Figure 1)

$$A = xy.$$

Thus the area A is a function of two independent variables. Similarly, the selling price P of an antibiotic depends upon its cost, its quality, and the selling price of the competitor. Thus P is a function of several variables.

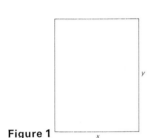

Figure 1

Definition 1. A real-valued function of two variables is a function f that has a set of ordered pairs of real numbers as its domain, and a subset of real numbers as its range.

In the case of functions of one variable, we write $f(x)$ to denote the value of the function at x. Similarly, if f is a function of two variables, we write $f(x, y)$ to denote the value of f at the ordered pair (x, y).

Example 1. Consider the relation

(1) $$z = f(x, y) = \sqrt{1 - x^2 - y^2}.$$

267

The relation (1) between x, y, and z determines a value of z or $f(x, y)$ corresponding to every ordered pair of numbers x, y such that $x^2 + y^2 \leq 1$.

For a real number b, $\sqrt{b}$ is real if and only if $b \geq 0$.

Denoting the pair of numbers (x, y) geometrically by a point on the plane, we see that the points (x, y) for which $x^2 + y^2 \leq 1$ lie on or within a circle with center at the origin and radius 1. Thus the domain of f is a region in the plane given by $\{(x, y) \mid x^2 + y^2 \leq 1\}$. The range of this function is obviously the closed interval $[0, 1]$.

Example 2. Describe the domain and the range of the function f, defined by

$$z = f(x, y) = e^{x^2 + y^2}.$$

Recall that $e^0 = 1$.

Solution. Domain of $f = \{(x, y) \mid x \in R, y \in R\}$, range of $f = \{z \mid z \geq 1\}$.

Example 3. Describe the domain of the function f, defined by

$$f(x, y) = \sqrt{(1 - x)(x - 2)} + \sqrt{(3 - y)(y - 5)}.$$

Solution. Since $(1 - x)(x - 2)$ is nonnegative if $1 \leq x \leq 2$ and $(3 - y)(y - 5)$ is nonnegative if $3 \leq y \leq 5$, the domain of $f = \{(x, y) \mid 1 \leq x \leq 2; 3 \leq y \leq 5\}$. This domain is a rectangle (see Figure 2) in the xy-plane bounded by the lines

$$x = 1, \quad x = 2, \quad y = 3, \quad y = 5.$$

We have seen in Chapter 2 that the graph of a function of one variable can be represented pictorially in the plane. In order to graph a function of two variables, we need a three-dimensional coordinate system. We draw three mutually perpendicular coordinate axes meeting at a common point, called the *origin* (see Figure 3). The *coordinate lines* are the x-axis, y-axis, and z-axis.

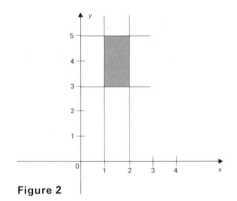

Figure 2

The y-axis and the z-axis are usually drawn in the plane of the paper (the y-axis horizontal and the z-axis vertical), and the x-axis is drawn to point toward the reader.

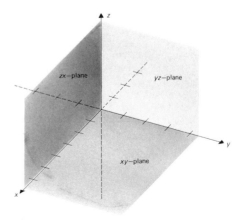

Figure 3

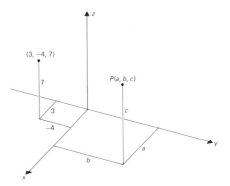

Figure 4

The positive directions of the axes are indicated by arrowheads. Each pair of coordinate axes determines a plane. Thus we have the *xy*-plane, *yz*-plane, and *zx*-plane. Any two of these three planes (each unlimited in extent) divide space into four equal parts, and all three of them divide space into eight equal parts known as *octants*.

Let *P* be any point in space. The position of the point *P* is described by (a, b, c), where a, b, and c are the directed distances from the three coordinate planes, as shown in Figure 4. For example, the point $(3, -4, 7)$ is located three units in front of the *yz*-plane, four units to the left of the *zx*-plane, and seven units above the *xy*-plane. Thus, given an ordered triple of real numbers (a, b, c), we can find a point in space which represents this ordered triple. Conversely, given a point *P* in space, we construct the three planes through *P* which are parallel to the three coordinate planes. The plane parallel to the *yz*-plane intersects the *x*-axis at the *x-coordinate*, the plane parallel to the *xz*-plane intersects the *y*-axis at the *y-coordinate*, and the plane parallel to the *xy*-plane intersects the *z*-axis at the *z-coordinate*. Thus, there is a unique correspondence between points in three-dimensional space and ordered triples of real numbers.

We observe that all points on the *x*-axis have their *y*- and *z*-coordinates both zero; that is, these points have the form $(x, 0, 0)$. Similarly, the points on the *y*-axis and *z*-axis have the forms $(0, y, 0)$ and $(0, 0, z)$ respectively. We also notice that all points in a plane perpendicular to the *z*-axis have the same value of their *z*-coordinates. The equation $z = c$ (where *c* is a real constant) is satisfied by every point (x, y, c) lying in a plane perpendicular to the *z*-axis and directed distance of *c* units from the *xy*-plane. Similarly,

$$y = c, \qquad x = c$$

represent planes perpendicular to the *y*-axis and *x*-axis respectively.

The following information will be helpful.

point	location of the point
$(0, 0, 0)$	origin
$(a, 0, 0)$	*x*-axis
$(0, b, 0)$	*y*-axis
$(0, 0, c)$	*z*-axis
$(a, b, 0)$	*xy*-plane
$(0, b, c)$	*yz*-plane
$(a, 0, c)$	*zx*-plane

We now prove

Theorem 1. The distance between the points $P(x_1, y_1, z_1)$ and $Q(x_2, y_2, z_2)$ is

Notice the similarity between (2) and the distance formula in two dimensions (Chapter 2).

(2) $d(P, Q) = \sqrt{(x_1 - x_2)^2 + (y_1 - y_2)^2 + (z_1 - z_2)^2}.$

Proof. See Figure 5. Draw PR and QS perpendicular to the xy-plane. The coordinates of R and S are $(x_1, y_1, 0)$ and $(x_2, y_2, 0)$. Also, draw PT perpendicular to QS. Then the coordinates of T are (x_2, y_2, z_1). From the figure we see that

$$d(R, S) = \sqrt{(x_1 - x_2)^2 + (y_1 - y_2)^2}.$$

In the rectangle $RPTS$, $d(P, T) = d(R, S)$. Furthermore,

$$d(T, Q) = |z_2 - z_1| = \sqrt{(z_1 - z_2)^2}.$$

By the Pythagorean theorem,

$$\begin{aligned} d(P, Q) &= \sqrt{[d(P, T)]^2 + [d(T, Q)]^2} \\ &= \sqrt{[d(R, S)]^2 + [d(T, Q)]^2} \\ &= \sqrt{(x_1 - x_2)^2 + (y_1 - y_2)^2 + (z_1 - z_2)^2}. \end{aligned}$$

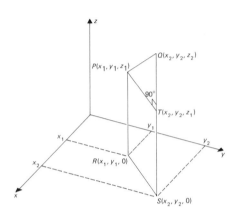

Figure 5

Example 4. Find the distance between the points

$$P(2, 3, -1) \quad \text{and} \quad Q(-1, 0, 2).$$

Solution. By direct substitution in (2) we obtain:

$$\begin{aligned} d(P, Q) &= \sqrt{(2 - (-1))^2 + (3 - 0)^2 + (-1 - 2)^2} \\ &= \sqrt{3^2 + 3^2 + 3^2} \\ &= \sqrt{27} = 3\sqrt{3}. \end{aligned}$$

Exercise 1

In problems 1 through 5, find the domain of f.

1. $f(x, y) = xy^2 + x^2y$

2. $f(x, y) = \sqrt{1 - \dfrac{x^2}{4} - \dfrac{y^2}{9}}$

3. $f(x, y) = \sqrt{(a - x)(x - b)} + \sqrt{(c - y)(y - d)}$, where a, b, c, and d are constants, $a < b$, and $c < d$.

4. $f(x, y) = \dfrac{1}{\ln x + \ln y}$

5. $f(x, y) = \dfrac{x^3 + y^3}{x^2 + y^2}$

6. Plot the following points.

(a) (1, 0, 0)   (b) (0, 2, 0)
(c) (0, 0, 3)   (d) (1, 2, 0)
(e) (1, 2, 3)   (f) (2, 1, 2)
(g) (−1, 2, 0)   (h) (−1, 0, −1)
(i) (2, 2, 2)   (j) (−1, −2, −3)

7. Describe the set of all points (x, y, z) that satisfy the given conditions.

(a) $x = 0$ (b) $y = 0$
(c) $z = 3$ (d) $x = 0$ and $y = 0$
(e) $x = 1$ and $y = 0$ (f) $z = -2$
(g) $x = y = z$ (h) $x = 1, y = -1$, and $z = 2$

8. Calculate the distance between the points.

(a) (0, 0, 0), (a, b, c)
(b) (1, 0, −1), (0, 2, −3)
(c) (1, 2, 3), (1, −2, 0)
(d) (3, 0, 1), (−1, 0, −2)
(e) (0, 4, 5), (0, −8, 0)
(f) (1, 2, 3), (−4, −5, −6)

2. Graphs

We recall from Chapter 2 that the graph of a function f of one variable is the set of all points (x, y) such that $y = f(x)$.

Definition 2. **The graph of a function f of two variables is the set of all points (x, y, z) such that $z = f(x, y)$.**

For a function of one variable, the graph of $y = f(x)$ is generally a curve in the plane. However, in general, the graph of a function f given by $z = f(x, y)$ is a surface in three dimensions. Obtaining the shape, or even a rough sketch, of the graph of a function of two variables is a difficult task. Attempting to imitate the case of a function of one variable, by plotting a reasonable number of points (x, y, z) in three dimensions and connecting them by a smooth surface, will not succeed. Even if we are able to plot many points in space (which itself is quite difficult), we may not be able to recognize the shape of the surface.

In this section we shall discuss some of the simplest surfaces.

The proof of the following theorem is beyond the scope of this book, and is therefore omitted.

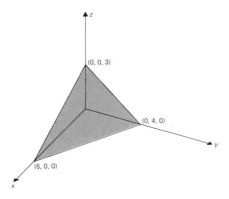

Figure 6

In other words, we draw the curve G given by z = f(x) on the zx-plane, and the graph of z = f(x) in three-dimensional space is generated by moving a horizontal line along the curve G.

Theorem 2. The graph of the linear equation

(3) $ax + by + cz = d$ (*a, b, c* not all equal to zero)

is a plane in three-dimensional space.

Example 1. Discuss the plane

(4) $2x + 3y + 4z = 12.$

Solution. It is known that three noncollinear points in space determine a unique plane. We can find points on (4) by choosing two coordinates and solving for the third. If we let $x = 0$ and $y = 0$, we obtain $z = 3$. Consequently, the point $(0, 0, 3)$ is on the plane given by (4). Similarly, $(0, 4, 0)$ and $(6, 0, 0)$ are also points on the plane. In fact, these are the points where the plane intersects the axes. Figure 6 is a sketch of the plane.

Suppose we wish to graph the function given by

(5) $z = f(x)$

in three-dimensional space. We notice that one of the variables, namely y, is missing; i.e., equation (5) is independent of y. Consequently, the point $(x_0, y_0, f(x_0))$ is on the graph if and only if the point $(x_0, 0, f(x_0))$ is on the graph. It follows that the graph of the surface given by (5) can be obtained by sliding the graph of $z = f(x)$ on the *zx*-plane parallel to the *y*-axis. Similar results hold for equations with x or z missing.

Example 2. Sketch the graph of

(6) $2x + y = 4$

in three-dimensional space.

Solution. Notice that in equation (6) the variable z is missing. From the above discussion, we draw the graph of (6) in the *xy*-plane, which we know is a straight line. We now slide this line parallel to the *z*-axis, and obtain the required graph, which is a plane, as shown in Figure 7. In other words, we generate the graph by moving a vertical line along the straight line given by (6) in the *xy*-plane.

Example 3. Sketch the graph of

(7) $z = x^2$

in three-dimensional space.

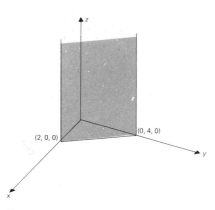

Figure 7

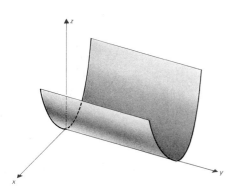

Figure 8

Solution. We know that the graph of $z = x^2$ in the zx-plane is a parabola. Sliding this parabola parallel to the y-axis gives the graph of (7) in three-dimensional space, as sketched in Figure 8.

As we pointed out earlier, graphing a surface given by $z = f(x, y)$ is very difficult. However, if we hold y constant, say at $y = b$, then $z = f(x, b)$ is a function of one variable—namely, x. The graph of $z = f(x, b)$ lies on the plane $y = b$; in fact, it is the intersection of the graph of $z = f(x, y)$ with the plane $y = b$. Such an intersection is called a *cross section* of $z = f(x, y)$. Drawing several cross sections $z = f(x, b_1)$, $z = f(x, b_2)$, . . . , yields a fairly good idea of the shape of the surface $z = f(x, y)$. We can find cross sections by holding x constant also.

The cross sections of a given surface obtained by the intersection of the surface with the coordinate planes are called the *traces* of the surface. For example, the trace of the surface $z = x^2 - y$ in the xy-plane is $x^2 - y = 0$. The trace in the xy-plane is obtained by letting $z = 0$. Similarly, the traces in the yz- and zx-planes are obtained by letting $x = 0$ and $y = 0$ respectively, in the equation $z = f(x, y)$.

Exercise 2

1. Sketch the graphs of the following planes.
   (a) $x + y + z = 1$ (b) $x - 2y + 3z = 6$
   (c) $2x + y + z = 4$ (d) $-x - y + z = 1$
2. Sketch the following graphs in three-dimensional space.
   (a) $x + y = 1$ (b) $x + 2z = 2$
   (c) $2y + 3z = 6$ (d) $x^2 - y = 0$
   (e) $z = \sqrt{1 - x^2}$ (f) $x^2 = 4y - 2$
3. Sketch the traces of the following surfaces.
   (a) $x^2 + y^2 + z^2 = 1$
   (b) $x^2 + y^2 - z = 1$
4. Sketch the graph of

$$z = \sqrt{4 - x^2 - y^2}$$

   by drawing several cross sections.

3. Partial Differentiation

Before we discuss the notion of partial differentiation, we shall mention the ideas of limit and continuity of a function of two variables very briefly.

Intuitively, we say

$$\lim_{(x,y)\to(a,b)} f(x, y) = L$$

provided that $f(x, y)$ is "close" to the number L, whenever (x, y) is close to (a, b).

To state the idea of limit formally, we first need

Definition 3. **Let δ be any positive real number. For a given point (a, b) in the plane, let**

$$N = \{(x, y) \mid a - \delta < x < a + \delta; b - \delta < y < b + \delta\}.$$

Then the set N is called the δ-*neighborhood* of the point (a, b).

Notice (see Figure 9) that the set N determines a square in the plane bounded by the lines

$$x = a - \delta, \qquad x = a + \delta, \qquad y = b - \delta, \qquad y = b + \delta.$$

The center of the square is the point (a, b). The sides of the square are not included in N. To emphasize this point, some authors call it an open neighborhood.

We now give the formal definition of the limit.

The quantity δ depends on ε.

Definition 4. **If, for every $\epsilon > 0$, there exists a δ-neighborhood N of (a, b) such that**

$$|f(x, y) - L| < \epsilon$$

for every $(x, y) \in N$, then

$$\lim_{(x,y)\to(a,b)} f(x, y) = L.$$

The above definition can be used to prove the usual results about limits, similar to the ones for a function of one variable.

We state (without proof)

Theorem 3. **Let**

$$\lim_{(x,y)\to(a,b)} f(x, y) = L \qquad \text{and} \qquad \lim_{(x,y)\to(a,b)} g(x, y) = M.$$

Then

(i) $$\lim_{(x,y)\to(a,b)} f([x, y) \pm g(x, y)] = L \pm M$$

$$= \lim_{(x,y)\to(a,b)} f(x, y) + \lim_{(x,y)\to(a,b)} g(x, y).$$

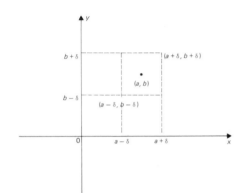

Figure 9
δ-neighborhood of (a, b)

The limit of the sum (or difference) is the sum (or difference) of the limits.

(ii) $\lim\limits_{(x,y)\to(a,b)} [f(x, y) \cdot g(x, y)] = L \cdot M$.

(iii) $\lim\limits_{(x,y)\to(a,b)} \dfrac{f(x, y)}{g(x, y)} = \dfrac{L}{M}$, provided $M \neq 0$.

The reader should write (ii) and (iii) above in words.

Simple limits can be calculated using the above theorem, as in the following example.

Example 1. Calculate $\lim\limits_{(x,y)\to(1,2)} \dfrac{x^3 + 3xy^2}{x^2 - y}$.

By Theorem 3(iii)

Solution. $\lim\limits_{(x,y)\to(1,2)} \dfrac{x^3 + 3xy^2}{x^2 - y} = \dfrac{\lim\limits_{(x,y)\to(1,2)} (x^3 + 3xy^2)}{\lim\limits_{(x,y)\to(1,2)} (x^2 - y)}$

By Theorem 3(i)

$$= \frac{\lim\limits_{(x,y)\to(1,2)} x^3 + \lim\limits_{(x,y)\to(1,2)} (3xy^2)}{\lim\limits_{(x,y)\to(1,2)} x^2 - \lim\limits_{(x,y)\to(1,2)} y}$$

$$= \frac{1^3 + 3(1)(2)^2}{1^2 - 2}$$

$$= \frac{1 + 12}{-1}$$

$$= -13.$$

The definition of continuity is similar to the definition given for one variable.

The function f of Example 1 is continuous at (1, 2), since

$$f(1, 2) = \frac{1^3 + 3(1)(2)^2}{1^2 - 2} = -13.$$

Definition 5. Let (a, b) be a point in the domain of f. Then the function f is said to be **continuous** at (a, b) if

$$\lim\limits_{(x,y)\to(a,b)} f(x, y) = f(a, b).$$

A function is said to be continuous on a set S if it is continuous at every point of S.

A. PARTIAL DIFFERENTIATION

Remark. When taking a partial derivative with respect to x (or y), we are essentially considering the behavior of the trace of the function on the zx- (or yz-) plane.

Let $z = f(x, y)$ be a function of two variables. If y is held constant or *fixed*, then the function $z = f(x, y)$ becomes a function of x alone, and its derivative may be found by the rules discussed in

Chapter 5. This derivative is called the *partial derivative of z with respect to x*, and is denoted by

$$\frac{\partial z}{\partial x}, \quad \frac{\partial f}{\partial x} \quad \text{or} \quad f_x.$$

The partial derivative with respect to y has a similar meaning.

Formally, we have

Definition 6. Let $z = f(x, y)$ be a function of two variables. The partial derivatives of z with respect to x and with respect to y are defined by

$$(8) \qquad \frac{\partial z}{\partial x} = \frac{\partial f}{\partial x} = f_x = \lim_{h \to 0} \frac{f(x + h, y) - f(x, y)}{h},$$

$$(9) \qquad \frac{\partial z}{\partial y} = \frac{\partial f}{\partial y} = f_y = \lim_{k \to 0} \frac{f(x, y + k) - f(x, y)}{k},$$

provided the limits in (8) and (9) exist.

Definition 6 can be easily extended to functions of more than two variables. For example, if

$$w = f(x, y, u, v),$$

then

$$\frac{\partial w}{\partial u} = \lim_{h \to 0} \frac{f(x, y, u + h, v) - f(x, y, u, v)}{h}$$

provided the limit exists.

Notations such as $(\partial z / \partial x)_{(a,b)}$ and $f_y(a, b)$ are used to indicate the value of the partial derivative at the point (a, b).

Example 2. Calculate $\partial z / \partial x$ and $\partial z / \partial y$, where

$$z = x^2 y + xy^{-3}.$$

Also, find the value of each partial derivative at $(1, 2)$.

Solution. Treating y as a constant, we have

$$\frac{\partial z}{\partial x} = 2xy + y^{-3}.$$

Thus

$$\left(\frac{\partial z}{\partial x} \right)_{(1,2)} = 2(1)(2) + 2^{-3} = 4 + \tfrac{1}{8} = \tfrac{33}{8}.$$

Similarly, treating x as a constant, we obtain

$$\frac{\partial z}{\partial y} = x^2 - 3xy^{-4}$$

and

$$\left(\frac{\partial z}{\partial y} \right)_{(1,2)} = (1)^2 - 3(1)(2)^{-4} = 1 - \tfrac{3}{16} = \tfrac{13}{16}.$$

B. SECOND PARTIAL DERIVATIVES

The partial derivatives themselves are functions of two variables. We can therefore calculate the partial derivatives of $\partial z/\partial x$ and $\partial z/\partial y$, just as we calculated the partial derivatives of z. Thus we have four *second partial derivatives* (with different notations):

(i) $$\frac{\partial}{\partial x}\left(\frac{\partial z}{\partial x}\right) = \frac{\partial^2 z}{\partial x^2} = \frac{\partial^2 f}{\partial x^2} = f_{xx},$$

(ii) $$\frac{\partial}{\partial y}\left(\frac{\partial z}{\partial x}\right) = \frac{\partial^2 z}{\partial y\,\partial x} = \frac{\partial^2 f}{\partial y\,\partial x} = f_{xy},$$

(iii) $$\frac{\partial}{\partial x}\left(\frac{\partial z}{\partial y}\right) = \frac{\partial^2 z}{\partial x\,\partial y} = \frac{\partial^2 f}{\partial x\,\partial y} = f_{yx},$$

(iv) $$\frac{\partial}{\partial y}\left(\frac{\partial z}{\partial y}\right) = \frac{\partial^2 z}{\partial y^2} = \frac{\partial^2 f}{\partial y^2} = f_{yy}.$$

In the last equations of (ii) and (iii), notice that the order of the symbols is reversed:

$$\frac{\partial^2 f}{\partial y\,\partial x} = f_{xy},$$

since $\dfrac{\partial^2 f}{\partial y\,\partial x}$ stands for $\dfrac{\partial}{\partial y}\left(\dfrac{\partial f}{\partial x}\right)$, and f_{xy} stands for $(f_x)_y$. Both symbols indicate that the partial derivative is first calculated with respect to x and then with respect to y.

The derivatives f_{xx} and f_{yy} are called the second partial derivatives with respect to x and y respectively, while the derivatives f_{xy} and f_{yx} are called the *mixed partial derivatives*.

The mixed partial derivatives f_{xy} and f_{yx} are distinguished by the order in which f is successively differentiated with respect to x and y. The following theorem (we omit its proof) shows that they are equal in most cases of interest.

Theorem 4. Let f, f_x, f_y, f_{xy}, and f_{yx} all be continuous at (a, b). Then

$$f_{xy}(a, b) = f_{yx}(a, b).$$

Example 3. Calculate all the second partial derivative of

$$f(x, y) = x^3 y^4 + x^2 e^y.$$

Solution.

$$f_x = 3x^2 y^4 + 2xe^y, \quad f_{xx} = 6xy^4 + 2e^y, \quad f_{xy} = 12x^2 y^3 + 2xe^y,$$
$$f_y = 4x^3 y^3 + x^2 e^y, \quad f_{yy} = 12x^3 y^2 + x^2 e^y, \quad f_{yx} = 12x^2 y^3 + 2xe^y.$$

Notice that $f_{xy} = f_{yx}$, as asserted in Theorem 4.

Exercise 3

1. Evaluate the following limits and comment on the continuity of these functions at the indicated points.

(a) $$\lim_{(x,y)\to(2,1)} \frac{x + y}{x - y}$$

(b) $\lim\limits_{(x,y)\to(1,1)} \dfrac{x^3 y^2 + x^2 + y}{x^2 + y^2}$

(c) $\lim\limits_{(x,y)\to(2,2)} \dfrac{x^3 - y^3}{x^2 - y^2}$

(d) $\lim\limits_{(x,y)\to(1,1)} \dfrac{x - y}{\sqrt{x} - \sqrt{y}}$

2. Calculate f_x and f_y.

(a) $f(x, y) = x^2 y - xy^2$

(b) $f(x, y) = (xy)^3$

(c) $f(x, y) = \dfrac{x}{y}$

(d) $f(x, y) = \dfrac{x + y}{x - y}$

3. Calculate f_x and f_y.

(a) $f(x, y) = xe^y$

(b) $f(x, y) = e^{x^2 + y^2}$

(c) $f(x, y) = e^{x+y} \ln(x^2 + y^2)$

(d) $f(x, y) = x^3 + y^2 \ln x + xe^{xy^2 + x^3 y}$

4. In problem 2, calculate all second partial derivatives at $(1, 2)$.

5. Let $z = f(x, y, u, v) = (x^2 + y^3) \cdot e^{u+v}$. Calculate

(a) $f_x, f_y, f_u, f_v,$

(b) $f_{xy}, f_{xu}, f_{uv},$

(c) $f_{xyu}, f_{uyx}, f_{uxy}.$

*6. State a theorem similar to Theorem 4 for functions of three variables.

4. Some Applications

One application of partial derivatives is obvious from its counterpart in functions of a single variable. The partial derivative f_x represents the rate of change of f with respect to x, when the variable y is kept fixed. A similar comment applies to f_y.

Example 1. *Ecology.* Suppose a nuclear power plant and a chemical company are situated on a large lake. Let the pollution index P of the lake be a function of x and y, where x measures the quantity of heat entering the lake due to the power plant and y measures the amount of waste chemical pouring into the lake. We can write $P = f(x, y)$. Then $(\partial P / \partial x)_{(a,b)}$ simply means that if y is held constant at value *b*, and *x* alone is allowed to vary, then the pollution index *P* is changing $(\partial P/\partial x)_{(a,b)}$ units for each unit change in *x* at the instant *x* takes the value *a*. Similar interpretation is given for $(\partial P/\partial y)_{(a,b)}$.

Example 2. *Economics.* Suppose the productivity *P* of a company is a function of the size of the labor force *x*, and the amount of capital invested *y*: $P = f(x, y)$. Then $\partial P/\partial y$ is called the *marginal productivity of capital*, which means that for $x = a$, $y = b$ the productivity of the company increases approximately $(\partial P/\partial y)_{(a,b)}$ (if it is positive) if the capital is increased one more dollar. Similarly, $\partial P/\partial x$ is called the *marginal productivity of labor*, which means that for $x = a$, $y = b$ the productivity of the company increases approximately $(\partial P/\partial x)_{(a,b)}$ if one more person is added to the labor force.

Example 3. *Geometry.* Let $z = f(x, y)$. We mentioned earlier that geometrically this equation represents a surface. If *y* is held constant, say $y = b$, then we obtain a cross section of this surface, determined by the surface $z = f(x, y)$ and the plane $y = b$. We thus have a curve $z = f(x, b)$ in the plane $y = b$. We notice that *z* is now a function of the single variable *x*. The partial derivative $\partial z/\partial x$ is therefore the slope of the tangent to the curve of intersection of the surface and the plane $y = b$ (see Figure 10).

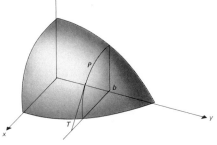

Figure 10

A. THE TOTAL DIFFERENTIAL

In Chapter 6, we defined the differential *dy* of the function $y = f(x)$ by

$$dy = f'(x)\, dx.$$

We shall define a similar term for a function of two variables.

Let $z = f(x, y)$ be a function of two variables *x* and *y*. Let (x, y) and $(x + \Delta x, y + \Delta y)$ be any two points, so that Δx and Δy are the changes in the variables *x* and *y*. Let Δz denote the consequent change in *z*.

We have

$$z + \Delta z = f(x + \Delta x, y + \Delta y)$$

or

(10)
$$\begin{aligned}
\Delta z &= f(x + \Delta x, y + \Delta y) - z \\
&= f(x + \Delta x, y + \Delta y) - f(x, y), \\
&= [f(x + \Delta x, y + \Delta y) - f(x, y + \Delta y)] \\
&\quad + [f(x, y + \Delta y) - f(x, y)].
\end{aligned}$$

We have added and subtracted $f(x, y + \Delta y)$.

The change Δz has been expressed as the sum of two differences. Considering the first pair of functions, we notice that f is evaluated at two points $(x + \Delta x, y + \Delta y)$ and $(x, y + \Delta y)$ which have the same y-value. Momentarily we think of f as a function of x alone. Thus, by the Mean Value Theorem, we have

Of course, we are assuming that the hypotheses of the Mean Value Theorem are satisfied.

$$f(x + \Delta x, y + \Delta y) - f(x, y + \Delta y) = f_x(c_1, y + \Delta y) \cdot \Delta x,$$

where c_1 is between x and $x + \Delta x$. Similarly,

$$f(x, y + \Delta y) - f(x, y) = f_y(x, c_2) \cdot \Delta y,$$

where c_2 is between y and $y + \Delta y$. Hence

(11) $$\Delta z = f_x(c_1, y + \Delta y) \cdot \Delta x + f_y(x, c_2) \cdot \Delta y.$$

If we assume that f_x and f_y are continuous, then when Δx and Δy are "small," c_1 and c_2 are "close" to x and y respectively. With these assumptions we write

(12) $$\Delta z \approx f_x(x, y) \, \Delta x + f_y(x, y) \, \Delta y.$$

Definition 7. Let $z = f(x, y)$ be a function of two independent variables x and y, and let f, f_x, f_y be continuous in some domain. We define

$$dx = \Delta x, \qquad dy = \Delta y,$$
$$dz = f_x \, \Delta x + f_y \, \Delta y = f_x \, dx + f_y \, dy.$$

We call dz the **total differential**.

Remark. The closer Δx and Δy are to zero, the closer the approximation dz is to the exact value of Δz.

Thus, as in the case of a single variable, dz serves as an approximation to Δz.

Example 4. Find the percentage error in the area of a rectangle, when an error of one percent is made in measuring the length and the width of the rectangle.

Solution. Let x, y, and A denote the length, width, and area of the rectangle, respectively. Since $A = xy$, the partial derivatives are

$$\frac{\partial A}{\partial x} = y$$

$$\frac{\partial A}{\partial y} = x.$$

Since we are given

$$dx = \frac{x}{100} \qquad \text{(one-percent error in length)}$$

$$dy = \frac{y}{100} \qquad \text{(one-percent error in width)},$$

we have

$$dA = \frac{\partial A}{\partial x}\, dx + \frac{\partial A}{\partial y}\, dy$$

$$= y\,\frac{x}{100} + x\,\frac{y}{100}$$

$$= \frac{2xy}{100}$$

$$= \frac{2A}{100}.$$

Therefore, there is a two-percent error in A.

Exercise 4

1. In Example 1, let the pollution index P be given by

$$P = cx^2 + dxy + xy,$$

where c and d are constants. Find $(\partial P/\partial x)_{(3,5)}$ and $(\partial P/\partial y)_{(2,3)}$. Interpret your results.

2. Repeat problem 1 for

$$P = 3e^{x+y+xy}.$$

3. In Example 2, suppose the productivity P of the ABC company is given by

$$P = \frac{e^{xy}}{x^2 + y^2 + y}.$$

(a) Find $\dfrac{\partial P}{\partial x}, \dfrac{\partial P}{\partial y}$.

(b) Find $\left(\dfrac{\partial P}{\partial x}\right)_{(200,5)}$ and $\left(\dfrac{\partial P}{\partial y}\right)_{(5,100)}$

(c) Interpret (b).

4. Repeat problem 3 for

$$P = (x^2 + e^y)^2.$$

5. A toothpaste company finds that sales per day are a function of x, the number of times that its commercial is shown on TV, and y, the number of people (in millions) that see the commercial. This function is

$$f(x, y) = 50xy + 1000.$$

(a) Find f_x, f_y.

(b) Interpret f_x when $y = 5$.

(c) Interpret f_y when $x = 2$.

6. Repeat problem 5 for

$$f(x, y) = x^2 + y^2 + 20xy + 500.$$

7. Find the total differential dz if
   (a) $z = xy^2 + xe^y$,
   (b) $z = x + y \ln x$.

8. Find the percentage error in the area of an ellipse when an error of one percent is made in the major and minor axes. *Hint:* If $2a$ and $2b$ are the lengths of the major and minor axes, then area $= \pi ab$.

9. Find the percentage error in the area of an ellipse when an error of one percent is made in measuring its major axis, and an error of two percent is made in measuring its minor axis.

10. In measuring a right circular cone, errors of two percent and one percent are made in the height and radius respectively. Find the percentage error in the volume.

11. In an experiment with a simple pendulum, the acceleration due to gravity g is calculated from the formula

$$T = 2\pi \sqrt{\frac{l}{g}},$$

where T is the period and l is the length of the pendulum. Calculate approximately the percentage error in g if l and T are measured within one percent of accuracy.

12. Find

$$3.001 \times 8.997$$

approximately. *Hint:* Let $f(x, y) = xy$. Then $f(3, 9) = 27$, $dx = 0.001$, $dy = -0.003$.

5. Maxima and Minima

The work in Chapter 6 regarding the maximum and minimum values of a function of one variable can be generalized to functions of two variables. We begin with

In the case of two variables, δ-neighborhoods are used instead of open intervals.

Definition 8. Let f be a function of two variables with domain D. We say that f has a relative maximum value at a point $P(a, b)$ if there exists a δ-neighborhood $N \subset D$ of P, such that for every $(x, y) \in N$,

$$f(a, b) \geq f(x, y).$$

Relative minimum value is defined analogously. We call $f(a, b)$ an *extreme value* if it is a maximum or a minimum value.

We recall that if a function of one variable has a relative extremum at a point and is differentiable there, then its derivative at that point is zero. A similar result for a function of two variables is given in

Theorem 5. Let $f(a, b)$ be a relative extreme value of a function f. If f_x and f_y both exist at (a, b), then

$$f_x(a, b) = 0 \quad \text{and} \quad f_y(a, b) = 0.$$

Proof. If $f(a, b)$ is a relative extreme value of the function f of two variables x and y, then clearly it is also a relative extreme value of the function $f(x, b)$ of one variable x for $x = a$ and thus (from Chapter 6) its derivative $f_x(a, b)$ for $x = a$ must necessarily be zero. Similarly, we can show that

$$f_y(a, b) = 0.$$

We remark that the conditions above are necessary but not sufficient; that is, the converse of Theorem 5 is not true. (See Example 2 below.)

The following theorem states a second derivative test which is used to decide whether or not the function has a relative maximum or a relative minimum.

Theorem 6. Let a function f of two variables have partial derivatives at all points of a δ-neighborhood of a point (a, b). Suppose that

$$f_x(a, b) = 0 \quad \text{and} \quad f_y(a, b) = 0.$$

Let

$$A = f_{xx}(a, b)$$
$$B = f_{xy}(a, b)$$
$$C = f_{yy}(a, b)$$

Then

(i) $f(a, b)$ is a relative maximum value if $AC - B^2 > 0$ and $A < 0$,

(ii) $f(a, b)$ is a relative minimum value if $AC - B^2 > 0$ and $A > 0$,

(iii) **f** has a *saddle point* at (**a**, **b**) if **AC − B² < 0,**

(iv) the case **AC − B² = 0** is doubtful and needs further considerations.

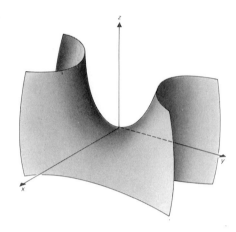

Figure 11
Saddle point

The proof of this theorem is beyond the scope of this book and is therefore omitted.

Working Rules:
Step 1. Find f_x and f_y.
Step 2. Solve $f_x = 0$ and $f_y = 0$ simultaneously.
Step 3. Find $A = f_{xx}$, $B = f_{xy}$, $C = f_{yy}$, and evaluate at the solution set of Step 2.
Step 4. Check $AC - B^2$ for its sign. Check A for its sign. Find the part of Theorem 6 that applies.

Example 1. Find the extreme values of

$$f(x, y) = x^2 + y^2 - 4x - 6y + 1.$$

Solution. We have

$$f_x = 2x - 4 \qquad f_y = 2y - 6.$$

Setting $f_x = 0$ and $f_y = 0$, we get $x = 2$ and $y = 3$. Thus (2, 3) is the only possible location for a relative extremum. Applying the second derivative test to the given function at (2, 3), we have

$$A = f_{xx} = 2$$
$$B = f_{xy} = 0$$
$$C = f_{yy} = 2.$$

Therefore,

$$AC - B^2 = 2 \cdot 2 - 0^2 = 4 > 0.$$

Since $A = 2 > 0$, then part (ii) of Theorem 6 applies and hence the function f has a relative minimum value at (2, 3).

Example 2. Find the extreme values of

$$f(x, y) = xy(3 - x - y).$$

Solution. $f(x, y) = 3xy - x^2y - xy^2$

$$f_x = 3y - 2xy - y^2 = y(3 - 2x - y)$$
$$f_y = 3x - x^2 - 2xy = x(3 - 2y - x)$$
$$A = f_{xx} = -2y$$
$$B = f_{xy} = 3 - 2x - 2y$$
$$C = f_{yy} = -2x.$$

We now solve the equations $\begin{cases} f_x = 0 \\ f_y = 0. \end{cases}$

We have a system of equations

$$\begin{cases} y(3 - 2x - y) = 0 \\ x(3 - 2y - x) = 0, \end{cases}$$

and we have to consider four pairs of equations

Recall that $ab = 0 \Rightarrow a = 0$ or $b = 0$.

(i) $y = 0,$ $x = 0$;
(ii) $3 - 2x - y = 0,$ $x = 0$
(iii) $y = 0,$ $3 - 2y - x = 0$
(iv) $3 - 2x - y = 0,$ $3 - 2y - x = 0.$

Solving these equations we find the following possible locations of the extreme values:

$$(0, 0), \ (0, 3), \ (3, 0), \ (1, 1).$$

For $(0, 0)$, we have $A = 0, B = 3, C = 0$, so that $AC - B^2 < 0$. Hence f has a saddle point at $(0, 0)$. For $(0, 3)$, we have $A = -6$, $B = -3, C = 0$, so that $AC - B^2 < 0$. Hence $(0, 3)$ is another saddle point of f. We can similarly show that $(3, 0)$ is also a saddle point of f. For $(1, 1)$, we have $A = -2, B = -1, C = -2$, so that $AC - B^2 > 0$. Since $A = -2 < 0$, part (i) of Theorem 6 applies. Consequently, $f(1, 1)$ is a relative maximum value.

Exercise 5

In problems 1 through 8, find all relative extrema and saddle points.
1. $f(x, y) = x^2 - y^2$
2. $f(x, y) = x^2 + y^2 - 1$
3. $f(x, y) = x^2 - y^2 + 2x - 4y + 3$
4. $f(x, y) = x^2 - xy$
5. $f(x, y) = x^2 - xy + y^2$
6. $f(x, y) = x^3 - 4y^2$
7. $f(x, y) = y^3 + x^2 - 3x$
8. $f(x, y) = xy(x - y)$
9. Find the points on the surface

$$z = \sqrt{xy + 1}$$

whose distance from the origin is minimum. *Hint:* Let $u = d[(0, 0, 0), (x, y, z)]^2 = x^2 + y^2 + z^2$. Find the minimum of u, when $z^2 = xy + 1$.

10. A rectangular tank is open at the top and holds 32 cubic feet. Find its dimensions so that the surface area of the tank is a minimum.

11. Find the dimensions of a rectangular box of maximum volume, if the box has no top and has surface area 108 square inches.

12. Consider all triangles with a given perimeter $2s$. Show that the one with the largest area is an equilateral triangle. *Hint:* If a, b, c are the sides of a triangle, then $a + b + c = 2s$, and the area A is given by

$$A = \sqrt{s(s - a)(s - b)(s - c)}.$$

13. The sum of three positive numbers is k. Show that their product is maximum when they are equal.

14. Suppose a rectangular box of volume 64 cubic feet is to be constructed from three different materials. The cost of the bottom is 50 cents per square foot, of the sides is 30 cents per square foot, and of the top is 40 cents per square foot. Find the dimensions of the most economical box.

15. Do problem 14 if the bottom costs 50 cents per square foot, and the sides and top cost 20 cents per square foot.

16. The cost of a day's production at ABC Company is given by

$$C = x^2 + y^2 + xy - 20x - 25y + 1500$$

where x is the labor force and y (in thousands of tons) is the amount of raw material. Find x and y that minimize the cost C.

17. Do problem 16 if

$$C = 3x^2 + 2y^2 + 3xy - 66x - 58y + 1600.$$

6. Extremal Problems with Constraints

In many maxima and minima problems, the functions defined are subject to certain restrictions (or conditions) called constraints. For example, suppose we wish to find the minimum value of the function

(13) $$z = f(x, y) = x^2 + y^2.$$

The solution is obvious, since $x^2 + y^2 \geq 0$ for all $x, y \in R$, and $x^2 + y^2 = 0$ if and only if $x = 0$ and $y = 0$. Consequently, the minimum value of f is 0, and it is attained at $(0, 0)$. However, we may be asked to find the minimum value of the function f defined by (13), subject to the constraint

(14) $$x + y - 1 = 0.$$

Obviously f cannot have minimum value at $(0, 0)$, since $(0, 0)$ does not satisfy (14). To solve the problem in this case, we solve equation (14) for y:

(15) $$y = 1 - x.$$

Substituting this value of y in (13), we have

(16) $$z = x^2 + y^2 = x^2 + (1 - x)^2 = x^2 + 1 - 2x + x^2$$
$$= 2x^2 - 2x + 1.$$

Since z is a function of one variable x, we can use the method of Chapter 6:

$$\frac{dz}{dx} = 4x - 2 = 0 \Rightarrow x = \tfrac{1}{2}.$$

Also,

$$\frac{d^2z}{dx^2} = 4 > 0.$$

Therefore $x = \tfrac{1}{2}$ gives the minimum value of z. Substituting $x = \tfrac{1}{2}$ in (15), we get $y = \tfrac{1}{2}$.

Thus, the minimum value of f is attained at $(\tfrac{1}{2}, \tfrac{1}{2})$.

$$f(\tfrac{1}{2}, \tfrac{1}{2}) = (\tfrac{1}{2})^2 + (\tfrac{1}{2})^2 = \tfrac{1}{4} + \tfrac{1}{4} = \tfrac{1}{2}.$$

The following example illustrates how to use a similar method when three variables are involved.

Example 1. Find the minimum value of

(17) $$u = x^2 + y^2 + z^2,$$

subject to the constraint

(18) $$x + y + z = 30.$$

Solution. Solving (18) for z, we obtain

(19) $$z = 30 - x - y.$$

We substitute this value of z in (17):

(20) $$u = x^2 + y^2 + (30 - x - y)^2.$$

Then

$$u_x = 2x + 2(30 - x - y)(-1) = 4x + 2y - 60,$$
$$u_y = 2y + 2(30 - x - y)(-1) = 2x + 4y - 60,$$
$$u_{xx} = 4, \qquad u_{xy} = 2, \qquad u_{yy} = 4.$$

Solving

$$\begin{cases} u_x = 0 \\ u_y = 0, \end{cases}$$

we find $x = 10$, $y = 10$. Also, $u_{xx} \cdot u_{yy} > (u_{xy})^2$, and $u_{xx} > 0$. Hence, u is minimum for $x = 10$, $y = 10$, and

$$z = 30 - x - y = 30 - 10 - 10 = 10,$$

and this minimum value is

$$u = 10^2 + 10^2 + 10^2 = 300.$$

The method above works, since the constraint equation (18) could be solved for one of the variables in terms of the other variables. Suppose equation (18) above was replaced by

(21) $x^3y^2 + 2xy^2 + y^5z^3 + 7xy + 8yz = \sqrt{71}$.

It would indeed be difficult to solve (21) for any one of the three variables. There is an alternate method developed by Lagrange, called *Lagrange's Method of Undetermined Multipliers*. We state this method in

Theorem 7. Let $f, g_1, g_2, \ldots, g_m$ be differentiable functions of the variables $x_1, x_2, \ldots, x_n$. Let f be subject to the constraints:

(22)
$$\begin{aligned} g_1(x_1, x_2, \ldots, x_n) &= 0 \\ g_2(x_1, x_2, \ldots, x_n) &= 0 \\ &\vdots \\ g_m(x_1, x_2, \ldots, x_n) &= 0. \end{aligned}$$

Then the points at which f attains an extreme value are found among the points $(x_1, x_2, \ldots, x_n)$ for which there exist constants $\lambda_1, \lambda_2, \ldots, \lambda_m$ (each λ_i is a Lagrange Multiplier) such that

(23)
$$\begin{cases} \dfrac{\partial f}{\partial x_1} - \lambda_1 \dfrac{\partial g_1}{\partial x_1} - \lambda_2 \dfrac{\partial g_2}{\partial x_1} - \ldots - \lambda_m \dfrac{\partial g_m}{\partial x_1} = 0 \\[2mm] \dfrac{\partial f}{\partial x_2} - \lambda_1 \dfrac{\partial g_1}{\partial x_2} - \lambda_2 \dfrac{\partial g_2}{\partial x_2} - \ldots - \lambda_m \dfrac{\partial g_m}{\partial x_1} = 0 \\[2mm] \dfrac{\partial f}{\partial x_n} - \lambda_1 \dfrac{\partial g_1}{\partial x_n} - \lambda_2 \dfrac{\partial g_2}{\partial x_n} - \ldots - \lambda_m \dfrac{\partial g_m}{\partial x_n} = 0 \\[2mm] g_1(x_1, x_2, \ldots, x_n) = 0 \\ g_2(x_1, x_2, \ldots, x_n) = 0 \\ g_m(x_1, x_2, \ldots, x_n) = 0. \end{cases}$$

The proof of this theorem is omitted. Suppose we consider a special case of this theorem, when the function f of two variables x and y is subject to a single constraint $g(x, y) = 0$. Then the theorem asserts that the points (x, y) at which the function f has an extreme value are among the points (x, y) for which there exists a real number λ such that

(24)
$$\begin{cases} \dfrac{\partial f}{\partial x} - \lambda \dfrac{\partial g}{\partial x} = 0 \\[2mm] \dfrac{\partial f}{\partial y} - \lambda \dfrac{\partial g}{\partial y} = 0 \\[2mm] g(x, y) = 0 \end{cases}$$

Example 2. Find the extreme values of

(25)
$$f(x, y) = xy,$$

subject to the constraint

(26)
$$g(x, y) = x^2 + y^2 - 1 = 0$$

Solution. Since

$$\frac{\partial f}{\partial x} = y, \qquad \frac{\partial f}{\partial y} = x,$$

$$\frac{\partial g}{\partial x} = 2x, \qquad \frac{\partial g}{\partial y} = 2y,$$

equations (24) become

(27)
$$\begin{cases} y - \lambda(2x) = 0 \\ x - \lambda(2y) = 0 \\ x^2 + y^2 - 1 = 0. \end{cases}$$

We solve the first equation in (27) for y and substitute in the second equation to obtain

(28)
$$y = \lambda(2x)$$

and

(29)
$$x(1 - 4\lambda^2) = 0.$$

From (29), we notice that either $x = 0$ or $1 - 4\lambda^2 = 0$. If $x = 0$, then from (28) we must have $y = 0$. But $x = 0$, $y = 0$ does not satisfy (26). Hence $1 - 4\lambda^2 = 0$ or

(30)
$$\lambda = \pm\tfrac{1}{2}.$$

Substituting these values of λ in (28), we get

(31) $\qquad\qquad y = x \qquad$ or $\qquad y = -x.$

For $y = \pm x$, we have from the third equation of (27)

$$x^2 + y^2 = 1$$
$$x^2 + (\pm x)^2 = 1$$
$$x^2 + x^2 = 1$$
$$2x^2 = 1.$$

Then $x = \pm 1/\sqrt{2}$ and $y = \pm x = \pm 1/\sqrt{2}$. Thus, the extreme values of f occur among

$$\left(\frac{1}{\sqrt{2}}, \frac{1}{\sqrt{2}}\right), \left(\frac{1}{\sqrt{2}}, -\frac{1}{\sqrt{2}}\right), \left(-\frac{1}{\sqrt{2}}, \frac{1}{\sqrt{2}}\right), \left(-\frac{1}{\sqrt{2}}, -\frac{1}{\sqrt{2}}\right).$$

We now compute the values of f at these points.

$$f\left(\frac{1}{\sqrt{2}}, \frac{1}{\sqrt{2}}\right) = \frac{1}{\sqrt{2}} \cdot \frac{1}{\sqrt{2}} = \frac{1}{2}$$

$$f\left(\frac{1}{\sqrt{2}}, -\frac{1}{\sqrt{2}}\right) = \left(\frac{1}{\sqrt{2}}\right)\left(-\frac{1}{\sqrt{2}}\right) = -\frac{1}{2}$$

$$f\left(-\frac{1}{\sqrt{2}}, \frac{1}{\sqrt{2}}\right) = \left(-\frac{1}{\sqrt{2}}\right)\left(\frac{1}{\sqrt{2}}\right) = -\frac{1}{2}$$

$$f\left(-\frac{1}{\sqrt{2}}, -\frac{1}{\sqrt{2}}\right) = \left(-\frac{1}{\sqrt{2}}\right)\left(\frac{1}{\sqrt{2}}\right) = \frac{1}{2}.$$

Thus the maximum value of f is $\frac{1}{2}$, which is attained at the points $(1/\sqrt{2}, 1/\sqrt{2})$ and $(-1/\sqrt{2}, -1/\sqrt{2})$. The minimum value of f is $-\frac{1}{2}$, attained at the points $(1/\sqrt{2}, -1/\sqrt{2})$ and $(-1/\sqrt{2}, 1/\sqrt{2})$.

Exercise 6

1. State Theorem 7 for a function f of three variables x, y, and z, subject to the constraint $g(x, y, z) = 0$.
2. State Theorem 7 for a function f of three variables x, y, and z, subject to two constraints $g_1(x, y, z) = 0$ and $g_2 = (x, y, z) = 0$.
3. Find the maximum and minimum values of $f(x, y) = x + y$, subject to $x^2 + y^2 = 1$.
4. Find the maximum and minimum values of $f(x, y) = x^2 - y^2$, subject to $x^2 + y^2 = 1$.
5. By Lagrange's method, find the maximum and minimum values of $f(x, y) = x^2 + y^2$, subject to $x + y = 1$.
6. Do Example 1, using Lagrange's method. *Hint:* Use problem 1.

In problems 7 through 11, use Lagrange's method.

7. Do problem 10 of Exercise 5.

8. Do problem 11 of Exercise 5.

9. Do problem 12 of Exercise 5.

10. Do problem 14 of Exercise 5.

11. Find the minimum value of $f(x, y, z) = x^2 + y^2 + z^2$, subject to $x + 2y + z = 1$ and $2x - y - 3z = 4$.

Pierre de Fermat

10

The

Trigonometric

Functions

Trigonometry was originally invented by the ancient Greeks to measure triangles; the word *trigonometry* means *triangle measurement*. Since then it has been used extensively in several fields, especially in surveying, astronomy, and navigation. Trigonometry is now the branch of mathematics concerned with the properties and the applications of the *circular* or *trigonometric functions*.

1. The Angle

Since angles play the central role in trigonometry, we shall discuss them first. We say that an angle is generated when a half-line (or a ray) rotates in a plane about its endpoint. The endpoint is called the *vertex* of the angle and we refer to the initial position of the half-line as the *initial side* and its final position as the *terminal side*.

For a given half-line in the plane, there are two possible directions of rotation from the initial position: clockwise rotation and counterclockwise rotation. We shall follow the usual convention that a clockwise rotation produces a negative angle, while counterclockwise rotation generates a positive angle. An angle is said to be in *standard position* if a rectangular coordinate system is associated with it such that the vertex of the angle is the origin and the initial side of the angle is the positive x-axis.

To measure an angle, a basic unit of measure is necessary. Since we are interested in measuring the amount of rotation by a half-line, the most natural unit to measure an angle is a complete rotation.

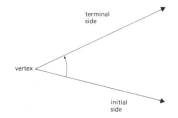

terminal side

vertex

initial side

Measure of an angle

293

Commonly, two kinds of units are used to express the measure of an angle: the *degree* and the *radian*.

Definition 1. An angle is said to have a measure of one *degree* if it is generated by a half-line rotating $\frac{1}{360}$th of a complete rotation about its endpoint. Thus one complete rotation measures 360 degrees.

The degree unit was introduced by the early Babylonians; it is believed that the number 360 was chosen to suit astronomical calculations. One solar year (the period during which the earth completes one revolution around the sun) is $365\frac{1}{4}$ days, while one lunar year (the period during which the moon completes twelve revolutions around the earth) is 354 days. Perhaps the number 360 was found to be a suitable approximation to these numbers! An advantage of the number 360 is that it is divisible by most of the "small" integers.

To denote the degree measure symbolically, we use the symbol $°$ written at the top to the right of the measure of an angle. For example, an angle with a measure of 45 degrees is denoted by 45°.

We thus have

$$360° = 1 \text{ complete rotation}$$
$$180° = \tfrac{1}{2} \text{ of a complete rotation}$$
$$90° = \tfrac{1}{4} \text{ of a complete rotation} = 1 \text{ right angle.}$$

One *minute* (written 1′) has one-sixtieth the measure of a degree, and one *second* (written 1″) has one-sixtieth the measure of a minute.

In summary,

$$1 \text{ complete rotation} = 4 \text{ right angles} = 360°$$
$$1 \text{ right angle} = 90°$$
$$1° = 60'$$
$$1' = 60''.$$

An angle whose measure is 27 degrees 47 minutes and 33 seconds is symbolically denoted by an angle 27°47′33″.

Although the degree-minute-second system is adequate for all practical purposes of angle measurement, the system best suited for the operations of calculus is "radian" measure.

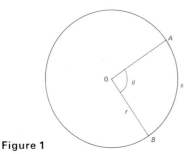

Figure 1

Definition 2. Let the vertex of the angle to be measured be placed at the center of a circle of radius *r* and let the length of the arc *AB* of the circle subtended by the two sides of the angle be *s*. Then the radian measure θ of the angle is given by $\theta = s/r$ (see Figure 1).

In Definition 2, it is easy to see that if $s = r$, then $\theta = 1$ radian. Consequently, one radian is the measure of the angle subtended by an arc whose length is equal to the radius of the circle.

It is easy to correlate degree measure with radian measure. We

Note. π (read "pie") is an irrational number whose approximate value is 3.14.

recall from geometry that the complete circle (of radius r) has arc length $2\pi r$. Thus the radian measure of the complete rotation at the center of the circle is $2\pi r/r = 2\pi$ radians. In the degree system this angle measures 360°. Consequently,

$$2\pi \text{ radians} = 360°$$

$$\pi \text{ radians} = 180°$$

$$1 \text{ radian} = \left(\frac{180}{\pi}\right)° \approx 57°17'44.8''$$

$$x \text{ radians} = \left(\frac{180x}{\pi}\right)°.$$

Similarly,

$$360° = 2\pi \text{ radians}$$

$$180° = \pi \text{ radians}$$

$$1° = \frac{\pi}{180} \text{ radians} \approx 0.017453 \text{ radians}$$

$$x° = \frac{\pi x}{180} \text{ radians}.$$

The following table gives the correspondence between degree and radian measures of some of the common angles.

Degrees	0°	15°	30°	45°	60°	90°	120°	150°	180°	270°	360°
Radians	0	$\frac{\pi}{12}$	$\frac{\pi}{6}$	$\frac{\pi}{4}$	$\frac{\pi}{3}$	$\frac{\pi}{2}$	$\frac{2\pi}{3}$	$\frac{5\pi}{6}$	π	$\frac{3\pi}{2}$	2π

Example 1. Convert (a) 36°15′25″ into radians,
(b) 1.46 radians into degrees.

Solution. (a) $36°15'25'' = 36 + \frac{15}{60} + \frac{25}{3600}$ degrees
$$= 36 + 0.25 + 0.00694$$
$$= 36.25694 \text{ degrees.}$$

Since 1 degree $= \dfrac{\pi}{180}$ radians,

$$36.25694 \text{ degrees} = \frac{\pi}{180} \times 36.25694 \text{ radians}$$

$$\approx \frac{3.14 \times 36.25694}{180}$$

$$\approx 0.638 \text{ radians.}$$

(b) Since 1 radian $= \dfrac{180}{\pi}$ degrees,

$$1.46 \text{ radians} = \frac{180}{\pi} \times 1.46$$

$$\approx \frac{180 \times 1.46}{3.14}$$

$$\approx 83.694 \text{ degrees}$$

$$\approx 83°41'38.4''.$$

Example 2. Suppose the center of the sun is 93,000,000 miles away from the earth and the diameter of the sun subtends an angle of 32′ at a point on the earth. Find approximately the diameter of the sun.

Solution. Let $d(A, B) = \ell$ be the length of the diameter of the sun with center C, and let O be the point on the earth (see Figure 2). Then $d(O, C) = 93,000,000$.

Earth

Figure 2

$$\text{The measure of the angle at } O = 32' = \frac{32}{60} = \left(\frac{8}{15}\right)^\circ$$

$$= \frac{8}{15} \times \frac{\pi}{180} \text{ radians}$$

$$= \frac{2\pi}{675} \text{ radians.}$$

Since d(O, C) is very large compared to ℓ, this is a good approximation.

We take the diameter ℓ of the sun to be approximately equal to the length of the arc of a circle with center at O and radius equal to $d(O, C)$. Consequently,

Recall the formula $\theta = \dfrac{s}{r}$, where θ is measured in radians.

$$\ell \approx d(O, C) \times \text{(the angle at } O \text{ in radians)}$$

$$= \frac{93,000,000 \times 2\pi}{675}$$

$$\approx 865,200 \text{ miles.}$$

Example 3. A hollow right circular cone is made from a piece of paper in the form of a sector of a circle of radius 50 cm and angle 216° by joining the straight edges of the sector. Find

    (a) the radius of the circle of the right circular cone,

    (b) the surface area of the cone.

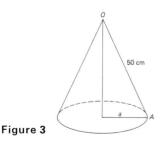

Figure 3

Solution. (a) Let a be the radius of the base circle of the right circular cone (see Figure 3). Then

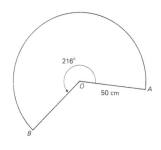

Figure 4

$$2\pi a = \text{circumference of the circle}$$
$$= \text{length of the arc } AB \text{ in Figure 4}$$
$$= d(O, A) \times (\text{angle } AOB \text{ in radians})$$
$$= 50 \cdot \frac{216\pi}{180}$$
$$= 60\pi.$$

Therefore $2\pi a = 60\pi$, which implies $a = 30$ cm.

(b) The surface area of the cone is equal to the area of the sector of the circle in Figure 4. Recall that the area of a circle with radius r is πr^2. If we divide this by 2π, we get the area of the sector which makes an angle of 1 radian at the center:

$$\frac{\pi r^2}{2\pi} = \frac{r^2}{2}.$$

Thus the area of the sector which makes an angle of θ radians at the center is $\theta r^2/2$. In this case,

$$\theta = \frac{216\pi}{180} = \frac{6\pi}{5} \text{ radians.}$$

Thus the area of the given sector

$$= \text{the surface area of the cone}$$
$$= \frac{6\pi}{5} \cdot \frac{r^2}{2} = \frac{3\pi r^2}{5} \text{ sq cm}$$

(see Figure 4).

Exercise 1

1. Convert to radian measure.
   (a) 27°   (b) 260°
   (c) 310°   (d) 450°
   (e) 47°20′   (f) 85°15′30″

2. Convert to degree measure.
   (a) 3.4 radians   (b) 2.7 radians
   (c) $\frac{9\pi}{7}$ radians   (d) $\frac{7\pi}{12}$ radians
   (e) $\frac{\pi}{8}$ radians   (f) 2.43 radians

3. If the angles of a triangle are $x, x + 1, x + 2$ (in radians), find x.

4. In an isoceles triangle the base angles are each equal to 1 radian. Find the third angle in degrees.

5. In an amusement park, a wheel has a radius of 7 ft and makes 30 revolutions per minute. Find its speed in ft per sec. (Choose $\pi = \frac{22}{7}$, and recall that speed is distance traveled divided by time.)

6. Express in radian measure the angle of a regular hexagon.
7. A satellite is launched into a circular orbit about the earth. If the distance of the satellite from the center of the earth is 6000 miles, how far does it travel while sweeping an angle of 45° at the center of the earth?
8. An astronomer who is 240,000 miles from the moon observes that its diameter subtends an angle of 30'. Find approximately the diameter of the moon.
9. The area of a circle is 81 sq ft. Find the length of its arc subtending an angle of 80° at the center and the area of the corresponding sector. (Recall that the area of a circle of radius r is πr^2.)
10. The area of a circle is 36 sq ft. Find the area of a sector of this circle determined by an angle of $5\pi/3$ radians at the center.

2. The Trigonometric Functions

The classical definition of the trigonometric functions is given for acute angles θ by the ratios of the lengths of the sides in a right triangle as follows (see Figure 5):

$$\text{the sine of } \theta = \sin \theta = \frac{a}{c}$$

$$\text{the cosine of } \theta = \cos \theta = \frac{b}{c}$$

$$\text{the tangent of } \theta = \tan \theta = \frac{a}{b}$$

$$\text{the cosecant of } \theta = \csc \theta = \frac{c}{a}$$

$$\text{the secant of } \theta = \sec \theta = \frac{c}{b}$$

$$\text{the cotangent of } \theta = \cot \theta = \frac{b}{a},$$

where a is the length of the side opposite angle θ
           b is the length of the side adjacent to angle θ
           c is the length of the hypotenuse.

Remark. Since similar triangles have proportional sides, the values of the six trigonometric functions of an angle θ depend only on the measure of the angle θ and not on the particular right triangle used for the computation of the trigonometric functions.

The approach given above defines the trigonometric functions for

The angle θ is said to be acute if

$$0 < \theta < \frac{\pi}{2}.$$

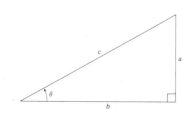

Figure 5

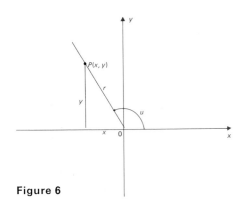

Figure 6

acute angles only. It cannot be used to compute trigonometric functions for obtuse angles (angles larger than 90° or $\pi/2$ radians), since a right triangle cannot have an obtuse angle (why?). In order to extend the results of calculus to the trigonometric functions, we employ a different approach based on analytic geometry.

Definition 3. Let *u* be a real number and let an angle whose radian measure is *u* be placed in standard position (see Figure 6). Let *P*(*x*, *y*) be a point on the terminal side of this angle and let $r = \sqrt{x^2 + y^2}$. Then the six trigonometric functions of *u* are defined by

$$\sin u = \frac{y}{r} \qquad \csc u = \frac{r}{y}$$

$$\cos u = \frac{x}{r} \qquad \sec u = \frac{r}{x}$$

$$\tan u = \frac{y}{x} \qquad \cot u = \frac{x}{y}.$$

It is easy to see that this definition of the trigonometric functions agrees with the right-triangle definition for acute angles. By measuring the angle clockwise if $u < 0$ and counterclockwise if $u \geq 0$, we can measure an angle of *u* radians for any real number *u*.

Since $r = \sqrt{x^2 + y^2}$, *r* is always positive. Consequently, the cosine and sine functions are defined for all real numbers *u*. In other words, the domain of these two functions is the set *R* of real numbers.

Domain of sine and cosine

It is also clear from the definition that the tangent and secant functions are not defined on the set $A = \{u \mid u = (2k + 1)\pi/2, k \in Z\}$, since for any $u \in A$, the related angle has its terminal side on the *y*-axis, and hence $x = 0$. Thus, the domain of the tangent and secant functions is the set $R - A$.

Domain of tangent and secant

For a similar reason, the cotangent and cosecant functions are not defined on the set $B = \{u \mid u = k\pi, k \in Z\}$. Consequently, the domain of the cotangent and cosecant functions is the set $R - B$.

Domain of cotangent and cosecant

Since $|x| \leq \sqrt{x^2 + y^2} = r$ and $|y| \leq \sqrt{x^2 + y^2} = r$, we have

Range of trigonometric functions

$$\left|\frac{x}{r}\right| \leq 1 \qquad \text{and} \qquad \left|\frac{y}{r}\right| \leq 1.$$

Thus

$$-1 \leq \frac{x}{r} \leq 1, \qquad -1 \leq \frac{y}{r} \leq 1.$$

Therefore the range of the sine and cosine functions is the set $I = \{t \mid -1 \leq t \leq 1\}$.

The energetic student can readily see that the range of the tangent and cotangent functions is R, and the range of the secant and cosecant functions is $R - C$, where C is the open interval $(-1, 1)$.

We observe from Definition 3 that we need only define the sine and cosine functions explicitly. The other four trigonometric functions can be defined in terms of these functions as follows:

$$\tan u = \frac{y}{x} = \frac{\dfrac{y}{r}}{\dfrac{x}{r}} = \frac{\sin u}{\cos u}, \qquad \cos u \neq 0;$$

$$\cot u = \frac{x}{y} = \frac{\dfrac{x}{r}}{\dfrac{y}{r}} = \frac{\cos u}{\sin u}, \qquad \sin u \neq 0;$$

$$\sec u = \frac{r}{x} = \frac{\dfrac{r}{r}}{\dfrac{x}{r}} = \frac{1}{\cos u}, \qquad \cos u \neq 0;$$

and

$$\csc u = \frac{r}{y} = \frac{\dfrac{r}{r}}{\dfrac{y}{r}} = \frac{1}{\sin u}, \qquad \sin u \neq 0.$$

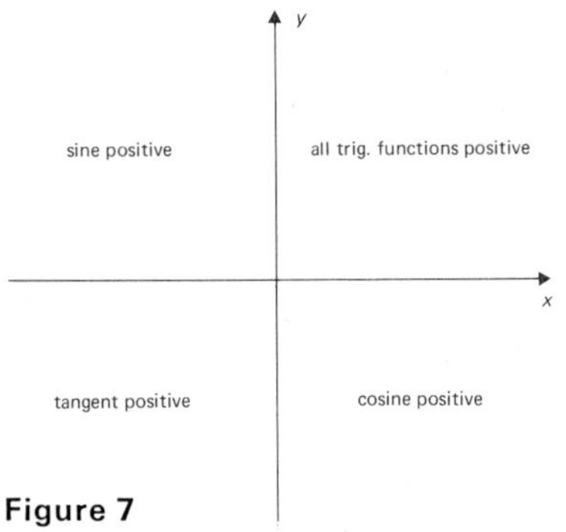

sine positive | all trig. functions positive

tangent positive | cosine positive

Figure 7

Note. $\cot u = \dfrac{\cos u}{\sin u} = \dfrac{1}{\tan u}.$

A. SIGNS OF THE TRIGONOMETRIC FUNCTIONS

Since r is always positive, it is clear from the definitions of the trigonometric functions that their signs depend upon x and y. We know that the signs of x and y depend upon the quadrant in which the terminal side of the angle belongs. For example, if the terminal side of the angle is in the second quadrant, then x is negative and y is positive. Therefore the sine and cosecant are positive, while the remaining functions are all negative. For convenience, we write the signs of the trigonometric functions in tabular form. In Table 1 we write the signs of the sine, cosine, and tangent functions only. Since the functions cosecant, secant, and cotangent are reciprocals of the functions sine, cosine, and tangent, respectively, their signs can be easily determined from the table or Figure 7.

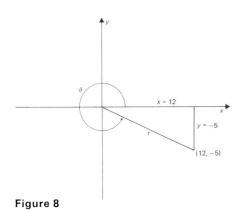

Figure 8

Table 1

Quadrant	I	II	III	IV
sine	+	+	−	−
cosine	+	−	−	+
tangent	+	−	+	−

Note that in the above table we have not given the sign of the trigonometric functions for the angles whose terminal sides lie along one of the axes, i.e., for the angles $u = 0$, $u = \pi/2$, $u = \pi$, $u = 3\pi/2$, $u = 5\pi/2$, etc.

Example 1. Let the positive angle θ be in standard position. Evaluate the six trigonometric functions of θ if its terminal side passes through the point $(12, -5)$.

Solution. From Figure 8, it is easy to see that

$$r = \sqrt{(12)^2 + (-5)^2} = \sqrt{169} = 13.$$

Thus

$$\sin \theta = \frac{y}{r} = \frac{-5}{13} \qquad \csc \theta = -\frac{13}{5}$$

$$\cos \theta = \frac{x}{r} = \frac{12}{13} \qquad \sec \theta = \frac{13}{12}$$

$$\tan \theta = \frac{y}{x} = \frac{-5}{12} \qquad \cot \theta = -\frac{12}{5}.$$

Example 2. Find the value of the six trigonometric functions of $\pi/2$.

Solution. For the angle $\theta = \pi/2$, we have

$$x = 0, \qquad y = r.$$

Therefore $\tan \pi/2$ and $\sec \pi/2$ are undefined, and

$$\sin \frac{\pi}{2} = \frac{y}{r} = 1 \qquad\qquad \cos \frac{\pi}{2} = \frac{x}{r} = \frac{0}{r} = 0$$

$$\csc \frac{\pi}{2} = \frac{r}{y} = \frac{r}{r} = 1 \qquad \cot \frac{\pi}{2} = \frac{x}{y} = \frac{0}{r} = 0.$$

Example 3. Suppose the terminal side of θ in standard position lies in the third quadrant and $\cos \theta = -\frac{3}{5}$. Find the values of the other trigonometric functions of θ.

Solution. In Figure 9, we note that

$$r^2 = x^2 + y^2.$$

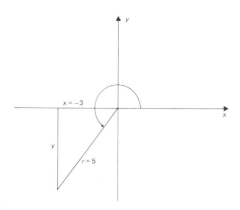

Figure 9

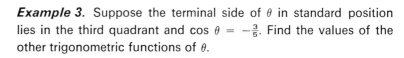

Thus $y^2 = r^2 - x^2 = (5)^2 - (-3)^2 = 25 - 9 = 16$, and $y = \pm 4$. Since y is negative in the third quadrant, we have

$$y = -4.$$

Hence the values of the remaining trigonometric functions are

$$\sin \theta = \frac{-4}{5} \qquad \tan \theta = \frac{4}{3} \qquad \cot \theta = \frac{3}{4}$$

$$\sec \theta = -\frac{5}{5} \qquad \csc \theta = -\frac{5}{4}.$$

Exercise 2

Throughout these problems, assume that the given angle is in standard position.

1. Identify the quadrant in which the terminal side of θ lies if
   (a) $\sin \theta < 0$, $\cos \theta > 0$
   (b) $\cot \theta < 0$, $\csc \theta > 0$
   (c) $\tan \theta > 0$, $\sec \theta > 0$
   (d) $\sec \theta < 0$, $\tan \theta > 0$.

2. In which quadrants may the terminal side of θ lie if
   (a) $\cot \theta > 0$
   (b) $\csc \theta > 0$
   (c) $\cos \theta > 0$?

3. What are the values of $\sin \theta$, $\cos \theta$, and $\tan \theta$ if the terminal side of θ passes through the point P?
   (a) $P(-7, 24)$
   (b) $P(-7, -12)$
   (c) $P(-3, 4)$

4. Evaluate $\sin \theta$, $\cot \theta$, and $\sec \theta$ if
   (a) $\cos \theta = \frac{10}{13}$
   (b) $\csc \theta = -\frac{4}{3}$.

5. Evaluate the six trigonometric functions of θ for
   (a) $\theta = 0$
   (b) $\theta = \pi$
   (c) $\theta = \dfrac{3\pi}{2}$.

6. If the terminal side of θ belongs to the fourth quadrant and $\sec \theta = \sqrt{2}$, show that

$$\frac{1 + \tan \theta + \csc \theta}{1 + \cot \theta - \csc \theta} = -1.$$

3. Some Special Angles

In Section 2, we showed that the sine and cosine functions are defined for all real numbers x, and each function has $I = \{y|\ -1 \leq y \leq 1\}$ as its range. It will be shown (in Section 8) that these two functions are differentiable everywhere (and hence continuous). Consequently, to sketch the graphs of these functions we need to know their values at several points. We first find the values of these functions at the so-called *standard angles*, namely, the angles 0, $\pi/6$, $\pi/4$, $\pi/3$, and $\pi/2$. Then we use the results to compute their values at several other angles.

From Section 2 we know that

$$\sin 0 = 0 \qquad \sin \frac{\pi}{2} = 1$$

$$\cos 0 = 1 \qquad \cos \frac{\pi}{2} = 0.$$

Let us now calculate the values of the trigonometric functions at $\pi/4$.

Since $\pi/4 = 45°$, the right triangle in Figure 10 is an isosceles triangle. Thus if we choose $x = 1$, then $y = 1$ and $r = \sqrt{1^2 + 1^2} = \sqrt{2}$. Consequently,

$$\sin \frac{\pi}{4} = \frac{1}{\sqrt{2}} \qquad \cos \frac{\pi}{4} = \frac{1}{\sqrt{2}}.$$

Knowing the sine and cosine of $\pi/4$, we can also calculate the values of the other trigonometric functions at $\pi/4$:

$$\tan \frac{\pi}{4} = \frac{\sin \dfrac{\pi}{4}}{\cos \dfrac{\pi}{4}} = 1 \qquad \cot \frac{\pi}{4} = \frac{1}{\tan \dfrac{\pi}{4}} = 1$$

$$\csc \frac{\pi}{4} = \frac{1}{\sin \dfrac{\pi}{4}} = \sqrt{2} \qquad \sec \frac{\pi}{4} = \frac{1}{\cos \dfrac{\pi}{4}} = \sqrt{2}.$$

We now wish to calculate the values of the trigonometric functions at $\pi/3$. Before we do so, let us consider the equilateral triangle in Figure 11. Each side of this triangle is 2 units long and each angle is $60° = \pi/3$ radians. If we divide this triangle into two right triangles (by dropping a perpendicular from the vertex to the base) as shown in Figure 11, then by the Pythagorean theorem we find the altitude of this triangle to be $\sqrt{3}$.

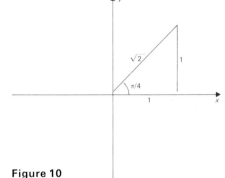

Figure 10

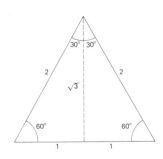

Figure 11

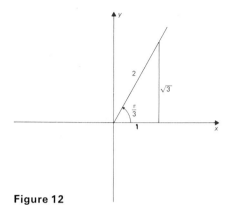

Figure 12

Thus, placing the angle $\pi/3$ in standard position as in Figure 12, and choosing $r = 2$, we have $x = 1$, $y = \sqrt{3}$. Therefore

$$\sin \frac{\pi}{3} = \frac{\sqrt{3}}{2} \qquad \csc \frac{\pi}{3} = \frac{2}{\sqrt{3}}$$

$$\cos \frac{\pi}{3} = \frac{1}{2} \qquad \sec \frac{\pi}{3} = 2$$

$$\tan \frac{\pi}{3} = \sqrt{3} \qquad \cot \frac{\pi}{3} = \frac{1}{\sqrt{3}}.$$

From the above argument, we can also deduce the values of the trigonometric functions at $\pi/6 = 30°$ (see Figure 13):

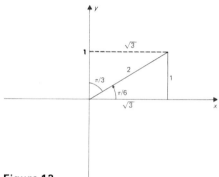

Figure 13

$$\sin \frac{\pi}{6} = \frac{1}{2} \qquad \csc \frac{\pi}{6} = 2$$

$$\cos \frac{\pi}{6} = \frac{\sqrt{3}}{2} \qquad \sec \frac{\pi}{6} = \frac{2}{\sqrt{3}}$$

$$\tan \frac{\pi}{6} = \frac{1}{\sqrt{3}} \qquad \cot \frac{\pi}{6} = \sqrt{3}.$$

We observe that the value of r is not a factor in the final values of the trigonometric functions. Therefore, we can use any value of r that we wish. Thus, from now on we shall use $r = 1$.

We shall presently see that it is sufficient to know the values of the trigonometric functions only for the acute angles (angles in the first quadrant). From these we shall be able to compute the values for angles in other quadrants. We first prove

The technique followed in Table 2 should be useful in memorizing the values of sine and cosine of

$$0, \frac{\pi}{6}, \frac{\pi}{4}, \frac{\pi}{3}, \frac{\pi}{2}.$$

angle	0	$\pi/6$	$\pi/4$	$\pi/3$	$\pi/2$
	0°	30°	45°	60°	90°
sine	$\sqrt{0}/4$	$\sqrt{1}/4$	$\sqrt{2}/4$	$\sqrt{3}/4$	$\sqrt{4}/4$
cosine	$\sqrt{4}/4$	$\sqrt{3}/4$	$\sqrt{2}/4$	$\sqrt{1}/4$	$\sqrt{0}/4$

Theorem 1. For θ, a real number,

(1) $$\sin(-\theta) = -\sin\theta$$

(2) $$\cos(-\theta) = \cos\theta.$$

Proof. Place the angles θ and $-\theta$ in standard position (see Figures 14 and 15). Then with center at O and radius 1 draw a circle. Thus $d(O, P) = 1$, and $d(O, Q) = 1$. Draw a straight line joining P and Q, and let it meet the x-axis at M. Then the triangles POM and QOM are congruent. Thus if the coordinates of P are (x, y), then the coordinates of Q are $(x, -y)$. Consequently,

$$\sin(-\theta) = \frac{-y}{1} = -\frac{y}{1} = -\sin\theta \qquad (\text{note } r = 1)$$

and

$$\cos(-\theta) = \frac{x}{1} = \cos\theta.$$

In Figures 14 and 15, we have taken θ such that its terminal side lies in the first and second quadrants respectively. The student may sketch diagrams to verify the result when the terminal side of θ lies in the third or fourth quadrant.

Theorem 2. For θ, a real number,

(3) $$\sin\left(\frac{\pi}{2} - \theta\right) = \cos\theta$$

(4) $$\cos\left(\frac{\pi}{2} - \theta\right) = \sin\theta.$$

Proof. We shall prove this theorem by considering the case where the terminal side of θ (in standard position) lies in the first quadrant (see Figure 16). The student may verify the validity of this theorem for θ in the remaining quadrants.

With center at O and radius 1, draw a circle. Let the terminal sides of θ and $(\pi/2 - \theta)$ intersect this circle at the points P and Q respectively. Draw PM and QN perpendicular to the x-axis. It is easy to see that the triangles POM and QON are congruent. Therefore

$$d(O, M) = d(Q, N)$$
$$d(P, M) = d(O, N).$$

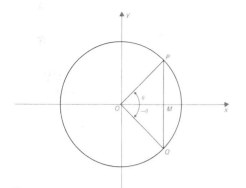

Figure 14

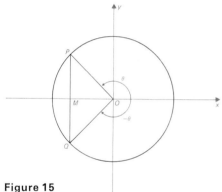

Figure 15

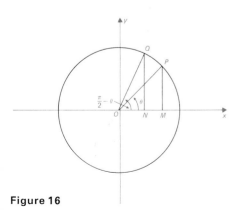

Figure 16

Thus, if the coordinates of P are (x, y) and the coordinates of Q are (x_1, y_1), then

$$x = y_1$$
$$y = x_1.$$

Consequently,

$$\sin \left(\frac{\pi}{2} - \theta \right) = \frac{y_1}{1} = \frac{x}{1} = \cos \theta$$

and

$$\cos \left(\frac{\pi}{2} - \theta \right) = \frac{x_1}{1} = \frac{y}{1} = \sin \theta.$$

We shall now use Theorems 1 and 2 to prove

Theorem 3. For θ, a real number,

(5) $\sin \left(\dfrac{\pi}{2} + \theta \right) = \cos \theta,$ (6) $\cos \left(\dfrac{\pi}{2} + \theta \right) = -\sin \theta,$

(7) $\sin (\pi + \theta) = -\sin \theta,$ (8) $\cos (\pi + \theta) = -\cos \theta,$

(9) $\sin (\pi - \theta) = \sin \theta,$ (10) $\cos (\pi - \theta) = -\cos \theta,$

(11) $\sin \left(\dfrac{3\pi}{2} + \theta \right) = -\cos \theta,$ (12) $\cos \left(\dfrac{3\pi}{2} + \theta \right) = \sin \theta,$

(13) $\sin \left(\dfrac{3\pi}{2} - \theta \right) = -\cos \theta,$ (14) $\cos \left(\dfrac{3\pi}{2} - \theta \right) = -\sin \theta,$

(15) $\sin (2\pi - \theta) = -\sin \theta,$ (16) $\cos (2\pi - \theta) = \cos \theta,$

(17) $\sin (2\pi + \theta) = \sin \theta,$ (18) $\cos (2\pi + \theta) = \cos \theta.$

Proof. We shall prove only (5), (10), and (13). The proofs of the remaining results are left for the student.

To prove (5) we write $\theta = -\alpha$:

$$\sin \left(\frac{\pi}{2} + \theta \right) = \sin \left(\frac{\pi}{2} - \alpha \right)$$

Theorem 2, (3)
$$= \cos \alpha$$

$$= \cos (-\theta)$$

Theorem 1, (2)
$$= \cos \theta.$$

(10): $\cos (\pi - \theta) = \cos \left(\dfrac{\pi}{2} + \dfrac{\pi}{2} - \theta \right)$

We write $\alpha = \dfrac{\pi}{2} - \theta.$

Theorem 3, (6)

We write $\alpha = \dfrac{\pi}{2} - \theta.$

Theorem 3, (7)

Theorem 2, (3)

$$= \cos\left(\frac{\pi}{2} + \alpha\right)$$

$$= -\sin\alpha$$

$$= -\sin\left(\frac{\pi}{2} - \theta\right)$$

$$= -\cos\theta.$$

(13): $\quad \sin\left(\dfrac{3\pi}{2} - \theta\right) = \sin\left(\pi + \dfrac{\pi}{2} - \theta\right)$

$$= \sin(\pi + \alpha),$$

$$= -\sin\alpha$$

$$= -\sin\left(\frac{\pi}{2} - \theta\right)$$

$$= -\cos\theta.$$

Since the other four trigonometric functions (namely, tangent, cotangent, secant, and cosecant) can be expressed in terms of sine and cosine functions, the results corresponding to those of Theorems 1, 2, and 3 for these functions can be deduced easily. For example,

$$\tan\left(\frac{\pi}{2} + \theta\right) = \frac{\sin\left(\frac{\pi}{2} + \theta\right)}{\cos\left(\frac{\pi}{2} + \theta\right)} = \frac{-\cos\theta}{\sin\theta} = -\cot\theta,$$

and

$$\sec\left(\frac{3\pi}{2} - \theta\right) = \frac{1}{\cos\left(\frac{3\pi}{2} - \theta\right)} = \frac{1}{-\sin\theta} = -\csc\theta.$$

A useful scheme for remembering these results is the following. Suppose we pair the six trigonometric functions:

$$\begin{cases}\sin\theta \\ \cos\theta\end{cases} \qquad \begin{cases}\tan\theta \\ \cot\theta\end{cases} \qquad \begin{cases}\sec\theta \\ \csc\theta.\end{cases}$$

These are called the *cofunctional pairs*. The second of each pair is called the cofunction of the first, and vice versa. With this nomenclature in mind, and denoting by $f(\theta)$ any one of the six trigonometric functions, we can state:

(19) $\qquad f\left(\theta \pm n\dfrac{\pi}{2}\right) = \begin{cases}\pm f(\theta), & n \text{ even} \\ \pm cof(\theta), & n \text{ odd.}\end{cases}$

The sign $\pm$ in the second member must be chosen to correspond to the sign of $f(\theta \pm n\pi/2)$. For example, if θ is an acute angle, then

$$\sec\left(\theta + \frac{3\pi}{2}\right) = \pm \csc \theta \qquad \text{(since } n = 3 \text{ is odd).}$$

Since $\theta + 3\pi/2$ has its terminal side in the fourth quadrant, the cosine and the secant are positive; thus

$$\sec\left(\theta + \frac{3\pi}{2}\right) = \csc \theta.$$

Example 1. Show that

$$\frac{4}{3}\cot^2\frac{\pi}{6} + 3\sin^2\frac{\pi}{3} - 2\csc^2\frac{\pi}{3} - \frac{3}{4}\tan^2\frac{\pi}{6} = \frac{10}{3}.$$

L.H.S. is an abbreviation for Left Hand Side; similarly R.H.S. denotes Right Hand Side.

Solution. By direct substitution, the L.H.S. of the above equation

$$= \frac{4}{3}(\sqrt{3})^2 + 3\left(\frac{\sqrt{3}}{2}\right)^2 - 2\left(\frac{2}{\sqrt{3}}\right)^2 - \frac{3}{4}\left(\frac{1}{\sqrt{3}}\right)^2$$

$$= \frac{4}{3}\cdot 3 + 3\cdot\frac{3}{4} - 2\cdot\frac{4}{3} - \frac{3}{4}\cdot\frac{1}{3}$$

$$= 4 + \frac{9}{4} - \frac{8}{3} - \frac{1}{4}$$

$$= \frac{10}{3}.$$

Example 2. Evaluate

(a) $\tan\left(-\frac{11\pi}{3}\right)$

(b) $\sec\left(\frac{11\pi}{4}\right)$

(c) $\sin\left(-\frac{15\pi}{4}\right).$

Solution. (a) $\tan\left(-\frac{11\pi}{3}\right) = -\tan\frac{11\pi}{3}$

$$= -\tan\left(3\pi + \frac{2\pi}{3}\right)$$

$$= -\tan\frac{2\pi}{3}$$

$$= -\tan\left(\frac{\pi}{2} + \frac{\pi}{6}\right)$$

$$= \cot\frac{\pi}{6}$$

$$= \frac{\cos\dfrac{\pi}{6}}{\sin\dfrac{\pi}{6}} = \sqrt{3}.$$

(b) $\sec\dfrac{11\pi}{4} = \dfrac{1}{\cos\dfrac{11\pi}{4}}$

$$= \frac{1}{\cos\left(2\pi + \dfrac{3\pi}{4}\right)}$$

$$= \frac{1}{\cos\dfrac{3\pi}{4}}$$

$$= \frac{1}{\cos\left(\pi - \dfrac{\pi}{4}\right)}$$

$$= \frac{1}{-\cos\dfrac{\pi}{4}}$$

$$= -\sqrt{2}.$$

(c) $\sin\left(-\dfrac{15\pi}{4}\right) = -\sin\dfrac{15\pi}{4}$

$$= -\sin\left(4\pi - \dfrac{\pi}{4}\right)$$

$$= -\sin\left(-\dfrac{\pi}{4}\right)$$

$$= \sin\dfrac{\pi}{4}$$

$$= \frac{1}{\sqrt{2}}.$$

Exercise 3   **1.** Find the sine and cosine of the following angles.

(a) 135°   (b) 330°

(c) 210°   (d) 405°
(e) 690°   (f) 1125°

2. Find the sines and cosines of the following real numbers.

(a) $\dfrac{3\pi}{4}$ (b) $\dfrac{16\pi}{3}$

(c) $-\dfrac{47\pi}{6}$ (d) $\dfrac{73\pi}{6}$

3. Express in terms of trigonometric functions of positive real numbers less than $\pi/2$.

(a) $\sin \dfrac{7\pi}{13}$ (b) $\csc \dfrac{19\pi}{7}$

(c) $\sec \dfrac{101\pi}{3}$ (d) $\cot\left(-\dfrac{13\pi}{4}\right)$

(e) $\tan\left(-\dfrac{5\pi}{3}\right)$ (f) $\cos\left(-\dfrac{2\pi}{3}\right)$

In problems 4 through 8, show that the given equations are true.

4. $\sin \dfrac{\pi}{3} \cos \dfrac{\pi}{6} + \cos \dfrac{\pi}{3} \sin \dfrac{\pi}{6} = 1$

5. $\cos^2 \dfrac{\pi}{4} - \sin^2 \dfrac{\pi}{4} = \cos \dfrac{\pi}{2}$

6. $\tan^2 \dfrac{\pi}{3} + 4 \cos^2 \dfrac{\pi}{4} + 3 \cos^2 \dfrac{\pi}{3} = \dfrac{23}{4}$

7. $\frac{1}{2} \sin^2 30° + \frac{1}{4} \tan^2 45° + \frac{5}{4} \sec^3 60° = \frac{41}{4}$

8. $\tan^2 \dfrac{4\pi}{3} + \sin^2 \dfrac{2\pi}{3} + \cos^2 \dfrac{4\pi}{3} = \sec^2 \dfrac{\pi}{3}$

4. Graphs of Trigonometric Functions

The trigonometric functions belong to a large class of functions called *periodic* functions, in which there is a regular repetition of the values of the function over a certain interval. For example, we observe that as θ increases from 0 to 2π, the point $P(x, y)$, such that $x^2 + y^2 = 1$, completes one revolution around the unit circle. Then as θ continues to increase, the point $P(x, y)$ begins its second trip around the unit circle and the trigonometric functions begin to repeat their earlier behavior.

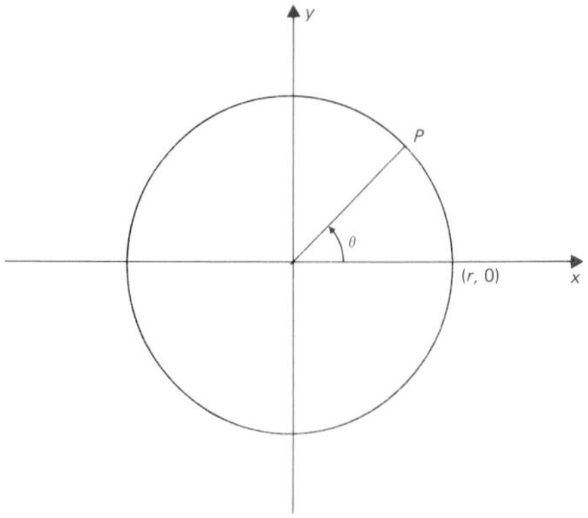

Figure 17

Definition 4. A function f is called *periodic* if there exists a nonzero fixed number p such that f(x + p) is defined and

(20) $$f(x + p) = f(x)$$

for all x in the domain of f. If p is the smallest positive number for which (20) holds, then p is called the *period* of the function f.

It is easy to see that if P (see Figure 17) represents the real number $\theta(0 \leq \theta < 2\pi)$ on the circumference of a circle of radius r, then P also represents all real numbers which are elements of the set

$$S = \{\alpha|\ \alpha = 2n\pi + \theta;\ \ n \in Z\}.$$

From the definition of the sine and cosine functions of a real number θ, we find that these functions depend on the coordinates (x, y) of the point P corresponding to the number θ. Since all real numbers of the set S are represented by the same point P, it follows that the values of the sine and cosine functions are equal for every element of S.

Consequently, for a real number θ and $n \in Z$,

$$\sin (\theta + 2n\pi) = \sin \theta$$

and

$$\cos (\theta + 2n\pi) = \cos \theta.$$

If $p = 2\pi n$, then

$$\sin (\theta + p) = \sin \theta$$
$$\cos (\theta + p) = \cos \theta.$$

It can be shown that the smallest value of $|p|$ $(p \neq 0)$ with the above property is 2π. Hence the sine and the cosine functions are periodic with period 2π.

Since $\sec \theta = 1/\cos \theta$ and $\csc \theta = 1/\sin \theta$, the secant and cosecant functions are also periodic functions with period 2π.

Since

$$\tan (\theta + \pi) = \frac{\sin (\theta + \pi)}{\cos (\theta + \pi)} = \frac{-\sin \theta}{-\cos \theta} = \tan \theta,$$

the tangent function and the cotangent function ($\cot \theta = 1/\tan \theta$) are periodic functions with period π.

Since the sine function is periodic with period 2π, the graph of this function repeats the basic shape obtained on the interval $[0, 2\pi]$. The following table summarizes the variation of $\sin \theta$ as θ varies from 0 to 2π.

Graph of y = sin θ

Table 3

θ varies	sin θ varies
0 to $\dfrac{\pi}{2}$	0 to 1
$\dfrac{\pi}{2}$ to π	1 to 0
π to $\dfrac{3\pi}{2}$	0 to −1
$\dfrac{3\pi}{2}$ to 2π	−1 to 0

Having noted a broad pattern of the variation of the values of $y = \sin\theta$, a good approximation of the graph is obtained by plotting a number of points $(\theta, \sin\theta)$ as θ takes values from 0 to 2π.

Table 4 gives the corresponding values of θ and $\sin\theta$, and Figure 18 shows the rough sketch of the graph of $y = \sin\theta$.

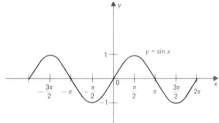

Figure 18

Table 4

θ	sin θ	θ	sin θ
0	0	π	0
$\dfrac{\pi}{6}$	$\dfrac{1}{2}$	$\dfrac{7\pi}{6}$	$-\dfrac{1}{2}$
$\dfrac{\pi}{4}$	$\dfrac{1}{\sqrt{2}}$	$\dfrac{5\pi}{4}$	$-\dfrac{1}{\sqrt{2}}$
$\dfrac{\pi}{3}$	$\dfrac{\sqrt{3}}{2}$	$\dfrac{4\pi}{3}$	$-\dfrac{\sqrt{3}}{2}$
$\dfrac{\pi}{2}$	1	$\dfrac{3\pi}{2}$	−1
$\dfrac{2\pi}{3}$	$\dfrac{\sqrt{3}}{2}$	$\dfrac{5\pi}{3}$	$-\dfrac{\sqrt{3}}{2}$
$\dfrac{3\pi}{4}$	$\dfrac{1}{\sqrt{2}}$	$\dfrac{7\pi}{4}$	$-\dfrac{1}{\sqrt{2}}$
$\dfrac{5\pi}{6}$	$\dfrac{1}{2}$	$\dfrac{11\pi}{6}$	$-\dfrac{1}{2}$
		2π	0

Graph of y = tan θ

The tangent function is periodic with period π. Thus the graph of $y = \tan\theta$ repeats the basic shape obtained on the interval $[0, \pi]$. The following table gives the variation of $y = \tan\theta$ as θ varies from 0 to π.

Table 5

θ varies	tan θ varies
0 to $\dfrac{\pi}{2}$	0 to $+\infty$
$\dfrac{\pi}{2}$ to π	$-\infty$ to 0

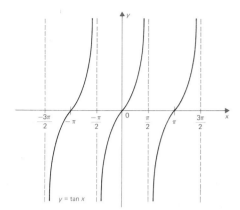

Figure 19

$\left(\dfrac{\pi}{2}\right)-$ *means θ approaches $\dfrac{\pi}{2}$ through values*

less than $\dfrac{\pi}{2}$, and $\left(\dfrac{\pi}{2}\right)+$ means θ approaches

$\dfrac{\pi}{2}$ *through values greater than $\dfrac{\pi}{2}$.*

Graph of $y = \sec \theta$

We note that tan $\pi/2$ is undefined, therefore the point $(\pi/2,$ tan $\pi/2)$ cannot be on the graph. In fact the graph has a *discontinuity* or break at $\theta = \pi/2$.

The following table gives the corresponding values of θ and tan θ as θ varies from 0 to π and Figure 19 shows the rough sketch of the graph of $y = $ tan θ.

Table 6

θ	tan θ	θ	tan θ
0	0	$\left(\dfrac{\pi}{2}\right)+$	$-\infty$
$\dfrac{\pi}{6}$	$\dfrac{1}{\sqrt{3}} = 0.58$	$\dfrac{2\pi}{3}$	$-\sqrt{3}$
$\dfrac{\pi}{4}$	1	$\dfrac{3\pi}{4}$	-1
$\dfrac{\pi}{3}$	$\sqrt{3} = 1.73$	$\dfrac{5\pi}{6}$	$-\dfrac{1}{\sqrt{3}}$
$\left(\dfrac{\pi}{2}\right)-$	$+\infty$	π	0

As indicated earlier, the secant function is periodic with period 2π, and it suffices to sketch the graph of $y = \sec \theta = 1/\cos \theta$ over $[0, 2\pi]$. The following table gives the values of θ and $\sec \theta$ as θ varies from 0 to 2π and Figure 20 shows the sketch of the graph.

Table 7

θ	sec θ	θ	sec θ
0	1	π	-1
$\dfrac{\pi}{6}$	$\dfrac{2}{\sqrt{3}}$	$\dfrac{7\pi}{6}$	$-\dfrac{2}{\sqrt{3}}$
$\dfrac{\pi}{4}$	$\sqrt{2}$	$\dfrac{5\pi}{4}$	$-\sqrt{2}$
$\dfrac{\pi}{3}$	2	$\dfrac{4\pi}{3}$	-2
$\left(\dfrac{\pi}{2}\right)-$	$+\infty$	$\left(\dfrac{3\pi}{2}\right)-$	$-\infty$
$\left(\dfrac{\pi}{2}\right)+$	$-\infty$	$\left(\dfrac{3\pi}{2}\right)+$	$+\infty$
$\dfrac{2\pi}{3}$	-2	$\dfrac{5\pi}{3}$	2
$\dfrac{3\pi}{4}$	$-\sqrt{2}$	$\dfrac{7\pi}{4}$	$\sqrt{2}$
$\dfrac{5\pi}{6}$	$-\dfrac{2}{\sqrt{3}}$	$\dfrac{11\pi}{6}$	$\dfrac{2}{\sqrt{3}}$
		2π	1

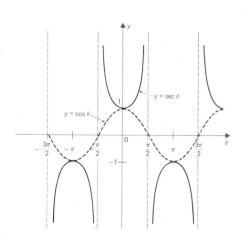

Figure 20

Example 1. Show that sin $k\theta$ is a periodic function of θ, and obtain the period of sin $k\theta$, where $k > 0$.

Solution. We know that

$$\sin (x + 2\pi) = \sin x.$$

Therefore

$$\sin (k\theta + 2\pi) = \sin k\theta$$

or

$$\sin k\left(\theta + \frac{2\pi}{k}\right) = \sin k\theta.$$

Thus, if the value of θ is increased by $2\pi/k$, the value of sin $k\theta$ remains unaltered. Consequently, sin $k\theta$ is periodic with period $2\pi/k$.

Example 2. Construct a cosine function whose period is 3.

Solution. Let the function cos kx, where k is to be determined, be such that

$$\cos kx = \cos k(x + 3).$$

We know that

$$\cos kx = \cos (kx + 2\pi).$$

Thus,

$$3k = 2\pi$$

or

$$k = \frac{2\pi}{3}.$$

Therefore, the required function is cos $(2\pi/3)x$.

Exercise 4

1. Sketch the graph of the function $f(x) = \cos x$.
2. Sketch the graph of the function $f(x) = \cot x$.
3. Sketch the graph of the function $f(x) = \csc x$.
4. Find the periods of
   (a) sin $3x$
   (b) cos $2x$
   (c) tan $5x$.
5. Construct a sine function of period 5.
6. Construct a tangent function of period $\dfrac{\pi}{4}$.
7. Find the period of tan $(2x + 3)$.

8. Find the period of $\sin(-2x)$.

*9. Show that $\sin x^2$ is not a periodic function of x.

*10. Show that $\cos\left(\dfrac{\pi x}{6}\right) + \csc\left(\dfrac{\pi x}{12}\right)$ is a periodic function of x with period 24.

5. The Pythagorean Identity

In the previous two sections, we came across some of the basic trigonometric identities. For example,

$$\tan \theta = \frac{\sin \theta}{\cos \theta}$$

$$\sec \theta = \frac{1}{\cos \theta}.$$

In this section, we shall prove and discuss some of the consequences of one of the most basic identities: the Pythagorean Identity. We state and prove this in

Theorem 4. **For any real number θ,**

(21) $$\sin^2 \theta + \cos^2 \theta = 1.$$

Proof. Let θ be any angle in standard position, and let $P(x, y)$ be a point on the terminal side of θ such that $d(O, P) = r$. Then by the Pythagorean theorem

(22) $$x^2 + y^2 = r^2.$$

Dividing both sides of (22) by $r^2 (r^2 \neq 0)$, we obtain

(23) $$\left(\frac{x}{r}\right)^2 + \left(\frac{y}{r}\right)^2 = 1.$$

But from the definition of the trigonometric functions,

$$\cos \theta = \frac{x}{r}, \quad \sin \theta = \frac{y}{r}.$$

Therefore from (23), we get

$$\cos^2 \theta + \sin^2 \theta = 1.$$

This proves Theorem 4.

Example 1. For any real number θ, prove

Why do we require sin $\theta \neq 0$?

(24) $\qquad\qquad 1 + \cot^2 \theta = \csc^2 \theta, \qquad \sin \theta \neq 0;$

(25) $\qquad\qquad 1 + \tan^2 \theta = \sec^2 \theta, \qquad \cos \theta \neq 0.$

Solution. Dividing both sides of (21) by $\sin^2 \theta$, we get

$$\left(\frac{\sin \theta}{\sin \theta}\right)^2 + \left(\frac{\cos \theta}{\sin \theta}\right)^2 = \left(\frac{1}{\sin \theta}\right)^2$$

or

$$1 + \cot^2 \theta = \csc^2 \theta.$$

Similarly, dividing both sides of (21) by $\cos^2 \theta$, we get

$$\left(\frac{\sin \theta}{\cos \theta}\right)^2 + \left(\frac{\cos \theta}{\cos \theta}\right)^2 = \left(\frac{1}{\cos \theta}\right)^2$$

or

$$\tan^2 \theta + 1 = \sec^2 \theta.$$

Example 2. If $\cot \theta = \frac{12}{5}$, $\pi < \theta < 3\pi/2$, find the values of $\sec \theta$ and $\sin \theta$.

Solution. By (24),

$$\begin{aligned}
\csc^2 \theta &= 1 + \cot^2 \theta \\
&= 1 + \left(\tfrac{12}{5}\right)^2 \\
&= 1 + \tfrac{144}{25} \\
&= \tfrac{169}{25}.
\end{aligned}$$

Thus

$$\csc \theta = \pm \tfrac{13}{5}.$$

Since $\pi < \theta < 3\pi/2$, $\csc \theta$ is negative. Therefore

$$\csc \theta = -\tfrac{13}{5},$$

and hence

$$\sin \theta = \frac{1}{\csc \theta} = -\frac{5}{13}.$$

We find $\sec \theta$ through a series of identities:

$$\begin{aligned}
\sec \theta &= \frac{1}{\cos \theta} \\
&= \frac{\sin \theta}{\cos \theta} \cdot \frac{1}{\sin \theta} \\
&= \tan \theta \cdot \csc \theta \\
&= \left(\tfrac{5}{12}\right)\left(-\tfrac{13}{5}\right) \\
&= -\tfrac{13}{12}.
\end{aligned}$$

Example 3. Prove that

$$\frac{\sin \theta}{1 + \cos \theta} + \frac{1 + \cos \theta}{\sin \theta} = 2 \csc \theta, \qquad \sin \theta \neq 0.$$

Solution.

$$\frac{\sin \theta}{1 + \cos \theta} + \frac{1 + \cos \theta}{\sin \theta} = \frac{\sin^2 \theta + (1 + \cos \theta)^2}{\sin \theta \, (1 + \cos \theta)}$$

$$= \frac{\sin^2 \theta + 1 + 2 \cos \theta + \cos^2 \theta}{\sin \theta (1 + \cos \theta)}$$

$sin^2 \theta + cos^2 \theta = 1$

$$= \frac{2 + 2 \cos \theta}{\sin \theta (1 + \cos \theta)}$$

$sin \theta \neq 0 \Rightarrow cos \theta \neq -1$

$$= \frac{2(1 + \cos \theta)}{\sin \theta (1 + \cos \theta)}$$

$$= \frac{2}{\sin \theta}$$

$$= 2 \csc \theta.$$

Example 4. Prove that

$$\frac{\tan \theta + \sec \theta - 1}{\tan \theta - \sec \theta + 1} = \frac{1 + \sin \theta}{\cos \theta}, \qquad \theta \neq n\pi + \frac{\pi}{2}.$$

Solution.

$$\frac{\tan \theta + \sec \theta - 1}{\tan \theta - \sec \theta + 1}$$

By (25), $1 = sec^2 \theta - tan^2 \theta.$

$$= \frac{\tan \theta + \sec \theta - (\sec^2 \theta - \tan^2 \theta)}{\tan \theta - \sec \theta + 1}$$

$$= \frac{(\tan \theta + \sec \theta) - (\sec \theta - \tan \theta)(\sec \theta + \tan \theta)}{\tan \theta - \sec \theta + 1}$$

$$= \frac{(\tan \theta + \sec \theta)(1 - \sec \theta + \tan \theta)}{\tan \theta - \sec \theta + 1}$$

$$= \tan \theta + \sec \theta$$

$$= \frac{\sin \theta}{\cos \theta} + \frac{1}{\cos \theta}$$

$$= \frac{1 + \sin \theta}{\cos \theta}.$$

*Example 5. For any nonzero real number x, show that

(26) $$\sin \theta = x + \frac{1}{x}$$

is never true.

Solution. Suppose (26) is true for some $x > 0$. Then (26) implies that

$$\sin \theta - 2 = x + \frac{1}{x} - 2$$

$$= \left(\sqrt{x} - \frac{1}{\sqrt{x}} \right)^2$$

$$\geq 0.$$

Thus

$$\sin \theta \geq 2,$$

which is false. Consequently, (26) cannot hold for any positive real number x.

If x is a negative real number, then replacing x by $-y$ (where y is positive), we obtain

$$\sin (-\theta) = y + \frac{1}{y} \qquad (y > 0),$$

and the above argument applies again.

Exercise 5

In problems 1 through 9, prove the given identities for suitable values of θ.

1. $\sin^4 \theta - \cos^4 \theta = \sin^2 \theta - \cos^2 \theta$

2. $\sin^6 \theta + \cos^6 \theta = 1 - 3 \sin^2 \theta \cos^2 \theta$

3. $\cot^4 \theta + \cot^2 \theta = \csc^4 \theta - \csc^2 \theta$

4. $\csc^6 \theta - \cot^2 \theta = 1 + 3 \csc^2 \theta \cot^2 \theta$

5. $\dfrac{1}{1 + \tan \theta} = \dfrac{\cot \theta}{1 + \cot \theta}$

6. $\dfrac{1}{1 + \cos \theta} + \dfrac{1}{1 - \cos \theta} = 2 \csc^2 \theta$

7. $(\sin \theta - \cos \theta)(\tan \theta + \cot \theta) = \sec \theta - \csc \theta$

8. $(\csc \theta - \sin \theta)(\sec \theta - \cos \theta) = \dfrac{1}{\tan \theta + \cot \theta}$

9. $\dfrac{1 - \tan^2 \theta}{1 + \tan^2 \theta} = 2 \cos^2 \theta - 1$

10. Suppose the terminal sides of two angles α and β are in the second and fourth quadrants respectively, and $\sin \alpha = -\sin \beta = \frac{5}{13}$. Find the values of $\cos \alpha$, $\cot \alpha$, $\sec \beta$, and $\csc \beta$.

11. If $\cot \theta + \csc \theta = 5$, show that $\cos \theta = \frac{12}{13}$.

12. Let α, β, and r be real numbers, and let

$$x = r \sin \alpha \cos \beta,$$
$$y = r \sin \alpha \sin \beta,$$
$$z = r \cos \alpha.$$

Show that $x^2 + y^2 + z^2 = r^2$.

13. Is the equation

$$2 \sin x + 3 \cos x = 8,$$

possible?

6. Addition Formulas

Let us consider a function f with domain D. Let x_1 and x_2 be any two elements of D such that $x_1 + x_2 \in D$. It is often of interest and importance to know how $f(x_1 + x_2)$ is related to $f(x_1)$ and $f(x_2)$. A formula which connects these quantities in some way may be called an *addition formula*. We see immediately that a linear function f, defined by $f(x) = mx$ (for some real number m), satisfies the relation

(27) $$f(x_1 + x_2) = f(x_1) + f(x_2).$$

It is easy to see that the cosine function does not satisfy (27). For example, let $x_1 = \pi/6$ and $x_2 = \pi/6$. Then

$$\cos\left(\frac{\pi}{6} + \frac{\pi}{6}\right) = \cos\left(\frac{\pi}{3}\right) = \frac{1}{2},$$

whereas

$$\cos\left(\frac{\pi}{6}\right) + \cos\left(\frac{\pi}{6}\right) = \frac{\sqrt{3}}{2} + \frac{\sqrt{3}}{2} = \sqrt{3},$$

and

$$\sqrt{3} \neq \tfrac{1}{2}.$$

Thus, in general,

$$\cos(x_1 + x_2) \neq \cos x_1 + \cos x_2.$$

We now prove

Theorem 5. For any real numbers α and β,

(28) $$\cos(\alpha - \beta) = \cos \alpha \cos \beta + \sin \alpha \sin \beta$$

(29) $$\cos(\alpha + \beta) = \cos \alpha \cos \beta - \sin \alpha \sin \beta.$$

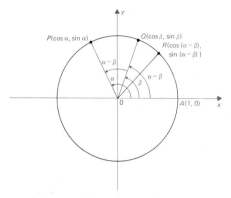

Figure 21

Proof. Let α and β be two angles (measured in radians) placed in standard position (see Figure 21). Place the angle $\alpha - \beta$ in standard position. With center at 0 and radius equal to 1 draw a circle. Let this circle intersect the terminal sides of the angles α, β, $\alpha - \beta$ at P, Q, and R respectively. Let A denote the point of intersection of this circle and the x-axis. By definition, we have

$$\cos \alpha = \frac{x \text{ coordinate of } P}{r} = \frac{x \text{ coordinate of } P}{1}$$

$$= x \text{ coordinate of } P$$

and

$$\sin \alpha = \frac{y \text{ coordinate of } P}{r} = \frac{y \text{ coordinate of } P}{1}$$

$$= y \text{ coordinate of } P.$$

Thus the coordinates of P are $(\cos \alpha, \sin \alpha)$. Similarly, the coordinates of Q are $(\cos \beta, \sin \beta)$ and the coordinates of R are

$$(\cos (\alpha - \beta), \quad \sin (\alpha - \beta)).$$

It is easy to see that the coordinates of A are (1, 0). In Figure 21, it is not difficult to show that $d(A, R) = d(P, Q)$. (Prove it!) Thus

(30) $$[d(A, R)]^2 = [d(P, Q)]^2,$$

or

$$(\cos (\alpha - \beta) - 1)^2 + (\sin (\alpha - \beta) - 0)^2$$
$$= (\cos \alpha - \cos \beta)^2 + (\sin \alpha - \sin \beta)^2$$
$$\cos^2 (\alpha - \beta) - 2 \cos (\alpha - \beta) + 1 + \sin^2 (\alpha - \beta)$$
$$= \cos^2 \alpha - 2 \cos \alpha \cos \beta + \cos^2 \beta$$
$$+ \sin^2 \alpha - 2 \sin \alpha \sin \beta + \sin^2 \beta$$

Recall that $\cos^2 \theta + \sin^2 \theta = 1$.

$$2 - 2 \cos (\alpha - \beta) = 2 - 2 \cos \alpha \cos \beta - 2 \sin \alpha \sin \beta$$
$$\cos (\alpha - \beta) = \cos \alpha \cos \beta + \sin \alpha \sin \beta.$$

This proves (28).

To prove (29), let us write $\beta = -\theta$. Then

$$\cos (\alpha + \beta) = \cos (\alpha - \theta)$$

By (28)

Replacing θ by $-\beta$

$\cos (-\beta) = \cos \beta$ and $\sin (-\beta) = -\sin \beta$

$$= \cos \alpha \cos \theta + \sin \alpha \sin \theta$$
$$= \cos \alpha \cos (-\beta) + \sin \alpha \sin (-\beta)$$
$$= \cos \alpha \cos \beta - \sin \alpha \sin \beta.$$

Therefore,

$$\cos (\alpha + \beta) = \cos \alpha \cos \beta - \sin \alpha \sin \beta.$$

We can now prove a similar theorem for the sine function.

Theorem 6. **For any real numbers α and β,**

(31) $\sin(\alpha - \beta) = \sin\alpha\cos\beta - \cos\alpha\sin\beta$

(32) $\sin(\alpha + \beta) = \sin\alpha\cos\beta + \cos\alpha\sin\beta.$

Proof. $\sin(\alpha - \beta) = \cos\left(\dfrac{\pi}{2} - (\alpha - \beta)\right)$

$$= \cos\left(\left(\frac{\pi}{2} - \alpha\right) + \beta\right)$$

$$= \cos\left(\frac{\pi}{2} - \alpha\right)\cos\beta - \sin\left(\frac{\pi}{2} - \alpha\right)\sin\beta$$

$$= \sin\alpha\cos\beta - \cos\alpha\sin\beta.$$

This establishes (31). We can obtain (32) from (31) by replacing $\beta = -\theta$ in (32) as follows:

$$\sin(\alpha + \beta) = \sin(\alpha - \theta)$$
$$= \sin\alpha\cos\theta - \cos\alpha\sin\theta$$
$$= \sin\alpha\cos(-\beta) - \cos\alpha\sin(-\beta)$$
$$= \sin\alpha\cos\beta + \cos\alpha\sin\beta.$$

For the tangent function, we have

Theorem 7. **For any real numbers α and β,**

(33) $\tan(\alpha - \beta) = \dfrac{\tan\alpha - \tan\beta}{1 + \tan\alpha\tan\beta}$

(34) $\tan(\alpha + \beta) = \dfrac{\tan\alpha + \tan\beta}{1 - \tan\alpha\tan\beta}.$

Proof.

$$\tan(\alpha - \beta) = \frac{\sin(\alpha - \beta)}{\cos(\alpha - \beta)}$$

$$= \frac{\sin\alpha\cos\beta - \cos\alpha\sin\beta}{\cos\alpha\cos\beta + \sin\alpha\sin\beta}$$

Dividing numerator and denominator by $\cos\alpha\cos\beta$ (of course, we are assuming $\cos\alpha\cos\beta \neq 0$)

$$= \frac{\dfrac{\sin\alpha\cos\beta}{\cos\alpha\cos\beta} - \dfrac{\cos\alpha\sin\beta}{\cos\alpha\cos\beta}}{\dfrac{\cos\alpha\cos\beta}{\cos\alpha\cos\beta} + \dfrac{\sin\alpha\sin\beta}{\cos\alpha\cos\beta}}$$

$$= \frac{\tan\alpha - \tan\beta}{1 + \tan\alpha\tan\beta}.$$

This proves (33). To prove (34), we write $\beta = -\theta$ in (34) to get

$$\tan (\alpha + \beta) = \tan (\alpha - \theta)$$

From (33)

$$= \frac{\tan \alpha - \tan \theta}{1 + \tan \alpha \tan \theta}$$

$$= \frac{\tan \alpha - \tan (-\beta)}{1 + \tan \alpha \tan (-\beta)}$$

tan $(-\beta) = -\tan \beta$ (Why ?)

$$= \frac{\tan \alpha + \tan \beta}{1 - \tan \alpha \tan \beta}.$$

Remark. All the formulas (28)–(34) are valid when α and β are multiples of certain other numbers. For example,

(i) if $\alpha = 2x$ and $\beta = 3y$, then

$$\cos (2x + 3y) = \cos 2x \cos 3y - \sin 2x \sin 3y ;$$

(ii) if $\alpha = \dfrac{x}{2}$ and $\beta = \dfrac{y}{5}$, then

$$\sin \left(\frac{x}{2} - \frac{y}{5} \right) = \sin \frac{x}{2} \cos \frac{y}{5} - \cos \frac{x}{2} \sin \frac{y}{5}.$$

From the above formulas, we can establish the following identities:

(35) $\cos 2\theta = \cos^2 \theta - \sin^2 \theta$ (36) $\cos 2\theta = 2 \cos^2 \theta - 1$

(37) $\cos 2\theta = 1 - 2 \sin^2 \theta$ (38) $\cos 2\theta = \dfrac{1 - \tan^2 \theta}{1 + \tan^2 \theta}$

(39) $\cos \theta = \cos^2 \dfrac{\theta}{2} - \sin^2 \dfrac{\theta}{2}$ (40) $\cos \theta = 2 \cos^2 \dfrac{\theta}{2} - 1$

(41) $\cos \theta = 1 - 2 \sin^2 \dfrac{\theta}{2}$ (42) $\cos \theta = \dfrac{1 - \tan^2 \dfrac{\theta}{2}}{1 + \tan^2 \dfrac{\theta}{2}}$

(43) $\sin 2\theta = 2 \sin \theta \cos \theta$ (44) $\sin 2\theta = \dfrac{2 \tan \theta}{1 + \tan^2 \theta}$

(45) $\sin \theta = 2 \sin \dfrac{\theta}{2} \cos \dfrac{\theta}{2}$ (46) $\sin \theta = \dfrac{2 \tan \dfrac{\theta}{2}}{1 + \tan^2 \dfrac{\theta}{2}}$

(47) $\tan 2\theta = \dfrac{2 \tan \theta}{1 - \tan^2 \theta}$ (48) $\tan \theta = \dfrac{2 \tan \dfrac{\theta}{2}}{1 - \tan^2 \dfrac{\theta}{2}}$

We shall prove only (38), (43), and (48) and leave the proof of the remaining identities for the students.

Proof of (38). $\cos 2\theta = \cos (\theta + \theta)$

$$= \cos \theta \cos \theta - \sin \theta \sin \theta$$

$$= \cos^2 \theta - \sin^2 \theta$$

Notice that this proves (35).

$$= \frac{\cos^2 \theta - \sin^2 \theta}{1}$$

$1 = \cos^2 \theta + \sin^2 \theta$

$$= \frac{\cos^2 \theta - \sin^2 \theta}{\cos^2 \theta + \sin^2 \theta}$$

Dividing numerator and denominator
by $\cos^2 \theta$ ($\cos \theta \neq 0$)

$$= \frac{\dfrac{\cos^2 \theta}{\cos^2 \theta} - \dfrac{\sin^2 \theta}{\cos^2 \theta}}{\dfrac{\cos^2 \theta}{\cos^2 \theta} + \dfrac{\sin^2 \theta}{\cos^2 \theta}}$$

$$= \frac{1 - \tan^2 \theta}{1 + \tan^2 \theta}.$$

Proof of (43).
$$\begin{aligned}
\sin 2\theta &= \sin (\theta + \theta) \\
&= \sin \theta \cos \theta + \cos \theta \sin \theta \\
&= 2 \sin \theta \cos \theta.
\end{aligned}$$

Proof of (48).
$$\tan \theta = \tan \left(\frac{\theta}{2} + \frac{\theta}{2} \right)$$

$$= \frac{\tan \dfrac{\theta}{2} + \tan \dfrac{\theta}{2}}{1 - \tan \dfrac{\theta}{2} \tan \dfrac{\theta}{2}}$$

$$= \frac{2 \tan \dfrac{\theta}{2}}{1 - \tan^2 \dfrac{\theta}{2}}.$$

Example 1. Evaluate $\sin 15°$.

Solution. We know that $15° = 45° - 30° = (\pi/4 - \pi/6)$ radians.
Thus

$$\sin 15° = \sin \left(\frac{\pi}{4} - \frac{\pi}{6} \right)$$

By (31)
$$= \sin \frac{\pi}{4} \cos \frac{\pi}{6} - \cos \frac{\pi}{4} \sin \frac{\pi}{6}$$

$$= \frac{1}{\sqrt{2}} \cdot \frac{\sqrt{3}}{2} - \frac{1}{\sqrt{2}} \cdot \frac{1}{2}$$

$$= \frac{\sqrt{3} - 1}{2\sqrt{2}}.$$

Example 2. Prove that $\dfrac{1 + \sin \theta}{1 - \sin \theta} = \tan^2 \left(\dfrac{\pi}{4} + \dfrac{\theta}{2} \right)$.

Solution.

$$\frac{1 + \sin \theta}{1 - \sin \theta} = \frac{1 - \cos \left(\dfrac{\pi}{2} + \theta \right)}{1 + \cos \left(\dfrac{\pi}{2} + \theta \right)}$$

$$= \frac{2 \sin^2 \left(\dfrac{\pi}{4} + \dfrac{\theta}{2} \right)}{2 \cos^2 \left(\dfrac{\pi}{4} + \dfrac{\theta}{2} \right)}$$

$$= \tan^2 \left(\frac{\pi}{4} + \frac{\theta}{2} \right).$$

Example 3. Prove that $\dfrac{\sin 3\theta}{\sin \theta} - \dfrac{\cos 3\theta}{\cos \theta} = 2$.

Solution. $\dfrac{\sin 3\theta}{\sin \theta} - \dfrac{\cos 3\theta}{\cos \theta} = \dfrac{\sin 3\theta \cos \theta - \cos 3\theta \sin \theta}{\sin \theta \cos \theta}$

$$= \frac{\sin (3\theta - \theta)}{\sin \theta \cos \theta}$$

$$= \frac{\sin 2\theta}{\sin \theta \cos \theta}$$

$$= \frac{2 \sin \theta \cos \theta}{\sin \theta \cos \theta}$$

$$= 2.$$

Exercise 6

1. Find sin 75°, cos 75°.

2. Suppose α and β are acute angles, $\sin \alpha = \frac{8}{17}$, and $\sin \beta = \frac{5}{13}$. Prove that
   (a) $\sin (\alpha - \beta) = \frac{21}{221}$
   (b) $\cos (\alpha + \beta) = \frac{140}{221}$.

3. Prove that $\cot (\alpha - \beta) = \dfrac{\cot \alpha \cot \beta + 1}{\cot \beta - \cot \alpha}$.

4. Prove that $\tan \left(\dfrac{\pi}{4} + \theta \right) \tan \left(\dfrac{\pi}{4} - \theta \right) = 1$.

5. Prove the identities (35) through (46) of this section.

6. Let $\cos \theta = \frac{7}{9}$, where $\pi < \theta < 2\pi$. Find $\sin \dfrac{\theta}{2}$ and $\cos \dfrac{\theta}{2}$.

7. Prove the following identities.
   (a) $\sin 3x = 3 \sin x - 4 \sin^3 x$
   (b) $\cos 3x = 4 \cos^3 x - 3 \cos x$

7. Inverse Trigonometric Functions

In Chapter 2, we have seen that a function f does not necessarily have an inverse function f^{-1}. We recall that f^{-1} exists if f is one-to-one. In Section 4 of this chapter, we noted that all the trigonometric functions are periodic. The period for the sine and cosine functions is 2π, whereas the period for the tangent function is π. Thus, for example, if

$$\sin x = \tfrac{1}{2},$$

then

$$x = \frac{\pi}{6} \pm 2n\pi \quad \text{or} \quad \frac{5\pi}{6} \pm 2n\pi, \qquad n \in Z.$$

Therefore, the sine function is not one-to-one and hence its inverse function does not exist.

However, we can restrict the domain (that is, choose a subset of the domain) over which the sine function is defined so that the restricted function is one-to-one.

To determine from the graph of a function $y = f(x)$ whether or not the function f has an inverse, we draw horizontal lines through each point in the range of f on the y-axis. If each such line cuts the graph of f at exactly one point, then f has an inverse.

We see from the graph of the sine function that there are many alternative ways of choosing an interval over which the sine function has an inverse. For example, some of these subintervals are

$$I_1 = \left\{ x \mid -\frac{\pi}{2} \le x \le \frac{\pi}{2} \right\}$$

$$I_2 = \left\{ x \mid -\frac{3\pi}{2} \le x \le -\frac{\pi}{2} \right\}$$

$$I_3 = \left\{ x \mid \frac{\pi}{2} \le x \le \frac{3\pi}{2} \right\}.$$

In such cases as these, it is customary to choose that interval
   (a) in which $|x|$ is the smallest value corresponding to a given value of the function, and
   (b) which maintains the full range for the values of the function.
   With (a) and (b) in mind, such an interval for the sine function is the set

$$\left\{ x \mid -\frac{\pi}{2} \le x \le \frac{\pi}{2} \right\}.$$

Hence the restricted sine function defined over the interval $\{x|\ -\pi/2 \le x \le \pi/2\}$ has an inverse function. Further, the domain of the inverse function is the set $\{y|\ -1 \le y \le 1\}$, and the range is the set $\{x|\ -\pi/2 \le x \le \pi/2\}$.

We formally have

Definition 5. The restricted sine function with the domain $[-\pi/2,\ \pi/2]$ has an inverse function called the *arcsine function*, defined by the relation

$$y = \arcsin x \quad \text{if, and only if,}$$

$$x = \sin y, \quad -\frac{\pi}{2} \le y \le \frac{\pi}{2}.$$

When considering the relation y = arc sin x, it is helpful to think "y is the angle whose sine is x."

A. GRAPH OF THE ARCSINE FUNCTION

Let $y = f(x) = \arcsin x$. The domain and the range of the arcsine function are $-1 \le x \le 1$ and $-\pi/2 \le y \le \pi/2$ respectively.

We can easily construct the following table to give us the values for the graph of the arcsine function (see Figure 22).

Table 8

x	$y = \arcsin x$
-1	$-\pi/2$
$-\sqrt{3}/2$	$-\pi/3$
$-1/\sqrt{2}$	$-\pi/4$
$-1/2$	$-\pi/6$
0	0
$1/2$	$\pi/6$
$1/\sqrt{2}$	$\pi/4$
$\sqrt{3}/2$	$\pi/3$
1	$\pi/2$

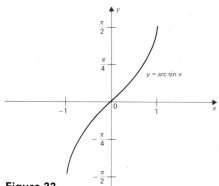

Figure 22

The graph of the arcsine function can also be constructed in an indirect fashion as follows:

In Chapter 2, we indicated that the graph of f^{-1} can be constructed from the graph of f by reflecting the graph of f through the line $y = x$. Consequently, we draw the graph of $y = \sin x$, $-\pi/2 \le x \le \pi/2$, and from this obtain the graph of $y = \arcsin x$, $-1 \le x \le 1$, as shown in Figure 23.

Since $\csc x = 1/\sin x$, $\sin x \ne 0$, both of these functions have the same period. By a similar argument we define the inverse cosecant function in

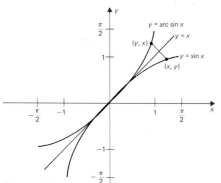

Figure 23

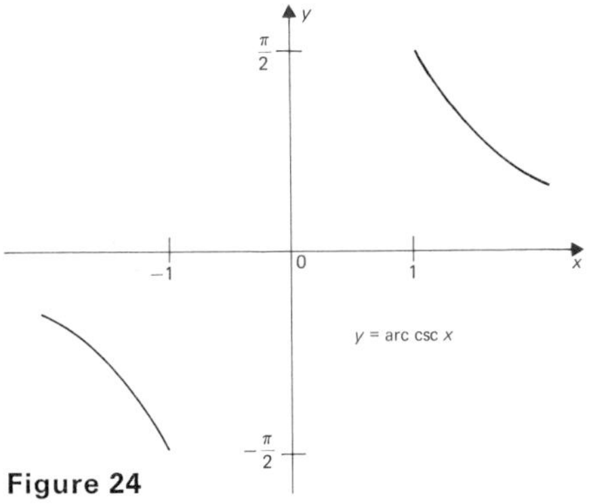

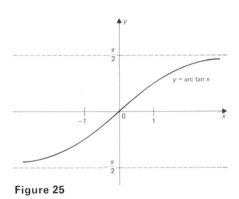

Figure 24

Figure 25

Definition 6. The *inverse cosecant function* is defined by the correspondence

$$y = \text{arccsc } x \quad \text{if, and only if,}$$

$$\text{csc } y = x, \ -\frac{\pi}{2} \le y \le \frac{\pi}{2}, \ y \ne 0.$$

The domain of the arccsc function is $\{x|\ 1 \le |x|\}$ and its range is $\{y|\ -\pi/2 \le y \le \pi/2, \ y \ne 0\}$.

The graph of this function can be obtained either by reflecting the graph of $y = \text{csc } x$, $-\pi/2 \le x \le \pi/2$, $x \ne 0$, through the line $y = x$; or by writing a table of its coordinates. Figure 24 is a rough sketch of the graph of $y = \text{arccsc } x$, $1 \le |x|$.

We now look for the definition of the *inverse tangent function*. The tangent function $y = \text{tan } x$ is defined with the domain $\{x|\ x \in R;\ x \ne (2n + 1)\pi/2, \ n \in Z\}$, and range $\{y|\ y \in R\}$. From the graph of the tangent function (Figure 19), it is easily seen that this function is not one-to-one and thus cannot have an inverse. However, if the domain is restricted to $\{x|\ -\pi/2 < x < \pi/2\}$, then the restricted tangent function is one-to-one. Consequently, we have

Definition 7. The *restricted tangent function* with the domain $\{x|\ -\pi/2 < x < \pi/2\}$ has an inverse function called the *arctan function*, defined by the relation

$$y = \text{arctan } x \quad \text{if, and only if,}$$

$$x = \text{tan } y, \ -\frac{\pi}{2} < y < \frac{\pi}{2}.$$

The domain of the arctan function is $\{x|\ x \in R\}$ and its range is $\{y|\ -\pi/2 < y < \pi/2\}$.

The graph of the arctangent function (Figure 25) can be obtained either by reflecting the graph of $y = \text{tan } x$, $-\pi/2 < x < \pi/2$, through the line $y = x$; or by constructing a table of its coordinates.

The definitions and graphs of arccosine, arcsecant, and arccotangent functions are left for the students. (See problem 1, Exercise 7.)

Example 1. Evaluate the following.

(a) $\text{arcsin } \dfrac{\sqrt{3}}{2}$

(b) $\cos (\text{arcsin } \frac{3}{5})$

(c) $\sin (\text{arctan } \frac{3}{4})$.

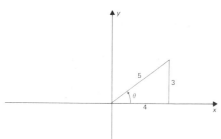

Figure 26

Solution. (a) Let $\theta = \arcsin \dfrac{\sqrt{3}}{2}$. Then, by definition

$$\sin \theta = \frac{\sqrt{3}}{2}, \qquad -\frac{\pi}{2} \le \theta \le \frac{\pi}{2}.$$

Thus $\theta = \pi/3$.

(b) Let $\theta = \arcsin \frac{3}{5}$, where $-\pi/2 \le \theta \le \pi/2$. Then

$$\sin \theta = \tfrac{3}{5}.$$

Since $\sin \theta$ is positive, θ is a positive acute angle. Therefore

$$\cos \theta = \sqrt{1 - \sin^2 \theta} = \sqrt{1 - \tfrac{9}{25}} = \sqrt{\tfrac{16}{25}} = \tfrac{4}{5},$$

that is,

$$\cos (\arcsin \tfrac{3}{5}) = \tfrac{4}{5}.$$

(c) Let $\theta = \arctan \frac{3}{4}$, where $-\pi/2 < \theta < \pi/2$. Then

$$\tan \theta = \tfrac{3}{4}.$$

Since $\tan \theta$ is positive, θ is a positive acute angle. If we construct the right triangle with θ in standard position as shown in Figure 26, we find by the Pythagorean theorem that the hypotenuse is 5. Consequently,

$$\sin \theta = \tfrac{3}{5},$$

that is,

$$\sin (\arctan \tfrac{3}{4}) = \tfrac{3}{5}.$$

Exercise 7

1. State why the following definitions are valid.
   (a) $\arccos x = y$ if, and only if, $x = \cos y$, $0 \le y \le \pi$.
   (b) $\operatorname{arcsec} x = y$ if, and only if, $x = \sec y$, $0 \le y < \pi$, $y \ne \dfrac{\pi}{2}$.
   (c) $\operatorname{arccot} x = y$ if, and only if, $x = \cot y$, $0 < y < \pi$.
2. Sketch the graph of
   (a) $\arccos x$
   (b) $\operatorname{arcsec} x$
   (c) $\operatorname{arccot} x$
   as defined in Problem 1.
3. Calculate.
   (a) $\arccos (\frac{1}{2}\sqrt{2})$ (b) $\arctan (\sqrt{3})$
   (c) $\operatorname{arccot} (0)$ (d) $\arcsin (-1)$
   (e) $\operatorname{arcsec} (-2)$ (f) $\operatorname{arccsc} (\sqrt{2})$
4. Calculate.

   (a) $\arcsin \left(\sin \dfrac{3\pi}{2} \right)$ (b) $\cos (\arcsin \tfrac{4}{5})$

(c) sin (arctan 1)   (d) sin (arccot $\frac{3}{4}$)

(e) arcsin $\left(\sin \dfrac{5\pi}{4} \right)$ (f) arccos $\left(\cos \dfrac{9\pi}{4} \right)$

(g) sin (arctan $\frac{5}{12}$)   (h) arctan $\left[\tan \left(-\dfrac{19\pi}{7} \right) \right]$

8. An Important Limit

In order to differentiate the trigonometric functions, we need to evaluate certain limits:

$$\lim_{h \to 0} \frac{\sin h}{h}, \quad \lim_{h \to 0} \frac{\cos h - 1}{h},$$

for example. In this section we shall prove that

$$\lim_{h \to 0} \frac{\sin h}{h} = 1$$

and use this result to evaluate several other limits.

Theorem 8.

Note. h is a real number, and sin h is the sine of angle h measured in radians.

(49) $$\lim_{h \to 0} \frac{\sin h}{h} = 1.$$

Proof. Case 1. Let *h* approach zero through positive values. Since we are interested in small positive values of *h*, it is sufficient to consider *h* such that

$$0 < h < \frac{\pi}{2}.$$

We place the angle *h* in standard position and draw a unit circle (circle with radius one) with center at the origin. We also construct the two right triangles *POM* and *TOA* as shown in Figure 27. It is obvious from Figure 27 that

(50)   area of $\triangle POA$ < area of sector *POA* < area of $\triangle TOA$.

It is easy to see that

$$d(O, A) = d(O, P) = 1$$
$$d(P, M) = \sin h$$
$$d(T, A) = \tan h = \frac{\sin h}{\cos h}.$$

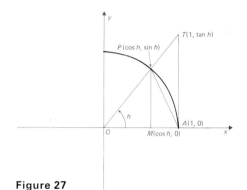

Figure 27

Area of sector POA $= \pi[d(O, A)]^2 \dfrac{h}{2\pi}$

Consequently, (50) is equivalent to

$$\tfrac{1}{2}d(O, A) \cdot d(P, M) < \tfrac{1}{2}[d(O, A)]^2 h < \tfrac{1}{2}d(O, A) \cdot d(T, A),$$

or

(51) $\qquad \dfrac{1}{2} \cdot 1 \cdot \sin h < \dfrac{1}{2} \cdot 1^2 \cdot h < \dfrac{1}{2} \cdot 1 \cdot \dfrac{\sin h}{\cos h}.$

Multiplying inequality (51) by $2/\sin h$ ($\sin h \neq 0$), we obtain

(52) $\qquad\qquad 1 < \dfrac{h}{\sin h} < \dfrac{1}{\cos h}.$

Taking reciprocals, we have

(53) $\qquad\qquad 1 > \dfrac{\sin h}{h} > \cos h.$

As $h \to 0+$, $\cos h$ approaches 1 ($\cos 0 = 1$). Thus $(\sin h)/h$ is "squeezed" between two functions, both of which approach 1 as $h \to 0+$. Therefore, $(\sin h)/h$ also approaches 1 as $h \to 0+$. This proves case 1.

Case 2. Let h approach zero through negative values, that is, $h \to 0-$.

If h is a negative number, we write $h = -k$, where k is positive.

Case 1

Then

$$\lim_{h \to 0-} \frac{\sin h}{h} = \lim_{-k \to 0-} \frac{\sin(-k)}{-k}$$

$$= \lim_{k \to 0+} \frac{-\sin k}{-k}$$

$$= \lim_{k \to 0+} \frac{\sin k}{k}$$

$$= 1.$$

Therefore,

$$\lim_{h \to 0+} \frac{\sin h}{h} = \lim_{h \to 0-} \frac{\sin h}{h} = 1.$$

Hence

$$\lim_{h \to 0} \frac{\sin h}{h} = 1.$$

Theorem 9.

$$\lim_{h \to 0} \frac{\cos h - 1}{h} = 0.$$

Proof. We shall prove this theorem by using Theorem 8.

$$\frac{\cos h - 1}{h} = \frac{\cos h - 1}{h} \cdot \frac{\cos h + 1}{\cos h + 1}, \qquad \cos h \neq -1.$$

$$= \frac{-\sin^2 h}{h(\cos h + 1)}$$

$$= (-1) \frac{\sin h}{h} \cdot \frac{\sin h}{\cos h + 1} .$$

Thus, $\displaystyle\lim_{h \to 0} \frac{\cos h - 1}{h} = \lim_{h \to 0} (-1) \frac{\sin h}{h} \cdot \frac{\sin h}{\cos h + 1}$

$$= (-1) \lim_{h \to 0} \frac{\sin h}{h} \cdot \lim_{h \to 0} \frac{\sin h}{\cos h + 1}$$

$$= (-1) \cdot 1 \cdot 0$$

$$= 0.$$

$$\lim_{h \to 0} \frac{\sin h}{\cos h + 1} = \frac{\displaystyle\lim_{h \to 0} \sin h}{\displaystyle\lim_{h \to 0} (\cos h + 1)} = \frac{0}{2} = 0$$

Example 1. Evaluate the following limits.

(a) $\displaystyle\lim_{x \to 0} \frac{\sin 5x}{x}$

(b) $\displaystyle\lim_{x \to 0} \frac{\sin mx}{\sin nx}$

(c) $\displaystyle\lim_{\theta \to 0} \frac{\sin^2 \dfrac{\theta}{2}}{\theta^2}$

Solution. (a) Let $h = 5x$. Then

$$\lim_{x \to 0} \frac{\sin 5x}{x} = \lim_{x \to 0} \frac{\sin 5x}{x} \cdot \frac{5}{5}$$

$$= \lim_{x \to 0} \frac{5 \sin 5x}{5x}$$

$$= \lim_{h \to 0} \frac{5 \sin h}{h}$$

$$= 5 \lim_{h \to 0} \frac{\sin h}{h}$$

$$= 5.$$

(b) $$\lim_{x \to 0} \frac{\sin mx}{\sin nx} = \lim_{x \to 0} \frac{\dfrac{\sin mx}{mx} \cdot mx}{\dfrac{\sin nx}{nx} \cdot nx}$$

$$= \lim_{x \to 0} \frac{m \cdot \dfrac{\sin mx}{mx}}{n \cdot \dfrac{\sin nx}{nx}}$$

$$= \frac{m \ \lim\limits_{mx \to 0} \dfrac{\sin mx}{mx}}{n \ \lim\limits_{nx \to 0} \dfrac{\sin nx}{nx}}$$

$$= \frac{m \cdot 1}{n \cdot 1}$$

$$= \frac{m}{n}.$$

(c)
$$\lim_{\theta \to 0} \frac{\sin^2 \dfrac{\theta}{2}}{\theta^2} = \lim_{\theta \to 0} \frac{\sin^2 \dfrac{\theta}{2}}{\left(\dfrac{\theta}{2}\right)^2 \cdot 4}$$

$$= \lim_{\theta \to 0} \frac{1}{4} \cdot \left(\frac{\sin \dfrac{\theta}{2}}{\dfrac{\theta}{2}} \right)^2$$

$$= \frac{1}{4} \lim_{\frac{\theta}{2} \to 0} \left(\frac{\sin \dfrac{\theta}{2}}{\dfrac{\theta}{2}} \right)^2$$

$$= \tfrac{1}{4}(1)^2$$

$$= \tfrac{1}{4}.$$

Example 2. Evaluate the following limits.

(a) $\lim\limits_{\theta \to 0} \dfrac{1 - \cos \theta}{\theta^2}$

(b) $\lim\limits_{\theta \to 0} \dfrac{\tan \theta - \sin \theta}{\sin^3 \theta}$

Solution.

An alternate solution to (a):

$$\lim_{\theta \to 0} \frac{1 - \cos \theta}{\theta^2} = \lim_{\theta \to 0} \frac{2 \sin^2 \dfrac{\theta}{2}}{\theta^2}$$

$$= \lim_{\theta \to 0} \frac{2 \sin^2 \dfrac{\theta}{2}}{\left(\dfrac{\theta}{2}\right)^2 \cdot 4}$$

$$= \frac{2}{4} \lim_{\theta \to 0} \left(\frac{\sin \dfrac{\theta}{2}}{\dfrac{\theta}{2}} \right)^2$$

$$= \frac{1}{2}$$

(a)
$$\lim_{\theta \to 0} \frac{1 - \cos \theta}{\theta^2} = \lim_{\theta \to 0} \frac{(1 - \cos \theta)}{\theta^2} \cdot \frac{1 + \cos \theta}{1 + \cos \theta}$$

$$= \lim_{\theta \to 0} \frac{1 - \cos^2 \theta}{\theta^2 (1 + \cos \theta)}$$

$$= \lim_{\theta \to 0} \frac{\sin^2 \theta}{\theta^2 (1 + \cos \theta)}$$

$$= \lim_{\theta \to 0} \frac{1}{1 + \cos \theta} \cdot \lim_{\theta \to 0} \left(\frac{\sin \theta}{\theta} \right)^2$$

$$= \tfrac{1}{2} \cdot 1$$
$$= \tfrac{1}{2}.$$

(b)
$$\frac{\tan \theta - \sin \theta}{\sin^3 \theta} = \frac{\dfrac{\sin \theta}{\cos \theta} - \sin \theta}{\sin^3 \theta}$$

$$= \frac{\sin \theta \left(\dfrac{1}{\cos \theta} - 1 \right)}{\sin^3 \theta}$$

$$= \frac{\sin \theta}{\sin^3 \theta} \frac{1 - \cos \theta}{\cos \theta}$$

$$= \frac{1 - \cos \theta}{\cos \theta \sin^2 \theta}$$

$$= \frac{1 - \cos \theta}{\cos \theta \sin^2 \theta} \cdot \frac{1 + \cos \theta}{1 + \cos \theta}$$

$$= \frac{1 - \cos^2 \theta}{\cos \theta \sin^2 \theta (1 + \cos \theta)}$$

$$= \frac{\sin^2 \theta}{\cos \theta \sin^2 \theta (1 + \cos \theta)}$$

$$= \frac{1}{\cos \theta (1 + \cos \theta)}.$$

Therefore,

$$\lim_{\theta \to 0} \frac{\tan \theta - \sin \theta}{\sin^3 \theta} = \lim_{\theta \to 0} \frac{1}{\cos \theta (1 + \cos \theta)}$$

$$= \frac{1}{\lim\limits_{\theta \to 0} \cos \theta (1 + \cos \theta)}$$

$$= \frac{1}{\lim\limits_{\theta \to 0} \cos \theta \lim\limits_{\theta \to 0} (1 + \cos \theta)}$$

$$= \frac{1}{1 \cdot 2}$$

$$= \tfrac{1}{2}.$$

Exercise 8

In problems 1 through 5, calculate the value (if any) of each of the limits.

1. $\lim\limits_{x \to 0} \dfrac{\sin 2x}{x}$

2. $\lim\limits_{x\to 0} \dfrac{\sin^2 x}{x^2}$

3. $\lim\limits_{x\to 0+} \dfrac{\sin x}{\sqrt{x}}$ *Hint:* $\dfrac{\sin x}{\sqrt{x}} = \sqrt{x}\,\dfrac{\sin x}{x}$

4. $\lim\limits_{x\to 0} \dfrac{\tan x}{x}$

5. $\lim\limits_{x\to 0} \dfrac{\tan^2 x - \sin^2 x}{x^4}$

Hint: $\dfrac{\tan^2 x - \sin^2 x}{x^4} = \dfrac{\tan x - \sin x}{x^3} \cdot \left(\dfrac{\tan x}{x} + \dfrac{\sin x}{x}\right)$

9. Derivatives of the Trigonometric Functions

We now use Theorems 8 and 9 to calculate the derivative of sin x.

Theorem 10 may be written in other notations:

$$y = \sin x, \frac{dy}{dx} = \cos x$$

$$\frac{d}{dx}(\sin x) = \cos x$$

$$D(\sin x) = \cos x$$

$$(\sin x)' = \cos x.$$

Formula (32)

Theorem 10. If $f(x) = \sin x$, then

(54) $$f'(x) = \cos x.$$

Proof. Setting up the difference quotient we have

$$\frac{f(x + h) - f(x)}{h} = \frac{\sin (x + h) - \sin x}{h}$$

$$= \frac{\sin x \cos h + \cos x \sin h - \sin x}{h}$$

$$= \sin x \left(\frac{\cos h - 1}{h}\right) + \cos x \left(\frac{\sin h}{h}\right).$$

Thus,

$$f'(x) = \lim_{h\to 0} \frac{f(x + h) - f(x)}{h}$$

$$= \sin x \lim_{h\to 0} \frac{\cos h - 1}{h} + \cos x \lim_{h\to 0} \frac{\sin h}{h}$$

Theorems 9 and 8

$$= (\sin x) \cdot 0 + (\cos x) \cdot 1$$

$$= \cos x.$$

Therefore, the derivative of sin x is cos x.

By applying the chain rule, we find a more general formula:

(55) $$\frac{d}{dx}(\sin u) = \cos u \cdot \frac{du}{dx},$$

where u is a differentiable function of x.

Example 1. Differentiate the following with respect to x.

(a) $\sin(3x^2 + 2x + 1)$

(b) $\sin(e^{x^2})$.

$u = 3x^2 + 2x + 1, \dfrac{du}{dx} = 6x + 2$

Solution. (a) Using (55), with $u = 3x^2 + 2x + 1$, we have

$$\frac{d}{dx}\sin(3x^2 + 2x + 1) = \frac{d}{dx}(\sin u)$$

$$= \cos u \cdot \frac{du}{dx}$$

$$= [\cos(3x^2 + 2x + 1)](6x + 2)$$

$$= 2(3x + 1)\cos(3x^2 + 2x + 1).$$

If $u = e^{x^2}$,

then $\dfrac{du}{dx} = \dfrac{d}{dx}(e^{x^2})$

$= e^{x^2}\dfrac{d}{dx}(x^2)$ *Chapter 8*

$= e^{x^2}(2x)$.

(b) Again using (55), with $u = e^{x^2}$, we have

$$\frac{d}{dx}[\sin(e^{x^2})] = \frac{d}{dx}\sin u$$

$$= \cos u \cdot \frac{du}{dx}$$

$$= \cos(e^{x^2}) \cdot e^{x^2} \cdot 2x$$

$$= 2xe^{x^2}\cos(e^{x^2}).$$

Theorem 11. If $f(x) = \cos x$, then

(56) $\qquad f'(x) = -\sin x.$

Recall Theorem 2:

$\sin\left(\dfrac{\pi}{2} - x\right) = \cos x$

$\cos\left(\dfrac{\pi}{2} - x\right) = \sin x$

Proof. Since

$$f(x) = \cos x = \sin\left(\frac{\pi}{2} - x\right),$$

we have

Apply (55).

$u = \dfrac{\pi}{2} - x \Rightarrow \dfrac{du}{dx} = -1$

$$\frac{d}{dx}\cos x = \frac{d}{dx}\sin\left(\frac{\pi}{2} - x\right)$$

$$= \cos\left(\frac{\pi}{2} - x\right)\frac{d}{dx}\left(\frac{\pi}{2} - x\right)$$

$$= \left[\cos\left(\frac{\pi}{2} - x\right)\right] \cdot (-1)$$

$$= -\cos\left(\frac{\pi}{2} - x\right)$$

$$= -\sin x.$$

Again, by the chain rule, we have

(57) $$\frac{d}{dx}(\cos u) = -\sin u \cdot \frac{du}{dx},$$

where u is a differentiable function of x.

Since the other trigonometric functions can be expressed in terms of the sine and cosine functions, we can find their derivatives by using the rules of differentiation established in Chapter 5. We have

Theorem 12.

(i) $\quad\dfrac{d}{dx}\tan x = \sec^2 x$

(ii) $\quad\dfrac{d}{dx}\cot x = -\csc^2 x$

(iii) $\quad\dfrac{d}{dx}\sec x = \sec x \tan x$

(iv) $\quad\dfrac{d}{dx}\csc x = -\csc x \cot x$

Proof. We shall prove only (i) and (iii).

(i) $\quad\dfrac{d}{dx}\tan x = \dfrac{d}{dx}\left(\dfrac{\sin x}{\cos x}\right)$

Recall: $\dfrac{d}{dx}\left(\dfrac{u}{v}\right) = \dfrac{v \cdot u' - u \cdot v'}{v^2}$

$$= \frac{\cos x \dfrac{d}{dx}(\sin x) - \sin x \dfrac{d}{dx}(\cos x)}{\cos^2 x}$$

$$= \frac{(\cos x)(\cos x) - (\sin x)(-\sin x)}{\cos^2 x}$$

$$= \frac{\cos^2 x + \sin^2 x}{\cos^2 x}$$

$$= \frac{1}{\cos^2 x}$$

$$= \sec^2 x.$$

(iii) $\quad\dfrac{d}{dx}\sec x = \dfrac{d}{dx}\left(\dfrac{1}{\cos x}\right)$

The derivative of a constant is zero.

$$= \frac{(\cos x)\dfrac{d}{dx}(1) - 1\dfrac{d}{dx}(\cos x)}{\cos^2 x}$$

The derivative of a constant is zero.

$$= \frac{(\cos x) \cdot 0 - (1)(-\sin x)}{\cos^2 x}$$

$$= \frac{\sin x}{\cos^2 x}$$

$$= \frac{1}{\cos x} \cdot \frac{\sin x}{\cos x}$$

$$= \sec x \tan x.$$

Applying the chain rule, we may write

$$(58) \qquad \frac{d}{dx} \tan u = \sec^2 u \cdot \frac{du}{dx}$$

$$(59) \qquad \frac{d}{dx} \cot u = -\csc^2 u \cdot \frac{du}{dx}$$

$$(60) \qquad \frac{d}{dx} \sec u = \sec u \tan u \cdot \frac{du}{dx}$$

$$(61) \qquad \frac{d}{dx} \csc u = -\csc u \cot u \cdot \frac{du}{dx}.$$

Example 2. If $y = \tan^3 (2x + 5)$, find $\dfrac{dy}{dx}$.

Solution. Let $u = \tan (2x + 5)$ and $v = 2x + 5$. Then

$$y = u^3 \qquad u = \tan v \qquad v = 2x + 5,$$

and

$$\frac{dy}{du} = 3u^2 \qquad \frac{du}{dv} = \sec^2 v \qquad \frac{dv}{dx} = 2.$$

By the chain rule,

$$\frac{dy}{dx} = \frac{dy}{du} \cdot \frac{du}{dv} \cdot \frac{dv}{dx}$$

$$= 3u^2 \cdot \sec^2 v \cdot 2$$

$$= 6u^2 \sec^2 v$$

$$= 6 \tan^2 (2x + 5) \sec^2 (2x + 5).$$

Knowing the derivatives of the trigonometric functions, it is easy to calculate the derivatives of their inverse functions.

Theorem 13. If $y = \arcsin x$ $\left(-\dfrac{\pi}{2} < y < \dfrac{\pi}{2}\right)$, then

(62)
$$\frac{dy}{dx} = \frac{1}{\sqrt{1 - x^2}}.$$

Proof. For $y = \arcsin x$, we have

$$x = \sin y \quad \left(\frac{-\pi}{2} < y < \frac{\pi}{2}\right).$$

We calculate the derivative on both sides with respect to x, treating y as a function of x. Thus,

$$\frac{d}{dx}(x) = \frac{d}{dx}(\sin y)$$

Using implicit differentiation

$$1 = \cos y \cdot \frac{dy}{dx}$$

$$\frac{dy}{dx} = \frac{1}{\cos y} \quad (\cos y \neq 0).$$

Since $-\pi/2 < y < \pi/2$, $\cos y > 0$. Thus $\cos y = \sqrt{1 - \sin^2 y}$. Therefore

$$\frac{dy}{dx} = \frac{1}{\cos y} = \frac{1}{\sqrt{1 - \sin^2 y}} = \frac{1}{\sqrt{1 - x^2}}.$$

Similar calculations yield

Theorem 14.

(63) If $y = \arccos x$ $(0 < y < \pi)$, then $\dfrac{dy}{dx} = -\dfrac{1}{\sqrt{1 - x^2}}$.

(64) If $y = \arctan x$ $\left(-\dfrac{\pi}{2} < y < \dfrac{\pi}{2}\right)$, then $\dfrac{dy}{dx} = \dfrac{1}{1 + x^2}$.

(65) If $y = \text{arccot } x$ $(0 < y < \pi)$, then $\dfrac{dy}{dx} = -\dfrac{1}{1 + x^2}$.

The proof of this theorem is left as an exercise for the students.

Example 3. Find the derivative of $\arcsin\left(\dfrac{2x}{1 + x^2}\right)$.

Solution. Let $y = \arcsin\left(\dfrac{2x}{1 + x^2}\right)$, and $u = \dfrac{2x}{1 + x^2}$. Then

$$y = \arcsin u \qquad u = \frac{2x}{1 + x^2}$$

$$\frac{dy}{du} = \frac{1}{\sqrt{1 - u^2}} \qquad \frac{du}{dx} = \frac{(1 + x^2)2 - 2x(2x)}{(1 + x^2)^2}$$

$$= \frac{2 + 2x^2 - 4x^2}{(1 + x^2)^2}$$

$$= \frac{2(1 - x^2)}{(1 + x^2)^2}.$$

Alternate method: Let $x = \tan\theta$. Then

$$y = \arcsin\left(\frac{2\tan\theta}{1 + \tan^2\theta}\right)$$

$$= \arcsin\,(\sin 2\theta)$$

$$= 2\theta.$$

Thus,

$$y = 2\theta \quad \text{and} \quad \theta = \arctan x$$

$$\frac{dy}{d\theta} = 2 \quad \text{and} \quad \frac{d\theta}{dx} = \frac{1}{1 + x^2}.$$

By the chain rule,

$$\frac{dy}{dx} = \frac{dy}{d\theta} \cdot \frac{d\theta}{dx}$$

$$= 2 \cdot \frac{1}{1 + x^2}$$

$$= \frac{2}{1 + x^2}.$$

By the chain rule,

$$\frac{dy}{dx} = \frac{dy}{du} \cdot \frac{du}{dx}$$

$$= \frac{1}{\sqrt{1 - u^2}} \cdot \frac{2(1 - x^2)}{(1 + x^2)^2}$$

$$= \frac{1}{\sqrt{1 - \dfrac{4x^2}{(1 + x^2)^2}}} \cdot \frac{2(1 - x^2)}{(1 + x^2)^2}$$

$$= \frac{(1 + x^2)}{\sqrt{(1 + x^2)^2 - 4x^2}} \cdot \frac{2(1 - x^2)}{(1 + x^2)^2}$$

$$= \frac{(1 + x^2)}{1 - x^2} \cdot \frac{2(1 - x^2)}{(1 + x^2)^2}$$

$$= \frac{2}{1 + x^2}.$$

Example 4. Find the maximum and minimum values of

$$4 \sin\theta + 3 \cos\theta \qquad (0 \le \theta \le 2\pi).$$

Solution. Let $f(\theta) = 4 \sin\theta + 3 \cos\theta$. Then

$$f'(\theta) = 4 \cos\theta - 3 \sin\theta.$$

Thus $f'(\theta) = 0$ if

$$4 \cos\theta - 3 \sin\theta = 0$$

$$4 \cos\theta = 3 \sin\theta$$

$$\tan\theta = \tfrac{4}{3}.$$

In the interval $[0, 2\pi]$, there are two values (in the first and third quadrants) α and β such that

$$\tan \alpha = \tfrac{4}{3} \quad \text{and} \quad \tan \beta = \tfrac{4}{3}.$$

It is easy to see that if α is in the first quadrant, then

$$\sin \alpha = \frac{4}{\sqrt{4^2 + 3^2}} = \frac{4}{5} \qquad \cos \alpha = \frac{3}{\sqrt{4^2 + 3^2}} = \frac{3}{5}.$$

Since β is in the third quadrant, $\sin \beta$ and $\cos \beta$ are negative; thus

$$\sin \beta = -\tfrac{4}{5} \qquad \cos \beta = -\tfrac{3}{5}.$$

We take the second derivative and evaluate it at α and β:

$$f''(\theta) = -4 \sin \theta - 3 \cos \theta$$
$$f''(\alpha) = -4 \cdot \tfrac{4}{5} - 3 \cdot \tfrac{3}{5} = -5, \qquad \text{a negative number}$$
$$f''(\beta) = -4(-\tfrac{4}{5}) - 3(-\tfrac{3}{5}) = 5, \qquad \text{a positive number.}$$

Therefore, $f(\alpha)$ is a maximum value and $f(\beta)$ is a minimum value of the function.

By substitution we find $f(\alpha) = 5$ and $f(\beta) = -5$, and these are respectively the maximum and the minimum values of f.

Exercise 9

In problems 1 through 14 differentiate with respect to x.

1. $x \sin x$

2. $\dfrac{4x + 1}{\cos x}$

3. $\dfrac{\sin x + \cos x}{\sin x - \cos x}$

4. $\dfrac{\sin (3x + 1)}{\cos (2x + 3)}$

5. $\tan^4 (7x^2 + 3x + 9)$

6. $\sin^2 (2x) \cos^3 (2x)$

7. $e^{\sin x - \tan x}$

8. $\ln (x^2 \sin x)$

9. $e^x \sin^2 (\cos x)$

10. $\arcsin (2x)$

11. $(x^2 + 1) \arctan (x)$

12. $\arcsin \sqrt{1 - x^2}$

13. $e^{\sin x + \arctan x}$

14. $\ln (x^2 + \tan x + \text{arccsc } x)$

15. Find $f'\left(\dfrac{\pi}{4}\right)$, if $f(x) = (1 + x) \tan x$.

16. A particle moves in a straight line with the equation of motion $x = a \cos(bt + c)$, where a, b, c are constants and x is the distance measured from a point 0 on the straight line. Show that the acceleration at any time t is proportional to the distance from 0.

17. Water is being poured into a conical vessel (which is held with its vertex down) at the rate of 2 cu ft per minute. The semivertical angle of the conical vessel is arctan $(\frac{1}{2})$. Find the rate at which the water is rising in the vessel when the depth of water is 2 ft.

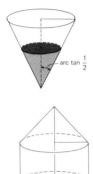

18. A bell-tent consists of a cylindrical portion resting on the ground and a conical portion on top of the cylinder. Prove that if the volume of the bell-tent is V and the circular base has radius a, then the amount of canvas used is a minimum when the semivertical angle of the cone is arccos $(\frac{2}{3})$.
Hint: A cone with radius a and height h has volume $= \frac{1}{3}\pi a^2 h$, and

$$\text{surface area} = \pi a \text{ length of the slant height}$$
$$= \pi a \cdot \sqrt{a^2 + h^2}.$$

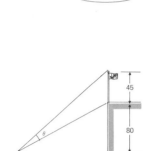

*19. A flagpole 45 ft high stands at the top of a building 85 ft high. If an observer's eye is 5 ft above the ground, how far should he stand from the foot of the building in order that the vertical angle (θ in the figure below) subtended by the flagpole at his eye be maximum?

10. Integration of the Trigonometric Functions

From the Fundamental Theorem of Calculus, it is clear that to each differentiation formula there corresponds an integration formula. The integrals corresponding to the derivatives in the previous section are

(66) $$\int \cos x \, dx = \sin x + C$$

(67) $$\int \sin x \, dx = -\cos x + C$$

(68) $$\int \sec^2 x \, dx = \tan x + C$$

(69) $$\int \csc^2 x \, dx = -\cot x + C$$

(70) $$\int \sec x \tan x \, dx = \sec x + C$$

(71) $$\int \csc x \cot x \, dx = -\csc x + C$$

(72) $$\int \frac{-1}{\sqrt{1 - x^2}} \, dx = \arcsin x + C, \; -1 < x < 1$$

(73) $$\int \frac{-1}{\sqrt{1 - x^2}} \, dx = \arccos x + C, \; -1 < x < 1$$

(74) $$\int \frac{1}{1 + x^2} \, dx = \arctan x + C$$

(75) $$\int \frac{-1}{1 + x^2} \, dx = \text{arccot } x + C.$$

Recall: $\int \frac{1}{x} \, dx = \ln |x| + C$

The integrals of tan x, cot x, sec x, and csc x can be obtained by the techniques of Chapter 8.

$u = \cos x$
$du = -\sin x \, dx$

(76) $$\int \tan x \, dx = \int \frac{\sin x}{\cos x} \, dx$$

$$= -\int \frac{1}{u} \, du$$

$$= -\ln |u| + C$$
$$= -\ln |\cos x| + C$$

$$= \ln \left| \frac{1}{\cos x} \right| + C$$

$$= \ln |\sec x| + C.$$

$u = \sin x$
$du = \cos x \, dx$

(77) $$\int \cot x \, dx = \int \frac{\cos x}{\sin x} \, dx$$

$$= \int \frac{1}{u} \, du$$

$$= \ln |u| + C$$
$$= \ln |\sin x| + C.$$

$u = \sec x + \tan x$
$du = (\sec^2 x + \sec x \tan x) \, dx$
$ = \sec x \, (\sec x + \tan x) \, dx$

(78) $$\int \sec x \, dx = \int \frac{\sec x (\sec x + \tan x)}{\sec x + \tan x} \, dx$$

$$= \int \frac{1}{u} \, du$$

$$= \ln |u| + C$$
$$= \ln |\sec x + \tan x| + C.$$

$u = \csc x + \cot x$

$du = (-\csc^2 x - \csc x \cot x)\, dx$

$\quad\ = -\csc s \,(\csc x + \cot x)\, dx$

(79)
$$\int \csc x \, dx = -\int \frac{-\csc x(\csc x + \cot x)}{\csc x + \cot x} \, dx$$

$$= -\int \frac{1}{u} \, du$$

$$= -\ln |u| + C$$

$$= -\ln |\csc x + \cot x| + C.$$

We shall now give some examples to illustrate how to solve simple integration problems involving trigonometric functions.

Example 1. Integrate $\int \sin^5 x \cos x \, dx$.

Solution. Let $u = \sin x$, then $du = \cos x \, dx$. Thus

To integrate $\int f (\sin x) \cos x \, dx$, we use the substitution $u = \sin x$; similarly, for $\int f (\cos x) \sin x \, dx$, we use $u = \cos x$.

$$\int \sin^5 x \cos x \, dx = \int u^5 \, du$$

$$= \frac{u^6}{6} + C$$

$$= \frac{\sin^6 x}{6} + C.$$

Example 2. Integrate.

(a) $\int \cos^5 x \, dx$

(b) $\int \cos^4 x \, dx$

Solution.

$u = \sin x, \, du = \cos x \, dx$

This technique is used to integrate $\sin^n x$ or $\cos^n x$ when n is odd.

(a)
$$\int \cos^5 x \, dx = \int \cos^4 x \cos x \, dx$$

$$= \int (1 - \sin^2 x)^2 \cos x \, dx$$

$$= \int (1 - u^2)^2 \, du$$

$$= \int (1 - 2u^2 + u^4) \, du$$

$$= u - \frac{2u^3}{3} + \frac{u^5}{5} + C$$

$$= \sin x - \tfrac{2}{3} \sin^3 x + \tfrac{1}{5} \sin^5 x + C.$$

(b) $\int \cos^4 x \, dx = \int (\cos^2 x)^2 \, dx$

$$= \int \left(\frac{1 + \cos 2x}{2}\right)^2 \, dx$$

$$= \frac{1}{4} \int (1 + 2 \cos 2x + \cos^2 2x) \, dx$$

$$= \frac{1}{4} \int dx + \frac{2}{4} \int \cos 2x \, dx + \frac{1}{4} \int \cos^2 2x \, dx$$

$$= \frac{x}{4} + \frac{2}{4} \cdot \frac{\sin 2x}{2} + \frac{1}{4} \int \left(\frac{1 + \cos 4x}{2}\right) \, dx$$

$$= \frac{x}{4} + \frac{1}{4} \sin 2x + \frac{1}{8} \int (1 + \cos 4x) \, dx$$

$$= \frac{x}{4} + \frac{1}{4} \sin 2x + \frac{1}{8} \left[x + \frac{\sin 4x}{4}\right] + C$$

$$= \frac{x}{4} + \frac{x}{8} + \frac{1}{4} \sin 2x + \frac{1}{32} \sin 4x + C$$

$$= \frac{3}{8} x + \frac{1}{4} \sin 2x + \frac{1}{32} \sin 4x + C.$$

$\cos 2x = 2 \cos^2 x - 1$
$\cos 2x = 1 - 2 \sin^2 x$

$\int \cos ax \, dx = \dfrac{\sin ax}{a} + C$, which can easily be
seen by letting $u = ax$, $du = a\,dx$.
This technique is used to evaluate $\sin^n x$ or
$\cos^n x$ when n is even.

Example 3. Integrate $\int x\sqrt{1 - x^2} \, dx$.

Solution. Let $x = \sin \theta$, then $dx = \cos \theta \, d\theta$.

$$\int x\sqrt{1 - x^2} \, dx = \int \sin \theta \sqrt{1 - \sin^2 \theta} \, \cos \theta \, d\theta$$

$$= \int \sin \theta \cos \theta \cos \theta \, d\theta$$

$$= \int \sin \theta \cos^2 \theta \, d\theta$$

$$= -\int \cos^2 \theta \, (-\sin \theta) \, d\theta$$

$$= -\int u^2 \, du$$

$$= -\frac{u^3}{3} + C$$

$$= -\frac{\cos^3 \theta}{3} + C$$

$$= -\frac{1}{3} \cos^3 (\arcsin x) + C.$$

Note. The following important substitutions
should be remembered.

(i)   If the integrand has $\sqrt{a^2 - x^2}$ as a factor,
let $x = a \sin \theta$ or $x = a \cos \theta$.

(ii) If the integrand has $\sqrt{a^2 + x^2}$ as a factor,
let $x = a \tan \theta$.

(iii) If the integrand has $\sqrt{x^2 - a^2}$ as a factor,
let $x = a \sec \theta$.

These substitutions are sometimes useful even
if the above factors are in the denominator.

Example 4. Integrate $\displaystyle\int \frac{dx}{\sqrt{7 + 6x - x^2}}$.

Solution. We have $7 + 6x - x^2 = 16 - (x - 3)^2 = (4)^2 - (x - 3)^2$.
Thus

Let $x - 3 = 4 \sin \theta$
$dx = 4 \cos \theta \, d\theta.$

$$\int \frac{dx}{\sqrt{7 + 6x - x^2}} = \int \frac{dx}{\sqrt{(4)^2 - (x - 3)^2}}$$

$$= \int \frac{4 \cos \theta \, d\theta}{\sqrt{(4)^2 - (4 \sin \theta)^2}}$$

$$= \int \frac{4 \cos \theta \, d\theta}{4\sqrt{1 - \sin^2 \theta}}$$

$$= \int \frac{\cos \theta \, d\theta}{\cos \theta}$$

$$= \int d\theta$$

$$= \theta + C.$$

Now

$$x - 3 = 4 \sin \theta \Rightarrow \frac{x - 3}{4} = \sin \theta$$

or

$$\theta = \arcsin \frac{x - 3}{4}.$$

Therefore, the given integral $= \theta + C$

$$= \arcsin \frac{x - 3}{4} + C.$$

Example 5. Integrate $\displaystyle\int \frac{dx}{e^x + e^{-x}}$.

Solution. We let $u = e^x$, then $du = e^x \, dx$.

$$\int \frac{1}{e^x + e^{-x}} \, dx = \int \frac{e^x}{e^{2x} + 1} \, dx$$

$$= \int \frac{du}{u^2 + 1}$$

$$= \arctan u + C$$

$$= \arctan e^x + C.$$

Exercise 10

In problems 1 through 30, integrate.

1. $\displaystyle\int \frac{\sec^2 x}{1 + \tan x} \, dx$ **2.** $\displaystyle\int \frac{\sin x}{1 + \cos x} \, dx$

3. $\int \cos x e^{\sin x} \, dx$

4. $\int \sec^2 x \tan^3 x \, dx$

5. $\int x \cos^2 (x^2 + 1) \, dx$

6. $\int e^{\sin x \cos x} \cos 2x \, dx$

7. $\int \dfrac{\sin x}{\cos^2 x} \, dx$

8. $\int \dfrac{1}{\cos^2 x (\tan x + 5)} \, dx$

9. $\int \sin^2 x \cos^2 x \, dx$

10. $\int x \cos x \, dx$

11. $\int \sec^3 x \, dx$

12. $\int x^2 \sin x \, dx$

13. $\int x \tan^2 x \, dx$

14. $\int_0^{\frac{\pi}{2}} \sin^2 x \, dx$

15. $\int_0^{\frac{\pi}{2}} \sin x \cos^2 x \, dx$

16. $\int_0^{\frac{\pi}{4}} \tan^5 x \, dx$

17. $\int_0^{\pi} x \sin x \cos^2 x \, dx$

18. $\int \sqrt{a^2 - x^2} \, dx$

19. $\int \dfrac{dx}{x^2 + 2x + 10}$

20. $\int x\sqrt{9 - x^2} \, dx$

21. $\int \arctan x \, dx$

22. $\int x \arctan x \, dx$

23. $\int x \operatorname{arcsec} x \, dx$

24. $\int \dfrac{dx}{(x + 1)(x^2 + 4)}$

25. $\int \dfrac{dx}{(x + 1)(x^2 + 4)^2}$

26. $\int \dfrac{1}{x^3 - 8} \, dx$

27. $\int \dfrac{x}{x^4 + x^2 + \frac{5}{4}} \, dx$

28. $\int \dfrac{1}{(x - 1)^2(x^2 + 4)}$

29. $\int \dfrac{x^2 + 6x - 8}{x^3 + 4x} \, dx$

30. $\int \dfrac{4x^2 - 3}{(x - 2)(x^2 + 2x + 5)} \, dx$

11. Applications

Business. In many manufacturing situations, it is found that the same basic process with slight adjustments produces two different products. If Company C produces the two products *A* and *B*, then both these products compete for the resources of Company C. Thus if we assume that Company C has a fixed amount of money to produce both these products, then if it wishes to increase the amount of production of product *A*, it will have to decrease the amount of production of product *B*. The amounts which can be produced (with the fixed resources of C) will satisfy a certain relation which usually can be placed in the form of an equation.

Example 1. The United Grain Co. produces two brands of cereals. The possible amounts *x* and *y* (in millions of cases) per year of brands *A* and *B* respectively that can be produced by the company are related by

$$(80) \qquad x^2 + y^2 + 2x + 4y = 31.$$

The company can sell brand *A* for $2 per case, and brand *B* for $1.50 per case. Find *x* and *y* so as to maximize the revenue of the company.

Solution. We can write (80) as

$$(x^2 + 2x + 1) + (y^2 + 4y + 4) = 31 + 1 + 4$$

or

$$(81) \qquad (x + 1)^2 + (y + 2)^2 = 6^2.$$

It is easy to see that (81) represents a circle (see Figure 28) with center at $A(-1, -2)$ and radius 6. If $P(x, y)$ is a point on the circle (81), then from Figure 28 we note that

$$(82) \qquad \sin \theta = \frac{d(P, Q)}{d(A, P)} = \frac{y + 2}{6}$$

and

$$(83) \qquad \cos \theta = \frac{d(A, Q)}{d(A, P)} = \frac{x + 1}{6}.$$

From (82) and (83), we get

$$(84) \qquad \begin{cases} x = 6 \cos \theta - 1 \\ y = 6 \sin \theta - 2. \end{cases}$$

Since brand *A* sells for $2 per case and brand *B* sells for $\frac{3}{2}$ per case, the revenue *R* is given by

$$(85) \qquad R = 2x + \tfrac{3}{2}y.$$

Substituting from (84), we obtain

$$(86) \qquad R = 2(6 \cos \theta - 1) + \tfrac{3}{2}(6 \sin \theta - 2)$$

or

$$(87) \qquad R = 12 \cos \theta + 9 \sin \theta - 5.$$

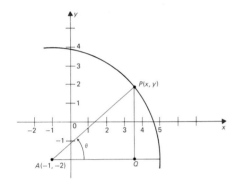

Figure 28

To maximize R, we differentiate R with respect to θ:

(88) $$R' = -12 \sin \theta + 9 \cos \theta,$$

and set $R' = 0$:

$$12 \sin \theta = 9 \cos \theta$$

or

(89) $$\tan \theta = \tfrac{3}{4}.$$

From (89), we get

$$\sec^2 \theta = 1 + \tan^2 \theta = 1 + \tfrac{9}{16} = \tfrac{25}{16}$$

or

$$\sec \theta = \tfrac{5}{4} \Rightarrow \cos \theta = \tfrac{4}{5}.$$

We also have

$$\sin \theta = \sqrt{1 - \cos^2 \theta} = \sqrt{1 - \tfrac{16}{25}} = \sqrt{\tfrac{9}{25}} = \tfrac{3}{5}.$$

If we substitute these values of $\sin \theta$ and $\cos \theta$ in

(90) $$R'' = -12 \cos \theta - 9 \sin \theta,$$

we see that R'' is negative. Therefore $\sin \theta = \tfrac{3}{5}$ and $\cos \theta = \tfrac{4}{5}$ gives the maximum revenue. Therefore

$$x = 6 \cos \theta - 1 = 6 \cdot (\tfrac{4}{5}) - 1 = \tfrac{24}{5} - 1 = \tfrac{19}{5}$$

and

$$y = 6 \sin \theta - 2 = 6 \cdot (\tfrac{3}{5}) - 2 = \tfrac{18}{5} - 2 = \tfrac{8}{5}$$

give the maximum revenue.

Physics. Recall that if a particle moves a distance $x = f(t)$ in time t, then dx/dt represents its instantaneous velocity and d^2x/dt^2 represents its instantaneous acceleration.

For a particle moving along a plane curve, we study its motion by resolving it along two perpendicular directions. For example, if the particle travels with a velocity u along a line which makes an angle α with the positive direction of the x-axis (see Figure 29), then we consider u as composed of two velocities: one along the x-axis: $u \cos \alpha$, the other along the y-axis: $u \sin \alpha$. This is exactly like the relationship between the lengths of the hypotenuse and legs of a right triangle.

Example 2. Projectile Problem. A projectile is fired from a cannon with a velocity u ft/sec making an angle θ ($0 < \theta < \pi/2$) with the horizontal. Assuming that gravity is the only force on the projectile, find the equation of the path of the projectile.

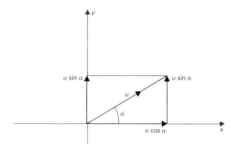

Figure 29

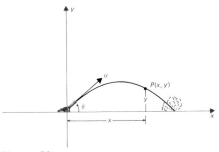

Figure 30

Acceleration due to gravity is denoted by g, and g is approximately 32 ft/sec².

In (91) we take minus g, since the accelerations $\dfrac{d^2y}{dt^2}$ and g are in opposite directions.

Solution. Let us take the cannon as the origin of the coordinate plane. After t seconds, if $P(x, y)$ is the position of the projectile, then x denotes the horizontal displacement of the projectile and y denotes its height above the cannon. As discussed earlier the initial velocities (at $t = 0$) along the x-axis and y-axis are

$$\left.\frac{dx}{dt}\right|_{t=0} = u \cos \theta$$

$$\left.\frac{dy}{dt}\right|_{t=0} = u \sin \theta.$$

Since gravity is the only force (we are ignoring air resistance, etc.) pulling on the projectile, there is no acceleration in the horizontal direction, while the acceleration in the vertical direction is g. Therefore,

(91)
$$\begin{cases} \dfrac{d^2x}{dt^2} = 0 \\ \dfrac{d^2y}{dt^2} = -g. \end{cases}$$

Integrating (91), we get

(92)
$$\begin{cases} \dfrac{dx}{dt} = c_1 \\ \dfrac{dy}{dt} = -gt + c_2. \end{cases}$$

Initially, when $t = 0$, $x = y = 0$, and $dx/dt = u \cos \theta$, $dy/dt = u \sin \theta$. Therefore,

$$u \cos \theta = c_1 \qquad u \sin \theta = c_2.$$

Thus, (92) becomes

(93)
$$\begin{cases} \dfrac{dx}{dt} = u \cos \theta \\ \dfrac{dy}{dt} = -gt + u \sin \theta. \end{cases}$$

Integrating (93), we get

(94)
$$\begin{cases} x = (u \cos \theta)t + c_3 \\ y = -\tfrac{1}{2}gt^2 + (u \sin \theta)t + c_4. \end{cases}$$

Again we apply the initial conditions that $t = 0$ when $x = 0$ and $y = 0$:

$$c_3 = 0 = c_4.$$

Thus

(95)
$$\begin{cases} x = (u \cos \theta)t \\ y = -\tfrac{1}{2}gt^2 + (u \sin \theta)t. \end{cases}$$

We eliminate t between the two equations of (95):

(96)
$$y = x \tan \theta - \frac{1}{2} \frac{gx^2}{u^2} \sec^2 \theta,$$

which represents a parabola. Thus the path of a projectile moving under gravity is a parabola.

Figure 31

Example 3. *Simple Harmonic Motion.* If a particle moves in a straight line with acceleration which is proportional to the distance of the particle from a fixed point and always directed toward that point, then the particle is executing *simple harmonic motion* (SHM). Write the differential equation of a particle in SHM and solve it.

Solution. Let us choose the origin to be the fixed point and the x-axis to be the straight line (see Figure 31). Then the differential equation of the particle in SHM is given by

(97)
$$\frac{d^2x}{dt^2} = -w^2x,$$

where w^2 is the constant of proportionality. Multiplying both sides of (97) by $2(dx/dt)$ and integrating, we obtain

(98)
$$\left(\frac{dx}{dt}\right)^{'} = -w^2x^2 + c.$$

Since the particle is undergoing negative acceleration, it has to momentarily rest at some point, say $x = a$. Then at $x = a$, $dx/dt = 0$. Consequently, from (98), we get

(99)
$$c = w^2a^2.$$

Substituting this value of c in (98), we get

(100)
$$\frac{dx}{dt} = \pm w\sqrt{a^2 - x^2}.$$

Solving the differential equation (100), we get

$$\arcsin\left(\frac{x}{a}\right) = wt + c_1$$

or

$$\arccos\left(\frac{x}{a}\right) = wt + c_2,$$

where c_1 and c_2 are the constants of integration. The solution therefore is

$$x = a \sin (wt + c_1)$$

or

$$x = a \cos (wt + c_2).$$

Recall: $\sin \left(\dfrac{\pi}{2} + \theta \right) = \cos \theta$.

If we write $c_1 = \pi/2 + c_2$, then both solutions can be written in the form

(101) $\qquad\qquad x = a \cos (wt + c_2).$

See Section 4.

From (101), we see that $2\pi/w$ is the period of the SHM. The number a is called the *amplitude*, and c_2 is called the *phase* of the SHM.

Exercise 11

1. The Slik Co. produces x million gidgets and y million gadgets a year. The amounts of x and y are related by

$$x^2 + y^2 = 169.$$

The Company makes a profit of 5¢ on each gidget and 12¢ on each gadget. Find x and y so as to maximize the company's profit.

2. In problem 1, find x and y if they are related by

$$x^2 + y^2 + 4x + 6y = 156.$$

3. The Bowry Co. produces x million gallons of Hangover wine and y million gallons of Delerium wine. The amounts x and y are related by

$$\frac{(x + 1)^2}{4} + \frac{(y - 2)^2}{9} = 1.$$

The Company sells the Hangover wine for \$4 per gallon, and the Delerium wine for \$2 per gallon. Find x and y so as to maximize the revenue of the Company.
Hint: Let $x + 1 = 2 \cos \theta$, $y - 2 = 3 \sin \theta$.

4. In problem 3, find x and y if they are related by

$$9x^2 + 4y^2 + 18x - 24y + 9 = 0.$$

5. For the projectile of Example 2, find
   (a) the maximum height
   (b) time of flight
   (c) horizontal range
   (d) θ for maximum horizontal range.

6. Show that the equation of the path of the projectile may be written as

$$y = (x \tan \theta) \left(1 - \frac{x}{R}\right),$$

where R is the horizontal range of the projectile.

7. A projectile is fired with velocity 192 ft/sec at an elevation of 30°. Find
   (a) the maximum height
   (b) time of flight
   (c) horizontal range.

8. If θ is the angle of projection of a projectile, show that the same horizontal range is obtained when the angle of projection is $\frac{\pi}{2} - \theta$.

9. An object projected with a velocity u ft/sec, inclined to the horizontal at an angle of 60°, just clears a wall 12 ft high which is $20\sqrt{3}$ ft from the point of projection. Find the velocity u of projection and the distance to the point at which it falls on the ground.

10. The horizontal range of a rifle bullet is 1728 yards when the angle of projection is 45°. If the rifle is fired at the same angle from a car, traveling 30 miles per hour (44 ft/sec) toward the target, show that the range is increased by 792 ft.

11. If the displacement x of a moving particle at time t is given by

    $$x = a \cos wt + b \sin wt, \ a \text{ and } b \text{ constants},$$

    show that the motion is SHM.

12. In Example 3, show that the speed of the particle at a distance x is given by $w\sqrt{a^2 - x^2}$. Where does the particle have maximum speed?

13. If the displacement, velocity, and acceleration at a particular instant of a particle moving with SHM are respectively 4 ft, 4 ft/sec, 4 ft/sec², find the greatest velocity of the particle and the period of motion.

14. A particle moving in SHM of period 8 sec oscillates through a distance of 4 ft on each side of its central position. Find its maximum speed. Also find the speed when the particle is 2 ft from the center.

Appendices

Appendix 1 **Mathematical Induction and the Binomial Theorem**

1. The $\sum$ Notation

One of the most important techniques for proving theorems for positive integers is mathematical induction. Let us first introduce the $\sum$ notation and its use in computation.

In this appendix, we shall prove some formulas such as

$$(1) \qquad 1 + 3 + 5 + \ldots + (2n - 1) = n^2$$

$$(2) \qquad 1 + 2 + 3 + \ldots + n = \frac{n(n + 1)}{2}$$

$$(3) \qquad 1^2 + 2^2 + 3^2 + \ldots + n^2 = \frac{n(n + 1)(2n + 1)}{6} \;;$$

each of the formulas is true for any natural number n. The three dots indicate the missing terms in the sum. We now introduce a short notation for sums.

The symbol $\sum$ is a capital sigma in the Greek alphabet and corresponds to our English S. Thus it reminds us of the word sum. The symbol

$$\sum_{i=1}^{5} x_i$$

(read "the sum of x_i as i assumes values from 1 to 5") means

$$\sum_{i=1}^{5} x_i = x_1 + x_2 + x_3 + x_4 + x_5.$$

The sum need not start at 1 and end at 5. Further, any letter can be used instead of i. The following examples indicate various possibilities.

$$\sum_{k=1}^{3} k^2 = 1^2 + 2^2 + 3^2 = 1 + 4 + 9 = 14.$$

$$\sum_{j=2}^{4} j(j + 3) = 2 \cdot 5 + 3 \cdot 6 + 4 \cdot 7 = 10 + 18 + 28 = 56.$$

$$\sum_{i=1}^{6} 1 = 1 + 1 + 1 + 1 + 1 + 1 = 6.$$

$$\sum_{i=4}^{7} x_i = x_4 + x_5 + x_6 + x_7.$$

$$\sum_{i=1}^{4} 3y_i = 3y_1 + 3y_2 + 3y_3 + 3y_4.$$

$$\sum_{n=2}^{6} na_n = 2a_2 + 3a_3 + 4a_4 + 5a_5 + 6a_6.$$

Example 1. Prove:

(a) $\displaystyle\sum_{i=1}^{4} x_i + \sum_{i=5}^{7} x_i = \sum_{i=1}^{7} x_i$

(b) $\displaystyle\sum_{i=1}^{4} (x_i + y_i) = \sum_{i=1}^{4} x_i + \sum_{i=1}^{4} y_i.$

(c) $\displaystyle\sum_{i=1}^{3} (x_i + y_i)^2 = \sum_{i=1}^{3} x_i^2 + 2\sum_{i=1}^{3} x_i y_i + \sum_{i=1}^{3} y_i^2.$

Solution. (a) $\displaystyle\sum_{i=1}^{4} x_i + \sum_{i=5}^{7} x_i$

$$= (x_1 + x_2 + x_3 + x_4) + (x_5 + x_6 + x_7)$$
$$= x_1 + x_2 + x_3 + x_4 + x_5 + x_6 + x_7$$
$$= \sum_{i=1}^{7} x_i.$$

(b) $\displaystyle\sum_{i=1}^{4} (x_i + y_i)$

$$= (x_1 + y_1) + (x_2 + y_2) + (x_3 + y_3) + (x_4 + y_4)$$
$$= (x_1 + x_2 + x_3 + x_4) + (y_1 + y_2 + y_3 + y_4)$$
$$= \sum_{i=1}^{4} x_i + \sum_{i=1}^{4} y_i.$$

(c) $\displaystyle\sum_{i=1}^{3} (x_i + y_i)^2 = (x_1 + y_1)^2 + (x_2 + y_2)^2 + (x_3 + y_3)^2$

$$= (x_1^2 + 2x_1y_1 + y_1^2) + (x_2^2 + 2x_2y_2 + y_2^2)$$
$$+ (x_3^2 + 2x_3y_3 + y_3^2)$$
$$= (x_1^2 + x_2^2 + x_3^2) + 2(x_1y_1 + x_2y_2 + x_3y_3)$$
$$+ (y_1^2 + y_2^2 + y_3^2)$$
$$= \sum_{i=1}^{3} x_i^2 + 2\sum_{i=1}^{3} x_iy_i + \sum_{i=1}^{3} y_i^2.$$

Example 2. Use the $\sum$ notation to rewrite formulas (1), (2), and (3).

Solution. With the $\sum$ notation, (1), (2), and (3) become

(1*) $$\sum_{k=1}^{n} (2k - 1) = n^2$$

(2*) $$\sum_{i=1}^{n} i = \frac{n(n + 1)}{2}$$

(3*) $$\sum_{j=1}^{n} j^2 = \frac{n(n + 1)(2n + 1)}{6}.$$

Exercise 1

In problems 1 through 5 write out each of the sums term by term.

1. $\displaystyle\sum_{k=1}^{5} k^2$ 2. $\displaystyle\sum_{i=3}^{8} (x_i + y_i)$ 3. $\displaystyle\sum_{k=1}^{5} 7x^k$

4. $\displaystyle\sum_{i=1}^{9} 6$ 5. $\displaystyle\sum_{i=2}^{4} (x_i + 1)^2$

In problems 6 to 10 write each expression in sigma notation.

6. $1 + 2 + 3 + \ldots + 50$

7. $1 \cdot 2 + 2 \cdot 3 + 3 \cdot 4 + \ldots + 16 \cdot 17$

8. $2 + 4 + 8 + 16 + \ldots + 2^n$

9. $x_1^2 + x_2^2 + x_3^2 + \ldots + x_{50}^2$

10. $\dfrac{x_1}{1} + \dfrac{x_2}{2} + \dfrac{x_3}{3} + \ldots + \dfrac{x_n}{n}$

2. Mathematical Induction

Suppose we wish to find the sum of the first n odd positive integers.

Direct computation shows that

$$
\begin{aligned}
&\text{if } n = 1, \ 1 = 1 && = 1^2 \\
&\text{if } n = 2, \ 1 + 3 = 4 && = 2^2 \\
&\text{if } n = 3, \ 1 + 3 + 5 = 9 && = 3^2 \\
&\text{if } n = 4, \ 1 + 3 + 5 + 7 = 16 && = 4^2 \\
&\text{if } n = 5, \ 1 + 3 + 5 + 7 + 9 = 25 && = 5^2 \\
&\text{if } n = 6, \ 1 + 3 + 5 + 7 + 9 + 11 = 36 = 6^2.
\end{aligned}
$$

These cases lead us to believe that the sum is always equal to the square of the number of terms in the sum. We observe that 1 is the first term in the sum, and it is easy to see that the nth term of the sum is $2n - 1$. Thus, we can express our assertion symbolically as

(4) $$1 + 3 + 5 + \ldots + (2n - 1) = n^2,$$

or using the $\sum$ notation, we claim that

(5) $$\sum_{k=1}^{n} (2k - 1) = n^2.$$

By direct computation, we have seen that equation (4) (or (5)) is true for $n = 1, 2, 3, 4, 5$, and 6. Does this mean that equation (4) is true for any positive integer n? Can we settle this by continuing our numerical computation? Certainly not! No matter how many cases we check, we can never prove that (4) is always true, because there are infinitely many cases and no amount of computation can settle them all. Thus, we need some *logical* argument that will prove that (4) is always true.

We use the symbol $p(n)$ (read "p of n") to denote some proposition which depends on the positive integer n. For example, $p(n)$ might denote the proposition that (4) is true. The proof that $p(n)$ is true for all positive integers n can be given in two steps and these two steps form the

Principle of Mathematical Induction. If for a given proposition $p(n)$ we can prove that

1. the proposition is true for $n = 1$,
2. if $p(n)$ is true for $n = k$, then it is also true for $n = k + 1$,

then the proposition is true for all positive integers n.

Note that the principle of mathematical induction can be used to prove a general proposition for positive integers only. Also the result to be proved should be surmised beforehand.

Example 1. Prove that equation (4) holds for every positive integer n.

Solution. 1. We have already seen that equation (4) is true for $n = 1$.

  2. We now assume that $p(n)$ is true for $n = k$:

(6) $$1 + 3 + 5 + \ldots + (2k - 1) = k^2.$$

To obtain the sum of the first $k + 1$ odd integers we merely add the next odd one, $2k + 1$, to both sides of (6):

(7) $$(1 + 3 + 5 + \ldots + (2k - 1)) + (2k + 1)$$
$$= k^2 + (2k + 1)$$
$$= (k + 1)^2.$$

Equation (7) is precisely equation (4) when $n = k + 1$, hence we have shown that if $p(n)$ is true for $n = k$, then it is also true for $n = k + 1$.

By the principle of mathematical induction this completes the proof that equation (4) is true for every positive integer n.

Example 2. Prove that for every positive integer n,

(8) $$1^2 + 2^2 + 3^3 + \ldots + n^2 = \frac{n(n + 1)(2n + 1)}{6}.$$

Solution. 1. For $n = 1$, the assertion of equation (8) is

$$1^2 = \frac{1(1 + 1)(2 + 1)}{6}$$
$$1 = 1,$$

which is certainly true.

  2. We assume that (8) is true for $n = k$:

(9) $$1^2 + 2^2 + 3^2 + \ldots + k^2 = \frac{k(k + 1)(2k + 1)}{6}.$$

To obtain the left-hand side of (8) for $n = k + 1$, we must add $(k + 1)^2$ to the left-hand side of (9). This gives

(10) $$1^2 + 2^2 + 3^2 + \ldots + k^2 + (k + 1)^2$$
$$= \frac{k(k + 1)(2k + 1)}{6} + (k + 1)^2$$
$$= \tfrac{1}{6}[k(k + 1)(2k + 1) + 6(k + 1)^2]$$
$$= \tfrac{1}{6}(k + 1)[k(2k + 1) + 6(k + 1)]$$
$$= \tfrac{1}{6}(k + 1)(2k^2 + 7k + 6)$$
$$= \tfrac{1}{6}(k + 1)(k + 2)(2k + 3)$$
$$= \tfrac{1}{6}(k + 1)((k + 1) + 1)(2(k + 1) + 1).$$

This is equation (8) when $n = k + 1$. Hence by the principle of mathematical induction, formula (8) is true for every positive integer n.

Exercise 2

In problems 1 through 9, prove that the given assertion is true for all positive integers n.

1. The nth positive even integer is n.

2. $1 + 2 + 3 + \ldots + n = \dfrac{n(n + 1)}{2}$

3. $1 \cdot 2 + 2 \cdot 3 + 3 \cdot 4 + \ldots + n(n + 1) = \dfrac{n(n + 1)(n + 2)}{3}$

4. $1 + 2^1 + 2^2 + 2^3 + \ldots + 2^n = 2^{n+1} - 1$

5. $1^3 + 2^3 + 3^3 + \ldots + n^3 = \dfrac{n^2(n + 1)^2}{4}$

6. $\dfrac{1}{1 \cdot 2} + \dfrac{1}{2 \cdot 3} + \dfrac{1}{3 \cdot 4} + \ldots + \dfrac{1}{n(n + 1)} = \dfrac{n}{n + 1}$

7. $1 \cdot 3 + 3 \cdot 5 + 5 \cdot 7 + \ldots + (2n - 1)(2n + 1)$
$$= \tfrac{1}{3}n(2n^2 + 6n + 1)$$

8. $\dfrac{1}{1 \cdot 3} + \dfrac{1}{3 \cdot 5} + \dfrac{1}{5 \cdot 7} + \ldots + \dfrac{1}{(2n - 1)(2n + 1)} = \dfrac{n}{2n + 1}$

9. $\dfrac{1}{2 \cdot 5} + \dfrac{1}{5 \cdot 8} + \dfrac{1}{8 \cdot 11} + \ldots + \dfrac{1}{(3n - 1)(3n + 2)} = \dfrac{n}{6n + 4}$

3. Binomial Theorem

Before we state and prove the binomial theorem, let us define some new notations.

A. THE FACTORIAL NOTATION

We define the symbol $n!$ (read "n factorial") where n is a nonnegative integer as

(11)
$$\begin{cases} 0! = 1 \\ (n + 1)! = (n + 1) \cdot n!. \end{cases}$$

For example, we have

$$0! = 1$$
$$1! = 1 \cdot 0! = 1 \cdot 1 = 1$$
$$2! = 2 \cdot 1! = 2 \cdot 1 = 2$$
$$3! = 3 \cdot 2! = 3 \cdot 2 = 6$$
$$4! = 4 \cdot 3! = 4 \cdot 6 = 24.$$

Thus for any positive integer n, $n!$ can be computed as the product of all positive integers less than or equal to n. For example, $5! = 5 \cdot 4 \cdot 3 \cdot 2 \cdot 1 = 120$.

B. THE SYMBOL $\binom{n}{r}$

Let n be a nonnegative integer and r a nonnegative integer less than or equal to n. The symbol $\binom{n}{r}$ is defined by

(12)
$$\binom{n}{r} = \frac{n!}{(n-r)!r!}.$$

By (12), we observe that

(13)
$$\binom{n}{0} = \frac{n!}{(n-0)!0!} = \frac{n!}{n! \cdot 1} = 1$$

$$\binom{0}{0} = \frac{0!}{(0-0)!0!} = 1.$$

The following Theorem shall be used in the proof of the Binomial Theorem.

Theorem 1. If k and r are integers such that $0 \leq r \leq k$, then

(14)
$$\binom{k}{r} + \binom{k}{r-1} = \binom{k+1}{r}.$$

Proof. $\binom{k}{r} + \binom{k}{r-1}$

$$= \frac{k!}{(k-r)!r!} + \frac{k!}{(k-r+1)!(r-1)!}$$

$$= \frac{k!(k-r+1)}{r!(k-r)!(k-r+1)} + \frac{k!r}{(k-r+1)!(r-1)!r}$$

$$= \frac{k!(k-r+1)}{r!(k-r+1)!} + \frac{k!r}{(k-r+1)!r!}$$

$$= \frac{k!(k+1)}{r!(k+1-r)!}$$

$$= \frac{(k+1)!}{(k+1-r)!r!}$$

$$= \binom{k+1}{r}.$$

C. THE BINOMIAL THEOREM

An expression which contains only two terms connected by a positive or a negative sign is called a *binomial* expression. For example, $a + b, 2x - 7, a + (1/a), x + 8y^2$ are all binomial expressions. If n is a positive integer, then $(x + a)^n$ is a power of the binomial expression $x + a$. We are familiar with the expansions of $(x + a)^2$, $(x + a)^3$, and $(x + a)^4$. These expansions can be written using the symbol $\binom{n}{r}$:

(15) $\qquad (x + a)^2 = x^2 + 2xa + a^2$

$$= \binom{2}{0} x^2 + \binom{2}{1} xa + \binom{2}{2} a^2,$$

since $\binom{2}{0} = 1$, $\binom{2}{1} = 2$, and $\binom{2}{2} = 1$.

(16) $\quad (x + a)^3 = x^3 + 3x^2a + 3xa^2 + a^3$

$$= \binom{3}{0} x^3 + \binom{3}{1} x^2a + \binom{3}{2} xa^2 + \binom{3}{3} a^3,$$

since $\binom{3}{0} = 1$, $\binom{3}{1} = 3$, $\binom{3}{2} = 3$, and $\binom{3}{3} = 1$.

(17) $\quad (x + a)^4$
$$= x^4 + 4x^3a + 6x^2a^2 + 4xa^3 + a^4$$

$$= \binom{4}{0} x^4 + \binom{4}{1} x^3a + \binom{4}{2} x^2a^2 + \binom{4}{3} xa^3 + \binom{4}{4} a^4,$$

since $\binom{4}{0} = 1$, $\binom{4}{1} = 4$, $\binom{4}{2} = 6$, $\binom{4}{3} = 4$, and $\binom{4}{4} = 1$.
From (15), (16), and (17) we make the following observations:

(i)　In successive terms the exponent of x decreases by 1 and the exponent of a increases by 1.

(ii)　The sum of the exponents of x and a remains the same in all the terms of the expansion.

(iii)　The coefficients of the successive terms follow a certain rule.

(iv)　The number of terms is one more than the exponent of $x + a$.

These facts lead us to state

Theorem 2. **The Binomial Theorem. For any positive integer** n,

$$(18) \quad (x + a)^n = \binom{n}{0} x^n + \binom{n}{1} x^{n-1}a + \binom{n}{2} x^{n-2}a^2$$

$$+ \dots + \binom{n}{r} x^{n-r}a^r + \dots + \binom{n}{n} a^n$$

$$= \sum_{r=0}^{n} \binom{n}{r} x^{n-r}a^r.$$

Proof. We use the principle of mathematical induction.
1. For $n = 1$, we have

$$(x + a)^1 = \binom{1}{0} x^1 + \binom{1}{1} a^1$$

$$= x + a.$$

which is certainly true.
2. We now assume that the theorem is true for $n = k$. Thus, we assume that

$$(19) \qquad (x + a)^k = \sum_{r=0}^{k} \binom{k}{r} x^{k-r}a^r.$$

Using equation (19) and the definition of $(x + a)^{k+1}$, we have

$$(x + a)^{k+1} = (x + a)(x + a)^k = (x + a) \sum_{r=0}^{k} \binom{k}{r} x^{k-r}a^r$$

$$= x \sum_{r=0}^{k} \binom{k}{r} x^{k-r}a^r + a \sum_{r=0}^{k} \binom{k}{r} x^{k-r}a^r.$$

Thus,

$$(20) \quad (x + a)^{k+1} = \sum_{r=0}^{k} \binom{k}{r} x^{k-r+1}a^r + \sum_{r=0}^{k} \binom{k}{r} x^{k-r}a^{r+1}$$

Collecting like terms on the right-hand side of (20), we find that the coefficient of $x^{k-r}a^{r+1}$ is

$$(21) \qquad \binom{k}{r + 1} + \binom{k}{r}.$$

By Theorem 1, this new coefficient is

$$(22) \qquad \binom{k}{r + 1} + \binom{k}{r} = \binom{k + 1}{r + 1}.$$

But the right-hand side of (22) has exactly the form required for the coefficient of $x^{k-r}a^{r+1}$ in the expansion of $(x + a)^{k+1}$. This argument holds if $r = 0, 1, 2, \ldots, k - 1$. The extreme terms, x^{k+1} and a^{k+1}, are not covered by the preceding discussion, but these terms obviously have 1 as coefficients, as required by Theorem 2.

Example 1. Expand $(x - 3y^2)^4$.

Solution. By Theorem 2, the coefficients are

$$\binom{4}{0} = \frac{4!}{4!0!} = 1, \quad \binom{4}{1} = \frac{4!}{3!1!} = 4, \quad \binom{4}{2} = \frac{4!}{2!2!} = 6,$$

$$\binom{4}{3} = \frac{4!}{1!3!} = 4, \quad \binom{4}{4} = \frac{4!}{0!4!} = 1.$$

We can write $(x - 3y^2)^4 = [x + (-3y^2)]^4$. Hence

$$
\begin{aligned}
(x - 3y^2)^4 &= x^4 + 4x^3(-3y^2) + 6x^2(-3y^2)^2 + 4x(-3y^2)^3 \\
&\quad + (-3y^2)^4 \\
&= x^4 - 12x^3y^2 + 54x^2y^4 - 108xy^6 + 81y^8.
\end{aligned}
$$

Example 2. Find the 6th term in the expansion of $(x^2 - 2y)^{11}$.

Solution. The $(r + 1)$th term in the expansion of $(x + a)^n$ by Theorem 2 is $\binom{n}{r}x^{n-r}a^r$. Thus the 6th term in the expansion of $(x^2 - 2y)^{11}$ is

$$
\begin{aligned}
\binom{11}{5}(x^2)^{11-5}(-2y)^5 &= \frac{11!}{6!5!}(x^2)^6(-32y^5) \\
&= \frac{11 \cdot 10 \cdot 9 \cdot 8 \cdot 7 \cdot 6!}{6! \cdot 5 \cdot 4 \cdot 3 \cdot 2 \cdot 1}x^{12}(-32y^5) \\
&= -32 \cdot 462x^{12}y^5 \\
&= -14784x^{12}y^5.
\end{aligned}
$$

Example 3. Find the term independent of x in the expansion of $(\sqrt{x} - 2/x^2)^{10}$.

Solution. The term independent of x means the term not containing x, and therefore the exponent of x must be zero.

The $(r + 1)$th term in the expansion of $(x^{1/2} - 2/x^2)^{10}$ is

$$
\begin{aligned}
\binom{10}{r}(x^{1/2})^{10-r}\left(-\frac{2}{x^2}\right)^r &= \binom{10}{r}x^{5-(r/2)}(-2)^r(x^{-2r}) \\
&= \binom{10}{r} \cdot (-2)^r \cdot x^{5-(5r/2)}.
\end{aligned}
$$

Since the term independent of x is desired, we must have

$$5 - \frac{5r}{2} = 0,$$

or

$$r = 2.$$

Therefore, the required term is

$$\binom{10}{2}(-2)^2 x^0 = \frac{10!}{8!\,2!} \cdot 4$$
$$= 45 \cdot 4$$
$$= 180.$$

Exercise 3

In problems 1 through 8, compute explicitly the given quantity.

1. $7!$

2. $10!$

3. $\binom{8}{3}$

4. $\binom{11}{7}$

5. $\binom{87}{86}\binom{2}{0}$

6. $\dfrac{16}{\binom{6}{4}}$

7. $\left(\dbinom{8}{\binom{4}{2}}\right)$

8. $\left(\dbinom{\binom{8}{4}}{2}\right)$

9. Show that $\binom{n}{r} = \binom{n}{n-r}$.

*10. Show that for $0 \le k \le n - 2$,

$$\binom{n+2}{k+2} = \binom{n}{k+2} + 2\binom{n}{k+1} + \binom{n}{k}.$$

Hint: Use (14) twice.

*11. Show that

$$\binom{n}{0} + \binom{n}{1} + \binom{n}{2} + \ldots + \binom{n}{n} = 2^n.$$

Hint: Let $x = a = 1$ in (18).

12. Expand the following.

(a) $(1 + x)^6$

(b) $(2x + 1)^5$

(c) $(x - 2)^5$

(d) $\left(x + \dfrac{1}{x}\right)^6$

13. Write the term indicated in the expansion of each of the following.

    (a) fourth term of $(x - 2y)^7$

    (b) seventh term of $\left(x^2 - \dfrac{1}{x} \right)^9$

    (c) ninth term of $(x - ay^2)^{12}$

    (d) fifteenth term of $\left(2 - \dfrac{1}{x} \right)^{19}$

14. Find the term independent of x in the following expressions.

    (a) $\left(2x^3 + \dfrac{1}{4x^8} \right)^{11}$ (b) $\left(2x - \dfrac{1}{x^2} \right)^6$

    (c) $\left(\sqrt{x} - \dfrac{2}{x^2} \right)^{10}$ (d) $\left(x^2 - \dfrac{1}{x} \right)^9$

Appendix 2   Complex Numbers

1. Complex Numbers

The equation $x^2 + 1 = 0$ has no real solution. In this appendix we are interested in defining new numbers which provide solutions to such equations. We shall use C to denote the set of objects called *complex numbers* and define

$$C = R \times R.$$

Thus, a complex number z is an ordered pair of real numbers (x, y) and we write

$$z = (x, y).$$

The first component x is called the *real part* of z ($x = \text{Re } z$), and the second component y is called the *imaginary part* of z ($y = \text{Im } z$).

    If $z_1 = (x_1, y_1)$ and $z_2 = (x_2, y_2)$ are two complex numbers, we define z_1 to be equal to z_2 ($z_1 = z_2$) if $x_1 = x_2$ and $y_1 = y_2$. The operations of addition and multiplication of complex numbers are defined as follows:

(1) $z_1 + z_2 = (x_1, y_1) + (x_2, y_2) = (x_1 + x_2, y_1 + y_2)$

(2) $z_1 \cdot z_2 = (x_1, y_1) \cdot (x_2, y_2) = (x_1 x_2 - y_1 y_2, x_1 y_2 + x_2 y_1)$

(3)   If $z_1 = (x_1, y_1)$ is a complex number and α is a real number, then $\alpha z_1 = \alpha(x_1, y_1) = (\alpha x_1, \alpha y_1)$.

It is clear from these definitions that the commutative, associative, and distributive laws hold. The *zero* complex number, denoted by 0, is defined by

$$0 = (0, 0),$$

and the *unit* complex number is the number $(1, 0)$. We see that

(4) $\qquad (0, 0) + (x, y) = (0 + x, 0 + y) = (x, y)$

(5) $\quad (1, 0) \cdot (x, y) = (1 \cdot x - 0 \cdot y, 1 \cdot y + x \cdot 0) = (x, y).$

There is a one-to-one correspondence between the real numbers x and the complex numbers $z = (x, 0)$. We find that if $(x_1, 0)$ and $(x_2, 0)$ are two complex numbers, then

(6) $\qquad\qquad (x_1, 0) + (x_2, 0) = (x_1 + x_2, 0)$

and

(7) $\qquad\qquad (x_1, 0) \cdot (x_2, 0) = (x_1 x_2, 0).$

Hence we identify the real number x with the complex number $(x, 0)$ and write

(8) $\qquad\qquad\qquad x = (x, 0).$

It is customary to denote $(0, 1)$ by i. We find that

(9) $\qquad\quad i^2 = (0, 1) \cdot (0, 1) = (-1, 0) = -1.$

Note that the complex number $z = (x, y)$ can be written in the form

(10) $\qquad\qquad z = (x, y) = (x, 0) + (0, y),$

and since

$$(0, y) = y(0, 1) = yi,$$

we have

(11) $\qquad\qquad z = (x, y) = x + yi.$

The advantage of the notation in (11) is that it simplifies finding the product of two complex numbers. Thus,

$$
\begin{aligned}
(2, 1) \cdot (-1, 5) &= (2 + i) \cdot (-1 + 5i) \\
&= -2 + 10i - i + 5i^2 \\
&= -2 + 9i - 5 \\
&= -7 + 9i \\
&= (-7, 9).
\end{aligned}
$$

If $z = (x, y) \neq (0, 0)$, there is a unique complex number called the *reciprocal* of z, denoted by z^{-1} (or $1/z$), defined by

(12) $\qquad\qquad z \cdot z^{-1} = 1 \quad (= (1, 0)).$

Let $z^{-1} = (u, v)$, then

$$(x, y) \cdot (u, v) = 1$$

or

(13) $$(xu - yv, xv + yu) = (1, 0).$$

Equating like components in (13) we obtain

$$xu - yv = 1$$
$$xv + yu = 0.$$

These equations have the solution

$$u = \frac{x}{x^2 + y^2}, \quad v = \frac{-y}{x^2 + y^2}, \quad (x^2 + y^2 > 0).$$

Thus

$$z^{-1} = \left(\frac{x}{x^2 + y^2}, \frac{-y}{x^2 + y^2} \right)$$

provided $z \neq 0$. This leads to the definition of the quotient z_1/z_2 where $z_2 \neq 0$:

$$z_1/z_2 = z_1 \cdot z_2^{-1} \quad \text{if } z_2 \neq 0.$$

Example 1. Evaluate the following.

(a) $(1, 3)/(2, -2)$

(b) $\dfrac{(2, -3) \cdot (0, 1)}{(1, 2)}$

Solution.

(a) $(1, 3)/(2, -2) = (1, 3) \cdot (2, -2)^{-1} = (1 + 3i) \cdot (\frac{1}{4} + \frac{1}{4}i)$
$= (-\frac{1}{2} + i) = (-\frac{1}{2}, 1).$

(b) $\dfrac{(2, -3) \cdot (0, 1)}{(1, 2)} = (2, -3) \cdot (0, 1) \cdot (1, 2)^{-1}$

$= (2 - 3i \cdot i \cdot (\frac{1}{5} - \frac{2}{5}i)$
$= (2i + 3) \cdot (\frac{1}{5} - \frac{2}{5}i)$
$= \frac{2}{5}i - \frac{4}{5}i^2 + \frac{3}{5} - \frac{6}{5}i$
$= \frac{7}{5} - \frac{4}{5}i$
$= (\frac{7}{5}, -\frac{4}{5}).$

The (x, y)-plane with the x-axis as the *real* axis and the y-axis as the *imaginary axis* is called the *complex plane*. Clearly, the complex numbers have a one-to-one correspondence with points in the complex plane.

If $z = 3 + 2i$, its mirror image in the x-axis is the point $3 - 2i$ (see Figure 1). This number is called the *complex conjugate* of z. In general, if $z = x + iy$, the complex conjugate, denoted by $\bar{z}$, is given by

$$\bar{z} = x - iy.$$

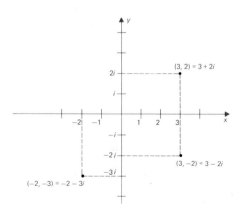

(3, 2) = 3 + 2i

(3, −2) = 3 − 2i

(−2, −3) = −2 − 3i

Figure 1

It is easy to verify that

(14) $$\overline{z_1 + z_2} = \overline{z_2} + \overline{z_1}$$

(15) $$\overline{z_1 z_2} = \overline{z_1}\, \overline{z_2}$$

(16) $$\overline{\left(\frac{z_1}{z_2}\right)} = \frac{\overline{z_1}}{\overline{z_2}}, \qquad z_2 \neq 0$$

(17) $$\overline{\overline{z}} = z.$$

The absolute value (*magnitude* or *modulus*) of the number $z = (x, y) = x + iy$, denoted by $|z|$, is given by

$$|z| = \sqrt{x^2 + y^2}.$$

Example 2. Find the absolute values of $(1, 1)$, $i/4$, and $3 - 4i$.

Solution.
$$|(1, 1)| = \sqrt{1^2 + 1^2} = \sqrt{2}$$
$$|i/4| = \sqrt{0^2 + (\tfrac{1}{4})^2} = \tfrac{1}{4}$$
$$|3 - 4i| = \sqrt{3^2 + (-4)^2} = \sqrt{25} = 5$$

It is noteworthy that the absolute value of a complex number obeys the same rules as the absolute value of a real number:

(18) $$|z| \geq 0$$

(19) $$|z_1 z_2| = |z_1|\, |z_2|$$

(20) $$|z_1 + z_2| \leq |z_1| + |z_2| \quad \text{(Triangle inequality)}$$

(21) $$|z_1| - |z_2| \leq |z_1 - z_2|.$$

Exercise 1

1. Write each of the following complex numbers in the form (x, y).
   (a) $3 - 2i$
   (b) $2(-2 - i)$
   (c) $(2 - 3i) + (-2 + 7i)$
   (d) $(-\sqrt{2}i + 5)/2i$
   (e) $3i - \dfrac{1}{i}$
   (f) $\dfrac{1}{2 - i}$
   (g) $(1 + 5i)/(3 + 4i)$
   (h) $3i/(i + 1)$

2. Write each of the following complex numbers in the form $x + iy$.
   (a) $(-1, -3)$
   (b) $-5(0, -\tfrac{1}{2})$
   (c) $(3, 1) - (2, -1)$
   (d) $(2, -2)^2$
   (e) $(0, 2)/(1, 1)$
   (f) $(2, -5)/(2, -3)$

3. Simplify.
   (a) i^3 (b) $(-i)^4$ (c) i^5 (d) i^7

4. Establish properties (14)–(17).

5. Establish properties (18)–(21).
6. Show that $z = i$ satisfies the equation $z^2 + 1 = 0$.
7. Show that $z = -1 - i$ satisfies the equation $z^2 + 2z + 2 = 0$.
8. Show that if z satisfies the equation $z^2 + az + b = 0$, where a and b are real numbers, then $\bar{z}$ also satisfies the equation.

In problems 9 through 12 find all complex numbers z which satisfy the given equation.

9. $z^2 + 4 = 0$
10. $z^2 + 2z + 2 = 0$
11. $z^2 + 2z + 4 = 0$
12. $3z^2 - z + 1 = 0$

Appendix 3   The Polynomial Functions

1. Polynomials

In this appendix we deal with a class of functions called polynomials. These are fairly simple functions and are significant for their use in approximating other more complicated functions.

Let n be a nonnegative integer and let $a_0, a_1, \ldots, a_n$ be real numbers with $a_n \neq 0$, then the function p defined by

$$(1) \qquad p(x) = a_n x^n + a_{n-1} x^{n-1} + \ldots + a_1 x + a_0$$

is called a *polynomial function of degree n* (or simply a *polynomial*). The numbers $a_0, a_1, \ldots, a_n$ are called the *coefficients* and a_n the *leading coefficient* of p.

The student has already encountered some special polynomials: the constant function, $p(x) = a_0$; the linear function, $p(x) = a_1 x + a_0$; and the quadratic function, $p(x) = a_2 x^2 + a_1 x + a_0$. These are polynomials of degree zero, one, and two respectively. No degree is assigned to the zero function, $p(x) \equiv 0$.

Polynomials may be added, subtracted, multiplied, or divided. Thus, if p and q are polynomials of degree m and n respectively, then

(i) $p \pm q$ is a polynomial of degree less than or equal to the maximum of m and n.
(ii) $p \cdot q$ is a polynomial of degree $m + n$.

Example 1. Let $p(x) = x^3 - 3x + 8$ and $q(x) = x^3 + 5x^2 - x + 1$. Then

(i) $\begin{aligned}[t] p(x) + q(x) &= (x^3 - 3x + 8) + (x^3 + 5x^2 - x + 1) \\ &= 2x^3 + 5x^2 - 4x + 9 \end{aligned}$

a polynomial of degree 3

(ii) $\begin{aligned}[t] p(x) - q(x) &= (x^3 - 3x + 8) - (x^3 + 5x^2 - x + 1) \\ &= -5x^2 - 2x + 7 \end{aligned}$

a polynomial of degree 2

(iii) $\begin{aligned}[t] p(x)q(x) &= (x^3 - 3x + 8)(x^3 + 5x^2 - x + 1) \\ &= x^3(x^3 + 5x^2 - x + 1) \\ &\quad - 3x(x^3 + 5x^2 - x + 1) \\ &\quad + 8(x^3 + 5x^2 - x + 1) \\ &= x^6 + 5x^5 - 4x^4 - 6x^3 + 43x^2 - 11x + 8 \end{aligned}$

a polynomial of degree 6

The concept of division involving polynomials is quite similar to that of integers. Thus if p and h are polynomials, then p is *divisible* by h if and only if there is a polynomial q such that

$$p = q \cdot h.$$

Example 2. Let $p(x) = x^3 - 1$ and $h(x) = x - 1$. Find a polynomial q such that $p = q \cdot h$.

Solution. Since

$$x^3 - 1 = (x - 1)(x^2 + x + 1),$$

for

$$q(x) = x^2 + x + 1$$

we have

$$p(x) = q(x)h(x).$$

Analogous to the division algorithm for integers is the division algorithm for polynomials.

Theorem 1. If p and h are polynomials and h is of degree greater than zero, then there exist unique polynomials q and r such that

(2) $p(x) = q(x) \cdot h(x) + r(x)$

where r is either a polynomial of degree less than the degree of h or the zero function.

The polynomial p is called the *dividend*, h the *divisor*, q the *quotient*, and r the *remainder*.

We illustrate this theorem in

Example 3. Let $p(x) = x^3 + 5x - 2$ and $h(x) = x + 1$. Find polynomials q and r such that

$$p(x) = q(x) \cdot h(x) + r(x).$$

Solution. The procedure is shown in the following sequence of steps (long division):

$$
\begin{array}{r}
x^2 - x + 6 \\
x + 1 \overline{\smash{)}\, x^3 + 0 \cdot x^2 + 5x - 2} \\
\underline{x^3 + x^2 } \\
-x^2 + 5x \\
\underline{-x^2 - x} \\
6x - 2 \\
\underline{6x + 6} \\
-8
\end{array}
$$

Thus, if we let $q(x) = x^2 - x + 6$ and $r(x) = -8$, then

$$x^3 + 5x - 2 = (x + 1)(x^2 - x + 6) - 8.$$

The student should observe how the term $0 \cdot x^2$ is used to fill in the missing term.

Theorem 2. **The Remainder Theorem.*** **Let p denote a polynomial of degree $n \leq 1$ and c a real number. Then there exists a polynomial q of degree $n - 1$ such that**

(3) $$p(x) = (x - c)q(x) + p(c)$$

for all numbers x.

Proof. We use Theorem 1 for polynomials $p(x)$ and $x - c$:

$$p(x) = q(x) \cdot (x - c) + r(x),$$

where $r(x)$ has degree less than the degree of $(x - c)$, which means that $r(x)$ has degree zero: $r(x) = r$, a constant. We evaluate p at c:

$$
\begin{aligned}
p(c) &= q(c) \cdot (c - c) + r \\
&= 0 + r \\
&= r.
\end{aligned}
$$

---

* Comparing the result of this theorem with division involving integers, it should become apparent to the student where the name "Remainder Theorem" originated.

Thus $p(x) = q(x) \cdot (x - c) + p(c)$, and it is clear that $q(x)$ must have degree $n - 1$. To find q, we let p and q be defined by

$$p(x) = a_n x^n + a_{n-1} x^{n-1} + \ldots + a_1 x + a_0$$
$$q(x) = b_{n-1} x^{n-1} + b_{n-2} x^{n-2} + \ldots + b_1 x + b_0.$$

Here $a_n, a_{n-1}, \ldots, a_0$ are known constants and we wish to find $b_{n-1}, b_{n-2}, \ldots, b_0$ and $p(c)$ such that equation (3) is satisfied. We observe that

(4) $$(x - c)q(x) = b_{n-1} x^n + (b_{n-2} - cb_{n-1}) x^{n-1} + \ldots$$
$$+ (b_0 - cb_1) x + (-cb_0).$$

Hence

(5) $$a_n x^n + a_{n-1} x^{n-1} + \ldots + a_1 x + a_0$$
$$= b_{n-1} x^n + (b_{n-2} - cb_{n-1}) x^{n-1} + \ldots$$
$$+ (b_0 - cb_1) x + (p(c) - cb_0).$$

We equate coefficients to find those for q:

$$b_{n-1} = a_n$$
$$b_{n-2} = a_{n-1} + cb_{n-1}$$
$$\vdots$$
$$b_0 = a_1 + cb_1$$
$$p(c) = a_0 + cb_0.$$

Thus for a given polynomial p, the coefficients of the polynomial q and $p(c)$ can be computed successively, and equation (3) can be obtained.

The student should observe carefully the pattern involved in determining the coefficients of q and $p(c)$. This provides us with a convenient method for dividing a polynomial p by the special polynomial $x - c$ (called the *monic polynomial*). The method is called *synthetic division*. For convenience, we arrange the work as follows:

c	a_n	a_{n-1}	a_{n-2}	$\ldots$	a_1	a_0
		$+$	$+$		$+$	$+$
		cb_{n-1}	cb_{n-2}	$\ldots$	$c \cdot b_1$	cb_0
				$\ldots$		
	b_{n-1}	b_{n-2}	b_{n-3}		b_0	$p(c)$

This gives an alternate method to long division involving polynomials when the divisor is a monic polynomial. We illustrate this procedure in

Example 4. Use synthetic division to divide $p(x) = 3x^4 - 8x^2 - 3x + 7$ by $x - 2$.

Solution.

$$
\begin{array}{r|rrrrr}
2 & 3 & 0 & -8 & -3 & 7 \\
  &   & 2\cdot3 & 2\cdot6 & 2\cdot4 & 2\cdot5 \\
\hline
  & 3 & 6 & 4 & 5 & \boxed{17}
\end{array}
$$

Hence $q(x) = 3x^3 + 6x^2 + 4x + 5$ and $p(2) = 17$; that is,

$$3x^4 - 8x^2 - 3x + 7 = (x - 2)(3x^3 + 6x^2 + 4x + 5) + 17.$$

We observe that

$$p(2) = 3(2)^4 - 8(2)^2 - 3(2) + 7 = 17.$$

The student will do well to work this problem using long division. If $p(c) = 0$, then from Theorem 2 we have

Theorem 3. **The Factor Theorem. Let p denote a polynomial and c a real number. Then $x - c$ is a factor of p (or p is divisible by $x - c$) if and only if $p(c) = 0$.**

Example 5. Show that $x + 1$ is a factor of

$$p(x) = x^3 - 6x^2 - 6x + 1.$$

Solution. By inspection we find that $p(-1) = 0$, hence $x + 1$ is a factor of p. Using synthetic division, we find

$$
\begin{array}{r|rrrr}
-1 & 1 & -6 & -6 & 1 \\
   &   & -1 & 7 & -1 \\
\hline
   & 1 & -7 & 1 & \boxed{0}
\end{array}
$$

Therefore

$$p(x) = (x + 1)(x^2 - 7x + 1).$$

Exercise 1

In problems 1 through 10, use long division to divide p by h and determine a quotient q such that the remainder r is a real number.

1. $p(x) = 4x^3 + 2x^2 + x + 3, \quad h(x) = 2x + 3$
2. $p(x) = 4x^3 - 5x^2 - 4x + 1, \quad h(x) = 3x - 2$
3. $p(x) = 6x^3 + x^2 + x + 4, \quad h(x) = 3x + 1$

4. $p(x) = 9x^3 - x^2 - 13x + 6$, $h(x) = x + 5$
5. $p(x) = x^3 - 75x^2 - 5x + 1$, $h(x) = x - 1$
6. $p(x) = 3x^4 - 6x^2 + 3x - 7$, $h(x) = x - 2$
7. $p(x) = -x^5 + 4x^3 + 2$, $h(x) = x + 2$
8. $p(x) = x^5 - 1$, $h(x) = x - 1$
9. $p(x) = x^6 + 2x^4 + x^3 + 5$, $h(x) = x + 1$
10. $p(x) = x^6 + 3x^2 + 10$, $h(x) = 2x + 3$

In problems 11 through 15 use synthetic division to divide p by h and find $p(c)$.

11. $p(x) = x^3 + 3x^2 - x + 12$, $h(x) = x + 2$
12. $p(x) = 2x^3 + 4x^2 + x - 15$, $h(x) = x - 3$
13. $p(x) = x^4 + 2x^2 - x - 7$, $h(x) = x + 1$
14. $p(x) = 2x^4 - 2x^3 + 9$, $h(x) = x - 1$
15. $p(x) = 3x^4 + 2x^3 + x + 1$, $h(x) = x - 2$
16. If $p(x) = 2x^2 - x + 4$, find a polynomial q such that for all x

$$p(x) = (x - 1)q(x) + p(1).$$

17. If $p(x) = x^3 + x^2 + x + 1$, find a polynomial q such that for all x

$$p(x) = (x + 2)q(x) + p(-2).$$

18. If $p(x) = x^2 - 3cx + c$, find c such that $x - 1$ is a factor of p.
19. If $p(x) = 2x^3 - cx^2 + x - c$, find c such that $x + 1$ is a factor of p.
20. If $p(x) = 2x^3 - x^2 - 4cx - 5c$, find c such that $x + 3$ is a factor of p.

2. Zeros of Polynomial Functions

We have seen from the Factor theorem that if p is a polynomial of degree $n \geq 1$ and c is a number, then $p(c) = 0$ implies that $x - c$ is a factor of p. The number c is called a *zero* (or *root*) of p. Geometrically, c represents the point where the graph of p intersects the x-axis.

Clearly, since a polynomial p of degree n can have no more than n factors which are monic polynomials, then p has at *most* n zeros. For the rational zeros of a polynomial we state

Theorem 4. If a/b, a rational number in lowest terms, is a zero of the polynomial

$$p(x) = a_n x^n + a_{n-1} x^{n-1} + \ldots + a_1 x + a_0$$

where the a_i's $(i = 0, 1, \ldots, n)$ are integers and $a_n \neq 0$, then a is an integral factor of a_0 and b is an integral factor of a_n.

Note that this theorem does not guarantee the existence of rational zeros of a polynomial. It merely enables us to identify the possible rational zeros. These are then checked using synthetic division or otherwise.

Example 1. Find all rational zeros of the polynomial

$$p(x) = 2x^3 + 5x^2 - 4x - 3.$$

Solution. If a/b is a rational zero of p, then by Theorem 4 a must be an integral factor of 3 and b must be an integral factor of 2. That is,

$$a \in \{-1, -3, 1, 3\}, \qquad b \in \{-1, -2, 1, 2\}$$

and the set of possible rational zeros of p is

$$\{-3, -\tfrac{3}{2}, -1, -\tfrac{1}{2}, \tfrac{1}{2}, 1, \tfrac{3}{2}, 3\}.$$

Using synthetic division we check each of the possible candidates:

$$
\begin{array}{r|rrrr}
-3 & 2 & 5 & -4 & -3 \\
   &   & -6 & 3 & 3 \\
\hline
   & 2 & -1 & -1 & 0
\end{array}
$$

From the Remainder theorem we see that -3 is a zero of p. For the remaining possible rational zeros we set up the following array:

	2	5	-4	-3
-3/2	2	2	-7	15/2
-1	2	3	-7	4
-1/2	2	4	-6	0
1/2	2	6	-1	-7/2
1	2	7	3	0

Thus far we have found three zeros of p: -3, $-\tfrac{1}{2}$, and 1. Since p is of degree 3 we cease our investigations of the remaining numbers. The set of all rational zeros of p is $\{-3, -\tfrac{1}{2}, 1\}$.

We now consider an alternate procedure. Having found that -3

is a zero of p, we use the Factor theorem and write

$$2x^3 + 5x^2 - 4x - 3 = (x + 3)(2x^2 - x - 1).$$

The quotient is a quadratic function whose zeros can be found by using the quadratic formula or simply by factoring. Thus,

$$2x^3 + 5x^2 - 4x = (x + 3)(2x + 1)(x - 1)$$

and the zeros of p are easily obtained.

Example 2. Find the rational zeros of

$$p(x) = x^3 + x^2 + 2x + 2.$$

Solution. The set of possible rational zeros of p is

$$\{-2, -1, 1, 2\}.$$

Using synthetic division we find

$$
\begin{array}{r|rrrr}
-2 & 1 & 1 & 2 & 2 \\
   &   & -2 & 2 & -8 \\
\hline
   & 1 & -1 & 4 & -6
\end{array}
$$

Thus -2 is not a zero of p. We try -1:

$$
\begin{array}{r|rrrr}
-1 & 1 & 1 & 2 & 2 \\
   &   & -1 & 0 & -2 \\
\hline
   & 1 & 0 & 2 & 0
\end{array}
$$

Hence -1 is a zero of p and by the Factor theorem we have

$$x^3 + x^2 + 2x + 2 = (x + 1)(x^2 + 2).$$

Since $x^2 + 2$ has no real zeros (use the quadratic formula), we find that -1 is the only rational zero of p.

It should be clear to the student that by the zeros of a polynomial p we mean the solutions of the equation $p(x) = 0$. In 1799, Gauss presented his doctoral dissertation in which he proved an important algebraic result which today is known as the Fundamental Theorem of Algebra:

Theorem 5. Let p denote the polynomial

$$p(z) = a_n z^n + a_{n-1} z^{n-1} + \ldots + a_1 z + a_0$$

of degree $n \geq 1$, where the a_i's ($i = 0, 1, \ldots, n$) and z are

complex numbers. Then p has at least one zero (real or complex).

In fact, as a consequence of this and the Factor theorem we can write p as

$$p(z) = a_n(z - c_1)(z - c_2) \ldots (z - c_n)$$

where $c_1, c_2, \ldots, c_n$ are complex numbers not necessarily all distinct. Thus the polynomial p of degree n has n zeros exactly. Since some of the factors may occur more than once, we refer to the corresponding zero as a *multiple zero*.

Example 3. Find all the zeros of

$$p(z) = 2z^4 - 4z^3 + 4z^2 - 4z + 2.$$

Solution. Using synthetic division we can write

$$p(z) = 2(z - 1)(z - 1)(z^2 + 1)$$

Hence the set of all zeros is $\{1, 1, i, -i\}$. Here we find that 1 is a multiple zero of p. We say that 1 is a zero of p of *multiplicity* 2.

Exercise 2

In problems 1 through 6 find all the rational zeros of p.
  1. $p(x) = x^3 - 2x^2 + x + 4$
  2. $p(x) = 3x^3 - x^2 - 4x + 2$
  3. $p(x) = 2x^3 - 3x^2 - 7x - 6$
  4. $p(x) = 2x^4 + x^3 + x^2 + x - 1$
  5. $p(x) = 5x^3 + 10x^2 - 2x - 8$
  6. $p(x) = x^4 + 4x^3 + 2x^2 - 4x - 3$

For each of the polynomials in 7 through 12, find the zeros and their multiplicity.
  7. $p(x) = x^4 - 8x^2 + 16$
  8. $p(x) = x^4 - 2x^2 - 15$
  9. $p(x) = 64x^4 - 1$
  10. $p(x) = x^4 + 2x^3 - 3x^2 - 4x + 4$
  11. $p(x) = x^4 - 2x^3 + 2x - 1$
  12. $p(x) = x^6 - 2x^4 + x^2$
  13. Show that the polynomial of degree 2 has at most two zeros. *Hint :* Assume that there are three distinct roots and show that this leads to a contradiction.
  14. If r_1 and r_2 are two zeros of $p(x) = ax^2 + bx + c$, show that

(a) $r_1 + r_2 = -\dfrac{b}{a}$ (b) $r_1 r_2 = \dfrac{c}{a}$.

15. Find the sum and product of the zeros of each of the following quadratic polynomials.
    (a) $p(x) = x^2 + 3x + 7$ (b) $p(x) = 2x^2 - 3x - 7$
    (c) $p(x) = 3x^2 + 4x - 2$ (d) $p(x) = x^2 - 10x + 19$
16. If r_1 and r_2 are two zeros of a quadratic polynomial p, find p given that
    (a) $r_1 + r_2 = 5$, $r_1r_2 = 9$ (b) $r_1 + r_2 = -2$, $r_1r_2 = 3$
    (c) $r_1 + r_2 = \frac{3}{2}$, $r_1r_2 = \frac{4}{3}$ (d) $r_1 + r_2 = \frac{5}{3}$, $r_1r_2 = 3$.
17. Find the values of c such that the graph of the given quadratic polynomial touches the x-axis at only one point.
    (a) $p(x) = x^2 + 3x + c$ (b) $p(x) = x^2 - 5x - c$
    (c) $p(x) = x^2 - 3cx + 4$
18. If $u + iv$ is a zero of the quadratic polynomial

$$p(z) = az^2 + bz + c,$$

    show that $u - iv$ is also a zero.
19. Find all the zeros of p if one of the zeros is the given number.
    (a) $p(z) = 3z^3 - 7z^2 + 27z - 63$; $-3i$
    (b) $p(z) = z^4 - 2z^3 + 6z^2 - 2z + 5$; $1 - 2i$
20. If $p(z) = a_3z^3 + a_2z^2 + a_1z + a_0$, where a_0, a_1, a_2, and a_3 are real numbers, show that p has at least one real zero.

Answers to Exercises

Chapter 1   **Exercise 1**

1. (a) P is a subset of Q.   (b) y is an element of A.
     (c) S is not a subset of T.   (d) x is not an element of Q.
     (e) the empty set
     (f) the set containing one element, 0.
2. (a) T (b) F (c) T (d) T (e) T (f) F
3. (a) $\{x \mid x$ is an even integer and $2 \leq x \leq 8\}$
     (b) $\{x^2 \mid x$ is an integer and $1 \leq x \leq 6\}$
     (c) $\{x \mid x$ is an integer, and $0 \leq x \leq 8\}$
     (d) $\{x \mid x$ is a prime number, and $2 \leq x \leq 19\}$
     (e) $\{x \mid x$ is an odd integer and $1 \leq x \leq 91\}$
     (f) $\{x \mid x$ is a positive odd integer$\}$
     (g) $\{x^2 \mid x$ is an integer and $5 \leq x \leq 9\}$
     (h) $\{x \mid x^2 = 1\}$
5. (a) $x = 2, y = 8; x = 4, y = 4$
     (b) $x = 0$ (c) $x = 10$ (d) $x = 2$
6. yes
10. $P, R, A, S \subset Q, R, A, S \subset P, S \subset R, S \subset A$
11. $\{a\}, \{b\}, \{c\}, \{d\}, \{a, b\}, \{a, c\}, \{a, d\}, \{b, c\}, \{b, d\}, \{c, d\},$
     $\{a, b, c\}, \{a, b, d\}, \{b, c, d\}, \{a, c, d\}, \{a, b, c, d\}, \phi$
12. 2^n
13. (a) $\{1\}, \{3\}, \{5\}, \{1, 3\}, \{1, 5\}, \{3, 5\}, \{1, 3, 5\}, \phi$
     (b) $\{0\}, \phi$ (c) $\{\phi\}, \phi$ (d) $\{\{\phi\}\}, \phi$
14. no

Exercise 2

1. (a) {1, 2, 3, 5}   (b) {4, 6, 8}
   (c) ϕ (d) ϕ
   (e) {1, 2, 3, 4, 5, 6, 7, 8, 9}
   (f) A (g) A
   (h) {1, 2, 3, 4, 5, 6, 8}   (i) {1, 2, 3, 4, 5, 6, 8}
2. (a) {3}   (b) ϕ
   (c) {{1, 2}, 1, 2, 3, 4}   (d) {{1, 2}, 1, 2, 3, 4}
3. (a) T (b) T (c) T (d) F
   (e) F (f) T (g) F (h) F
4. Let $P = \{a, e\}$ and let $Q = \{\{a, e\}, a, e, b\}$. Then $P = \{a, e\} \in Q$ and $P \subset Q$.
5. $A = \{2, 3, 4\}$, $B = \{2, 4, 5\}$, $C = \{1, 2, 3\}$
6. (a) ϕ (b) C
   (c) {1, 2, 4, 5, 7, 8, 10, . . . , 28, 29}
   (d) A
   (e) {1, 5, 7, 11, 13, 17, 19, 23, 25, 29}
   (f) same as in (e)   (g) $U - \{6, 12, 18, 24, 30\}$
   (h) same as in (g)   (i) $B \cup A$
8. (a)

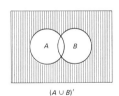 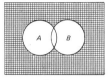

$(A \cup B)'$ $A' \cap B'$

9. (a)

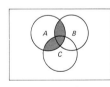 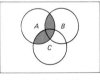

$A \cap (B \cup C)$ $(A \cap B) \cup (A \cap C)$

Exercise 4

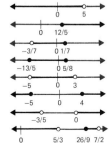

10. $x < 4$ **11.** $x < 5$
12. $x \geq \frac{12}{5}$
13. $-\frac{3}{7} < x \leq \frac{1}{7}$
14. $-\frac{13}{5} \leq x \leq \frac{5}{8}$
15. $-5 < x < 3$
16. $(-\infty, -5] \cup [4, +\infty)$
17. $(-\infty, -\frac{3}{5}) \cup (0, +\infty)$
18. $(-\infty, \frac{5}{3}) \cup [\frac{26}{9}, \frac{7}{2})$

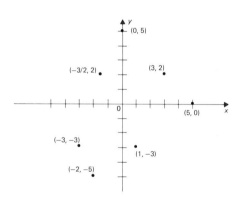

19. $(-\infty, 1) \cup [3, 5) \cup [7, +\infty)$

20. $\left(-\infty, \dfrac{1 - \sqrt{3}}{2}\right] \cup \left[\dfrac{11 + \sqrt{73}}{2}, +\infty\right)$

Exercise 5

1. $\left[-\dfrac{11}{3}, \dfrac{7}{3}\right]$

2. $\left(-\infty, \dfrac{2}{5}\right) \cup \left(\dfrac{12}{5}, +\infty\right)$

3. $\left[-\dfrac{7}{2}, \dfrac{7}{8}\right]$

4. $(-\infty, -2] \cup \left[\dfrac{1}{5}, +\infty\right)$

5. $(-\infty, -5) \cup \left(-5, \dfrac{1}{10}\right] \cup \left[\dfrac{11}{8}, +\infty\right)$

6. $\left[\dfrac{13}{9}, \dfrac{3}{2}\right) \cup \left(\dfrac{3}{2}, \dfrac{17}{11}\right]$

15. $\left(-\infty, -\dfrac{5}{2}\right) \cup \left(\dfrac{1}{2}, +\infty\right)$

16. $\left(-\dfrac{7}{2}, -1\right] \cup \left[2, \dfrac{9}{2}\right)$

Chapter 2

Exercise 1

1.

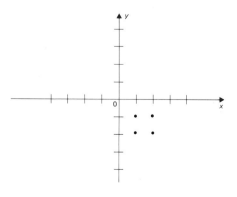

2. $\{(a, c), (b, c)\}$

3. $\{(a, b), (a, c), (a, d)\}$

4. $\{(1, -1), (1, -2), (2, -1), (2, -2)\}$

5. {(0, −1), (0, 0), (1, −1), (1, 0), (2, −1), (2, 0)}

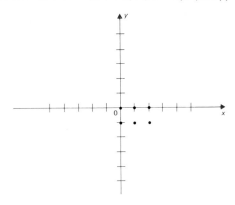

6. {(0, 0), (0, 1), (0, 2), (1, 0), (1, 1), (1, 2), (2, 0), (2, 1), (2, 2)}

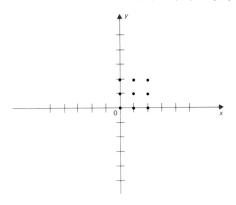

7. {(*a*, *b*), (*a*, *d*), (*a*, *e*), (*a*, *f*), (*b*, *b*), (*b*, *d*), (*b*, *e*), (*b*, *f*),
(*c*, *b*), (*c*, *d*), (*c*, *e*), (*c*, *f*)}

8. {(*x*, *y*)| −2 ≤ *x* ≤ 1, *y* = 3}

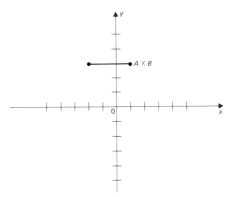

9. $\{(x, y) \mid |x - 2| < 3, y = 2\} = \{(x, y) \mid -1 < x < 5, y = 2\}$

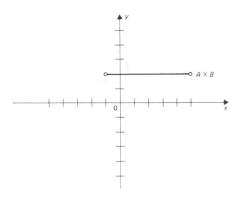

10. $\{(x, y) \mid -5 < x < 3, y = 3, 4\}$

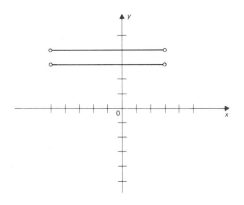

11. $\{(x, y) \mid -8 < x < 2, y > -2\}$

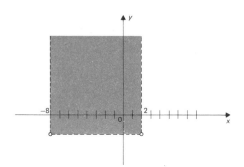

12. $\{(x, y) \mid x \geq -3, -1 < y < 2\}$

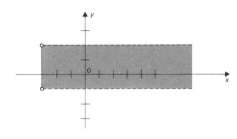

13. $\{(x, y) \mid -2 \leq x \leq 4, |y| > 1\}$

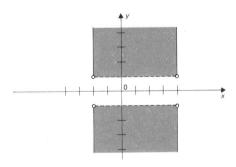

14. $\{(x, y) \mid x < -1 \text{ or } x > 4, y \geq -1\}$

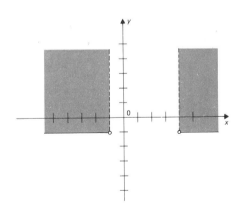

15. $A \times B$ has 20 elements as does $B \times A$.

16. (a) mn (b) m
(c) $m = 1, n = 4$; $m = 1, n = 7$; $m = 1, n = 9$.

17.
$$C_1 = (1, 4) \quad D_1 = (4, 4)$$
$$C_2 = (\tfrac{5}{2}, \tfrac{5}{2}) \quad D_2 = (\tfrac{5}{2}, -\tfrac{1}{2})$$
$$C_3 = (1, -2) \quad D_3 = (4, -2)$$

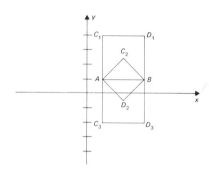

18. infinitely many

Exercise 2

1. $\sqrt{34}$, $(\tfrac{5}{2}, \tfrac{11}{2})$ **2.** $\sqrt{10}$, $(\tfrac{1}{2}, \tfrac{9}{2})$

3. $\sqrt{13}$, $(-2, -\tfrac{9}{2})$ **4.** $5\sqrt{2}$, $(-\tfrac{3}{2}, \tfrac{3}{2})$

5. $\sqrt{2}$, $(-\tfrac{5}{2}, \tfrac{1}{2})$ **6.** $\tfrac{1}{2}\sqrt{58}$, $(\tfrac{5}{4}, -\tfrac{5}{4})$

7. 4, $(-1, 5)$ **8.** 8, $(1, 2)$

9. $d(A, B) = \sqrt{73}$, $d(A, C) = \sqrt{73}$, $d(B, C) = \sqrt{146}$; $146 = 73 + 73$

10. $d(A, B) = 5$, $d(A, C) = 5$, $d(B, C) = 5\sqrt{2}$; $50 = 25 + 25$

11. $d(A, B) = \sqrt{13}$, $d(A, C) = \sqrt{52}$, $d(B, C) = \sqrt{65}$

12. $d(A, B) = \sqrt{10}$, $d(A, C) = 10$, $d(B, C) = 3\sqrt{10}$

13. $d(A, B) = 2\sqrt{2}$, $d(A, C) = 4$, $d(B, C) = 2\sqrt{2}$

14. $d(A, B) = \sqrt{5}$, $d(A, C) = \sqrt{10}$, $d(B, C) = \sqrt{5}$

15. $d(A, B) = \sqrt{17} = d(A, C)$, $d(B, C) = 5$, isosceles triangle

16. $d(A, B) = 5 = d(A, C)$, $d(B, C) = 5\sqrt{2}$, $d(B, C)^2 = 50 = 25 + 25 = d(A, B)^2 + d(A, C)^2$, isosceles right triangle

17. $d(A, B) = 2\sqrt{2} = d(A, C) = d(B, C)$, equilateral triangle

18. $d(A, B) = \sqrt{17} = d(A, C)$, $d(B, C) = 3\sqrt{2}$, isosceles triangle

19. (a) collinear   (b) not collinear   (c) collinear

20. $(-4, \pm\sqrt{7885})$

21. $d(A, B) = 17 = d(B, C) = d(C, D) = d(A, D)$, $d(A, C) = 17\sqrt{2} = d(B, D)$

22. $M(A, B) = (-\tfrac{3}{2}, \tfrac{7}{2})$, $M(A, C) = (2, \tfrac{9}{2})$, $M(B, C) = (-\tfrac{1}{2}, 6)$ length of median from A is $\tfrac{1}{2}\sqrt{73}$, from B is $\tfrac{1}{2}\sqrt{145}$, from C is $\tfrac{1}{2}\sqrt{130}$

23. (a)

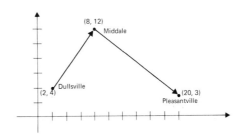

(b) $d(D, M) = 1000$ miles, $d(M, P) = 1500$ miles, total distance 2500 miles

(c) $d(D, P) = 500\sqrt{13}$ miles

Exercise 3

1. $\{x \mid x \in R\}$ **2.** $\{x \mid x \in R \text{ and } x \neq 0\}$

3. $\{t \mid t \in R\}$ **4.** $\{u \mid u \in R \text{ and } u \neq 0, 5\}$

5. $\{r \mid r \in R \text{ and } r \neq -2, 2\}$

6. $\{s \mid s \geq 0\}$ **7.** $\{s \mid s \in R \text{ and } s \neq -2, 1\}$

8. $\{x \mid x \leq 2\}$ **9.** $\{x \mid x \in R\}$

10. $\{z \mid -2 \leq z \leq 2\}$ **11.** $\{x \mid |x| \geq 4\}$

12. $\{t \mid t \in R \text{ and } t \neq 1\}$ **13.** $\{s \mid |s| < 1\}$

14. $\{t \mid t \geq 0\} \cup \{t \mid t \leq -4\}$

15. $\{t \mid t \leq -2\} \cup \{t \mid t \geq -1\}$ **16.** $\{x \mid x < -2\} \cup \{x \mid x > 3\}$

17.

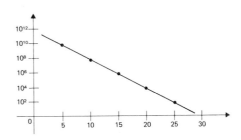

23. (a) 2   (b) 5   (c) 10   (d) $2a + 4$

(e) $2a + 6$ (f) $2a + 2$ (g) $2x + 2h + 4$

24. (a) 15   (b) 3   (c) $\frac{21}{8}$

(d) $2a^2 + 2a + 3$ (e) $4a$

25. (a) $-\frac{3}{2}$ (b) -3 (c) $\frac{1}{2}(c - 1)$

26. (a) ± 2 (b) $\pm\sqrt{7}$ (c) no solution

27. not in general, but if $ab = 0$ (i.e., $a = 0$ or $b = 0$) then
$f(a) + f(b) = f(a + b)$

28. (a) even   (b) odd   (c) even   (d) neither

Exercise 4

1. $(f + g)(x) = x^2 + 2x + 5$, $(f - g)(x) = -x^2 + 2x + 5$
$(fg)(x) = 2x^3 + 5x^2$, $(f/g)(x) = \dfrac{2x + 5}{x^2}$
domain $f + g$, $f - g$, $fg = R$, domain $f/g = \{x \mid x \neq 0\}$.

2. $(f + g)(x) = x^2 + \sqrt{x} + 1$, $(f - g)(x) = -x^2 + \sqrt{x} - 1$
$(fg)(x) = x^2\sqrt{x} + \sqrt{x}$, $(f/g)(x) = \dfrac{\sqrt{x}}{x^2 + 1}$
domain $f + g$, $f - g$, fg, $f/g = \{x \mid x \geq 0\}$.

3. $(f + g)(x) = x^2 + 2x + \sqrt{x^2 - 4} - 3$,
$(f - g)(x) = -x^2 - 2x + \sqrt{x^2 - 4} + 3$
$(fg)(x) = (x^2 + 2x - 3)\sqrt{x^2 - 4}$,
$(f/g)(x) = \dfrac{\sqrt{x^2 - 4}}{x^2 + 2x - 3}$
domain $f + g$, $f - g$, $fg = \{x \mid |x| \geq 2\}$,
domain $f/g = \{x \mid |x| \geq 2, x \neq -3\}$.

4. $(f + g)(x) = \dfrac{x^2 + 3x - 1}{(2x - 1)(x + 1)}$,
$(f - g)(x) = \dfrac{-x^2 + x - 1}{(2x - 1)(x + 1)}$
$(fg)(x) = \dfrac{x}{(2x - 1)(x + 1)}$, $(f/g)(x) = \dfrac{2x - 1}{x(x + 1)}$
domain $f + g$, $f - g$, $fg = \{x \mid x \neq -1, \frac{1}{2}\}$
domain $f/g = \{x \mid x \neq -1, 0, \frac{1}{2}\}$

6. $(f + g)(x) = \sqrt{x^2 - 1} + \sqrt[3]{x}$, $(f - g)(x) = \sqrt[3]{x} - \sqrt{x^2 - 1}$
$(fg)(x) = \sqrt[3]{x}\sqrt{x^2 - 1}$, $(f/g)(x) = \dfrac{\sqrt[3]{x}}{\sqrt{x^2 - 1}}$
domain $f + g$, $f - g$, $fg = \{x \mid |x| \geq 1\}$
domain $f/g = \{x \mid |x| > 1\}$

7. $y = x + 1$, $x \in R$ **8.** $y = \dfrac{x - 6}{4}$, $x \in R$

9. $y = \dfrac{5 - x}{2}$, $x \in R$ **10.** $y = \dfrac{4 - 2x}{3}$, $x \in R$

11. $y = \sqrt[3]{x}$, $x \in R$ **12.** $y = \sqrt[3]{\dfrac{x + 1}{2}}$, $x \in R$

13. $\{x \mid x \geq 0\}$ or $\{x \mid x \leq 0\}$

14. $(f \circ g)(x) = 2 + 3x$, domain $= \{x \mid x \in R\}$
$(g \circ f)(x) = 6 + 3x$, domain $= \{x \mid x \in R\}$

15. $(f \circ g)(x) = 3x^2 - 12x + 8$, domain $= \{x \mid x \in R\}$
$(g \circ f)(x) = 6 - 3x^2$, domain $\{x \mid x \in R\}$

16. $(f \circ g)(x) = x + 4$, domain $= \{x \mid x \geq 0\}$
$(g \circ f)(x) = \sqrt{(x^2 + 4)}$, domain $= \{x \mid x \in R\}$

17. $(f \circ g)(x) = \sqrt[3]{2x^3 + 2}$, domain $= \{x \mid x \in R\}$
$(g \circ f)(x) = 2x + 3$, domain $= \{x \mid x \in R\}$

18. $(f \circ g)(x) = \dfrac{1}{x}$, domain $= \{x \mid x \neq 0\}$

$(g \circ f)(x) = \dfrac{-x}{x + 1}$, domain $= \{x \mid x \neq -1\}$

19. $(f \circ g)(x) = \dfrac{x}{1 + 2x^2}$, domain $= \{x \mid x \neq 0\}$

$(g \circ f)(x) = \dfrac{x^2 + 2}{x}$, domain $= \{x \mid x \neq 0\}$

20. $(f \circ g)(x) = \dfrac{\sqrt{3x^4 + 8}}{x^2}$, domain $= \{x \mid x \neq 0\}$

$(g \circ f)(x) = \dfrac{2}{2x^2 + 3}$, domain $= \{x \mid x \in R\}$

21. $(f \circ g)(x) = \dfrac{1}{\sqrt{x - 2}}$, domain $= \{x \mid x > 2\}$

$(g \circ f)(x) = \dfrac{\sqrt{x} - 2x}{x}$, domain $= \{x \mid x > 0\}$

22. (a) 16　(b) −52　(c) −2　(d) 7　(e) 56　(f) 242

23. (a) undefined　　(b) $2(\sqrt{5} - \sqrt{10})$　　(c) undefined
(d) undefined　　(e) undefined　　　　(f) undefined

24. (a) x, domain $= \{x \mid x \neq 0\}$
(b) $9x + 8$, domain $= \{x \mid x \in R\}$

(c) $\sqrt{4 + \sqrt{4 + x}}$, domain $= \{x \mid x \geq -4\}$

(d) $\dfrac{x + 2}{2x + 5}$, domain $= \{x \mid x \neq -2, -\frac{5}{2}\}$

(e) $\dfrac{8x^4 + 9x^2 + 2}{2x^3 + x}$, domain $= \{x \mid x \neq 0\}$

25. $(g \circ h)(x)$, where $h(x) = x + 1$ and $g(x) = x^2$

26. $(g \circ h)(x)$, where $h(x) = 2x + 3$ and $g(x) = \sqrt{x}$

27. $(g \circ h)(x)$, where $h(x) = x + 1$ and $g(x) = 1 - 2/x$

28. $(g \circ h)(x)$, where $h(x) = x + 2$ and $g(x) = \dfrac{1}{x^2}$

29. $(g \circ h)(x)$, where $h(x) = 2x^2 - 3x + 1$ and $g(x) = x^{-1/3}$

30. $(g \circ h)(x)$, where $h(x) = x^2 + 1$ and $g(x) = \dfrac{\sqrt{x}}{x + 3}$

Exercise 5

1. $2x + 5y - 16 = 0$ **2.** $3x + y - 4 = 0$

3. $12x - 7y + 15 = 0$ **4.** $8x + y + 11 = 0$

5. $2x - 3y = 0$ **6.** $x + y - 5 = 0$

7. $4x + 7y + 29 = 0$ **8.** $2x + y - 3 = 0$

9. $3x - 12y + 2 = 0$ **10.** $3x - 6y + 17 = 0$

11. $2x + 3y - 11 = 0$ **12.** $x - y + 1 = 0$

13. $x + y + 6 = 0$ **14.** $2x + y - 3 = 0$

15. $2x - y - 6 = 0$ **16.** $y = 3$

17. $y = -3$ **18.** $-\frac{3}{4}, \frac{3}{2}, 2$

19. $\frac{1}{2}, 2, -4$ **20.** $\frac{4}{5}, -\frac{12}{5}, 3$

21. $-2, 0, 0$ **22.** (a) $3x + 5y - 15 = 0$

23. $-\frac{7}{2}$ **24.** -21

25. (a) $\$\frac{2600}{3}$ (b) $\$200$ (c) $\$\frac{2000}{3}$ (d) $\$\frac{100}{3}$ (e) $\$\frac{700}{3}$

26. (a) $\$350$ (b) $\$0.50$ (c) $\$600$ (d) $\$250$

27. (a) $p = 5t - 20$, $t \geq 8$

   (c) $20, 40, 70$ (d) 5

28. (a) $25t + 6p - 420 = 0$, where t is measured from 8:00 P.M.

   (b) $53\frac{1}{3}, 45, 36\frac{2}{3}$ (c) $-\frac{25}{6}$

Exercise 6

1. $\left(-\frac{5}{3}, -\frac{7}{3}\right)$ **2.** $\left(-\frac{11}{3}, \frac{37}{3}\right)$

3. $\left(-\frac{5}{7}, -\frac{1}{7}\right)$ **4.** $\left(-\frac{19}{6}, -\frac{17}{6}\right)$

5. $\left(-\frac{3}{16}, -\frac{7}{16}\right)$ **6.** $\left(\frac{50}{23}, \frac{112}{23}\right)$

7. $\left(-\frac{2}{11}, -\frac{10}{11}\right)$ **8.** $\left(\frac{5}{2}, \frac{7}{6}\right)$

9. $\left(2, \frac{5}{4}\right)$ **10.** $\left(\frac{44}{25}, \frac{16}{25}\right)$

11. parallel   **12.** perpendicular

13. perpendicular   **14.** parallel

15. parallel   **16.** parallel

17. perpendicular   **18.** parallel

19. (a) 0 (b) -3 (c) -8 (d) $-\frac{5}{2}$

20. (a) 4 (b) 2 (c) $\frac{19}{13}$ (d) no

21. (a) right triangle   (b) rectangle

   (c) right triangle   (d) rectangle

24. (a) $y = \frac{22}{35}x + 22{,}000$

(b)

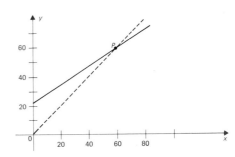

(c) $\approx \$59{,}230$ (d) $\approx \$7{,}714$

25. (a) $\$24{,}000$ (b) $\$16{,}667$ (c) $\$36{,}000$ (d) $\$2{,}667$

26. (a) $C = \frac{5}{9}F - \frac{160}{9}$ (b) $F = \frac{9}{5}C + 32$

(c) $-40°$

Exercise 7

1. lowest point $(0, -4)$; symm. $x = 0$; $x = \pm 2$
2. lowest point $(\frac{3}{2}, -\frac{1}{4})$; symm. $x = \frac{3}{2}$; $x = 1, 2$
3. lowest point $(\frac{5}{2}, -\frac{9}{4})$; symm. $x = \frac{5}{2}$; $x = 1, 4$
4. lowest point $(-\frac{1}{2}, -\frac{9}{4})$; symm. $x = -\frac{1}{2}$; $x = 1, -2$
5. highest point $(\frac{3}{2}, \frac{1}{4})$; symm. $x = \frac{3}{2}$; $x = 1, 2$
6. highest point $(\frac{5}{2}, \frac{9}{4})$; $x = \frac{5}{2}$; $x = 1, 4$
7. highest point $(-\frac{3}{4}, \frac{25}{8})$; symm. $x = -\frac{3}{4}$; $x = \frac{1}{2}, -2$
8. highest point $(-1, \frac{1}{2})$; symm. $x = -1$; $x = 0, -2$
9. lowest point $(0, 3)$; symm. $x = 0$; none
10. lowest point $(-6, -9)$; symm. $x = -6$; $x = 0, -12$
11. 5, 5   **12.** 5×10
13. (a) 100,000   (b) $\$250{,}000$
14. $x = \dfrac{36}{4 + \pi}$, $y = \dfrac{18}{4 + \pi}$
15. (b) $x = 90$ (c) $I = 44{,}100$
16. (b) 27 or 28   (c) $\$11{,}024$ (d) $\$218$, $\$222$

Exercise 8

1. circle, $C(0, 0)$, $r = 8$
2. parabola, $f(-\frac{3}{4}, 0)$, $V(0, 0)$, D: $x = \frac{3}{4}$
3. ellipse, $F_1(0, 2\sqrt{3})$, $F_2(0, -2\sqrt{3})$, $V_1(0, 4)$, $V_2(0, -4)$, $V_3(2, 0)$, $V_4(-2, 0)$

4. parabola, $f(0, -1)$, $V(0, 0)$, $D: y = 1$
5. circle, $C(3, 0)$, $r = 5$ 6. circle, $C(0, 0)$, $r = 3$
7. parabola, $f(0, \frac{1}{10})$, $V(0, 0)$, $D: y = -\frac{1}{10}$
8. ellipse, $F_1(\sqrt{15}, 0)$, $F_2(-\sqrt{15}, 0)$, $V_1(5, 0)$, $V_2(-5, 0)$, $V_3(0, \sqrt{10})$, $V_4(0, -\sqrt{10})$
9. parabola, $f(1, 0)$, $V(0, 0)$, $D: x = -1$
10. circle, $C(-2, 3)$, $r = 6$
11. degenerate case of circle: one point $(3, 4)$
12. ellipse, $F_1(-2, 0)$, $F_2(2, 0)$, $V_1(-\sqrt{10}, 0)$, $V_2(\sqrt{10}, 0)$, $V_3(0, -\sqrt{6})$, $V_4(0, \sqrt{6})$
13. $(x + 1)^2 + (y - 2)^2 = 16$
14. $(x - 3)^2 + (y + 4)^2 = 97$
15. $(x - 5)^2 + (y - 3)^2 = 8$
16. $(x + 1)^2 + (y - 3)^2 = 16$
17. $(x - 2)^2 + (y - 2)^2 = 9$
18. $(x - 1)^2 + (y - 2)^2 = 8$
19. $y^2 = -8x$ 20. $x^2 = 12y$ 21. $x^2 = -16y$
22. $3y^2 = 8x$ 23. $y^2 = -12x$ 24. $2x^2 = 3y$
25. $y^2 = 8x$
27. $x^2 = 12(y + 2)$, $V(0, -2)$, axis: $x = 0$
28. $(x + 1)^2 = -4(y - 1)$, $V(-1, 1)$, axis: $x = -1$
29. $(x - 2)^2 = 8(y - 1)$, $V(2, 1)$, axis: $x = 2$
30. $(x - 2)^2 = -4(y - 4)$, $V(2, 4)$, axis $x = 2$
31. $16x^2 + 9y^2 = 144$ 32. $16x^2 + 25y^2 = 400$
33. $25x^2 + 9y^2 = 225$ 34. $16x^2 + 11y^2 = 176$
35. $9x^2 + 16y^2 = 225$ 36. $x^2 + 2y^2 = 18$

Chapter 3   Exercise 1

1. $s_2 = 2$, $s_6 = 2$, $s_9 = 0$ 2. $s_2 = \frac{1}{4}$, $s_6 = \frac{1}{64}$, $s_9 = \frac{1}{512}$
3. $s_2 = \frac{1}{4}$, $s_6 = \frac{1}{36}$, $s_9 = -\frac{1}{81}$
4. $s_2 = 3$, $s_6 = 11$, $s_9 = 17$
5. $s_2 = \frac{3}{8}$, $s_6 = \frac{231}{1024}$, $s_9 = \frac{36465}{196608}$
6. $s_2 = s_6 = s_9 = 3$
8. $\{2^n\}$, $\{(n - 3)(n - 2)(n - 1) + 2^n\}$
9. $\{(-1)^{n-1}(4n - 3)\}$, $s_n = \begin{cases} n^2; & n \text{ odd} \\ -2n - 1; & n \text{ even} \end{cases}$
10. $\left\{\dfrac{n - 1}{n}\right\}$; $s_n = -\frac{1}{6}n^2 + n - \frac{5}{6}$.
11. $\frac{3}{5}, \frac{3}{10}, \frac{3}{20}, \frac{3}{40}, \frac{3}{80}$ 12. $1, 7, -5, 19, -29$
13. $2, 4, 8, 16, 32$ 14. $2, 5, -8, 62, -296$

15. $\frac{3}{5}\left(\frac{1}{2}\right)^{n-1}$, 2^n

16. $1, \frac{1}{2}, \frac{2}{3}, \frac{3}{5}, \frac{5}{8}, \frac{8}{13}, \frac{13}{21}, \frac{21}{34}$

17. $s_n = \dfrac{3}{10} + \dfrac{3}{10^2} + \ldots + \dfrac{3}{10^n}$

18. $s_8 = 3$, $s_{17} = 15$, $s_{20} = 7$, $s_{51} = 31$

19. $a_{n+1} = a_n(1 + 0.05) = a_n(1.05)$, $a_n = 1000(1.05)^n$

Exercise 2

15. $s_n = \{n\}$, $t_n = \{-n\}$

Exercise 3

1. 16

2. 10

3. $\frac{16}{7}$

4. does not converge

5. $\dfrac{1}{x}$ for $x > 0$ or $x < -2$

6. $\frac{1}{3}$

7. $\frac{172}{33}$

8. $\frac{15611}{495}$

9. $\frac{245}{999}$

10. $\frac{670069}{9999}$

12. 1140 ft

15. 104940

16. $\frac{1}{6}$

17. $\frac{1}{4}$

18. $\frac{1}{3}$

19. $\frac{2}{15}$

20. $\frac{1}{3}$

Exercise 4

3. 26

4. (a) $\frac{10}{9}(10^n - 1) - n$

(b) $\frac{7}{9}[\frac{10}{9}(10^n - 1) - n]$

5. 332,667

6. \$3828.84

7. \$3846.11

8. (a) 5.06%

(b) 12.68%

(c) 16.08%

9. \$3069.57

10. \$2214.70

11. \$8495.91

12. \$243.00

13. \$164.00

14. 28,440

15. $N = 4(8)^9 \cong 134,200,000$

Chapter 4

Exercise 1

1. (a) -8 (b) 1

2. (a) 0 (b) 0

3. (a) 12 (b) 22

4. (a) 5 (b) 7

5. (a) -2 (b) -2

6. (a) 4 (b) 4

7. (a) -3 (b) -3

8. (a) 0 (b) does not exist

9. (a) 3 (b) does not exist

10. (a) 0 (b) 0

11. 1

12. -1

13. 20

14. $-\frac{25}{13}$

15. $\frac{31}{17}$

16. $\frac{6}{5}$

17. 12

18. 12

19. $2\sqrt{2}$

20. 2

21. $\sqrt{\frac{4}{3}}$

22. -2

23. 5

24. $4x$

25. $2x - 3$

26. $\frac{1}{2}x^{-1/2}$

Exercise 2

1. -7	**2.** 9	**3.** -92	**4.** $\frac{16}{7}$
5. $-\frac{13}{2}$	**6.** $\frac{25}{67}$	**7.** $\frac{1}{9}$	**8.** $\sqrt{\frac{7}{2}}$
9. $\sqrt{2}$	**10.** 2	**11.** 0	**12.** $\sqrt{\frac{3}{2}}$
13. $\sqrt[3]{-7}$	**14.** $\sqrt[3]{\frac{1}{3}}$	**18.** -5	**19.** $\frac{1}{3}$
20. 0	**21.** 0	**22.** 1	**23.** -1

24. $-\infty$ **25.** $+\infty$ **26.** does not exist

27. does not exist   **28.** $+\infty$ **29.** $-\infty$

30. $-\infty$ **31.** $-\infty$ **32.** does not exist

Exercise 3

1. discont. **2.** discont. **3.** discont. **4.** discont.
5. cont. **6.** discont. **7.** cont. **8.** discont.
9. discont. **10.** cont. **11.** cont. **12.** cont.
13. cont. at 4, discont. at -4 **14.** cont. at -1 and discont. at 1
15. cont. **16.** cont. **17.** discont. **18.** cont.
19. cont. **20.** discont. **21.** cont. **22.** cont.
23. 3 **24.** $\frac{1}{4}$ **25.** -2 **26.** 12
27. no

Chapter 5

Exercise 1

1. $1, R$ **2.** $2x, R$ **3.** $10x, R$ **4.** $0, R$
5. $6x, R$ **6.** $2x + 3, R$

7. $-\dfrac{2}{x^2}, x \neq 0$ **8.** $-\dfrac{1}{(x+3)^2}, x \neq -3$

9. $1 - \dfrac{1}{x^2}, x \neq 0$ **10.** $\dfrac{1-x^2}{(1+x^2)^2}, R$

11. $-\dfrac{1}{2x\sqrt{x}}, x > 0$ **12.** $1 - \dfrac{1}{2\sqrt{x}}, x > 0$

13. $\dfrac{x}{\sqrt{x^2+1}}, R$ **14.** $3x^2 - \dfrac{1}{2x\sqrt{x}}, x > 0$

15. 0   **16.** 1
17. -1 **18.** none
19. 0   **20.** none

Exercise 2

1. (a) 100 ft/min   (b) 100 ft/min   (c) 2:02 P.M.

2. (a) 50 ft/min, 110 ft/min, 170 ft/min
   (b) 8:40 A.M. (c) 35,000 feet
3. (b) 7 ft/sec
4. (a) $4.00 (b) $2.00 (c) $2.00, $2.00
5. (a) $160.00 (b) $4x + 2h$
   (c) $200.00, $320.00
6. (a) $\dfrac{\sqrt{10}}{4}$ (b) 2500 gal, 625 gal
7. (a) $375,000 (b) $120, $140, $200
   (c) 1250
8. (a) 7×10^7, 6×10^7 (b) 6×10^7, 4×10^7
   (c) $t^* = 40$ minutes (d) 0

Exercise 3

1. $3x^2 - 14x - 9$ 2. $\frac{7}{2}t^{5/2} - 27t^2 + 3$
3. $\frac{2}{5}x + 2$ 4. -3
5. $505x^{100} + 34x$ 6. $2x + 6$
7. $2s - 2/s^3$ 8. $3x^2 + \frac{9}{2}x^{1/2} - \frac{3}{2}x^{-5/2}$
9. $3x^2 + 2(a + b + c)x + ab + ac + bc$
10. $\frac{25}{2}x^{3/2} + \frac{9}{2}x^{1/2} + 2x^{-1/2} + x^{-3/2}$
11. $\frac{5}{2}x^{3/2} + 3x^{-1/2} - \frac{27}{2}x^{-5/2}$
12. $12x^3 + 21x^2 + 26x + 2$
14. (a) 1 mph, $\frac{3}{4}$ mph, $\frac{1}{2}$ mph (b) $2\frac{1}{4}$ miles

   (c) $\dfrac{81\pi}{16}$

15. 216 17. 1,000
18. (a) 21,000 gal/min, 18,000 gal/min, 12,000 gal/min,
       6,000 gal/min (b) 40 min

Exercise 4

1. $2x + a + b$ 2. $3ax^2 + 2bx + c$
3. $15ax^4 + 4(7a + 3b)x^3 + 3(7b + 3c)x^2 + 2(8a + 7c)x + 8b$
4. $x^4(5 - 12x + 14x^2 - 32x^3)$

5. $\dfrac{ad - bc}{(cx + d)^2}$ 6. $\dfrac{4 + 10x}{7}$

7. $\dfrac{-(36x + 25)}{x^6}$ 8. $x^5(30 + 49x)$

9. $\dfrac{-2x}{(x^2 - 5)^2}$ 10. $\dfrac{-5}{(x - 7)^2}$

11. $\dfrac{-24x(x+1)}{(2x^3+3x^2+5)^2}$

12. $\dfrac{21x^2+24x+20}{(7x+4)^2}$

13. $\dfrac{ax^2+2ax+3a+b+2bx^{-1}}{(1+x+x^{-1})^2}$

14. $\dfrac{(c+1-a-b)x^2+2(c-ab)x+ca+cb-abc-ab}{[x^2+(c+1)x+c]^2}$

15. $\dfrac{-12x^{-1}}{(x^3-x^{-3})^2}$

16. $\dfrac{7x^2-10x-35}{x^2(x+7)^2}$

17. $\dfrac{-(12x^2+64x+69)}{(3x+1)^4}$

18. $\dfrac{3x^6-50x^4+54x^3-4x^2+60x+19}{(x^3-5x+6)^2}$

Exercise 5

1. $12x-y-16=0$

2. $x-4y+4=0$

3. $x+4y-4=0$

4. $7x-4y-2=0$

5. $y=-4,\ 4x-y-4=0$

6. $\left(-\frac{3}{2},\ -\frac{21}{4}\right)$

7. $24,12$

8. $(2,4),\ (-2,-4)$

9. $(5,5)$

10. $(2,60)$

11. $(2,0)$

12. $0<c<2$

13. 1

14. $\frac{4}{3}$

15. $\frac{3}{2}$

16. $-2+\sqrt{10}$

Exercise 6

1. (a) $6(3x+1)$

(b) $2\left(x-\dfrac{1}{x}\right)\left(1+\dfrac{1}{x^2}\right)$

(c) $-(x+1)^{-2}$

(d) $3(x+1)^2$

2. $16(2x+4)^7$

3. $15(5x^3+6x^2+8x+9)^{14}(15x^2+12x+8)$

4. $\frac{5}{2}(3-8x^2+9x^3)^{3/2}(27x^2-16x)$

5. $\frac{1}{3}(ax^2+bx+c)^{-2/3}(2ax+b)$

6. $(4x^2+x+10)(x^2+5)^{-1/2}$

7. $-\frac{1}{5}(x^3+ax^2+bx+c)^{-6/5}(3x^2+2ax+b)$

8. $2x-x(x^2+3)^{-3/2}$

9. $(x^2+1)^{-3/2}$

10. $(16x^3+6xb^2+10xa^2)(x^2+a^2)^2(x^2+b^2)^4$

11. $\frac{13}{2}(5x+11)^{-3/2}(3x+4)^{-1/2}$

12. $-\frac{1}{3}(5x^2+7x+3)^{-4/3}(x^2+2x+4)^{-2/3}(3x^2+34x+22)$

13. $\frac{1}{2}(2x+1)(x^2+x+1)^{-1/2}$

14. $\dfrac{1}{2(a-b)}\left[(x+a)^{-1/2} - (x+b)^{-1/2}\right]$

15. $-\frac{1}{3}\left[(x+3)^{-2/3} - (x-3)^{-2/3}\right]\left[(x+3)^{1/3} - (x-3)^{1/3}\right]^{-2}$

16. 80π ft²/sec, 200π ft²/sec, 800π ft²/sec

17. 25 in²/min   **18.** $\frac{1}{125}\pi$ in/hr, $\frac{4}{5}$ in²/hr

19. increasing, 15% per min   **20.** $\frac{8}{25}\pi$ ft/min

Exercise 7

1. $-\dfrac{x}{y}$ **2.** $\dfrac{9x^2}{2y}$ **3.** $-\dfrac{2}{y(x-2)^2}$

4. $-\dfrac{x^2}{y^2}$ **5.** $\dfrac{9x^2 + 10x}{2y}$ **6.** $\dfrac{1-3y}{3(y^2+x)}$

7. $\dfrac{x(a^2-y^2)}{y(x^2+a^2)}$ **8.** $-\dfrac{xb^2}{ya^2}$

9. $\dfrac{x(3xy^2 + 2y^3 - x^3)}{y(y^3 - 3x^2y - 2x^3)}$ **10.** $-\left(\dfrac{y}{x}\right)^{1/3}$

11. $\dfrac{8 + x(1-x^2)^{-1/2}}{8 - y(1-y^2)^{-1/2}}$ **12.** $\dfrac{5(x+4)^4}{6y^2}$

13. $\dfrac{6x^2 - 4xy - 1}{1 - 2x^2}$ **14.** $\dfrac{2x(x^2+7)^{-2/3} - 9y^2}{18xy}$

15. $x - y + 1 = 0$ **16.** $\sqrt{3}\,x - y + 4 = 0$

17. $x + y - 5 = 0$ **18.** $x + 8y - 5 = 0$

19. $2x + y - 5 = 0$ **20.** 8 in/sec

21. $\dfrac{85}{\sqrt{13}}$ knots

Exercise 8

1. (a) $6x$, 6   (b) $20x^3 + 6$, $60x^2$

(c) $6x + \frac{3}{2}x^{-1/2} - 48x^{-5}$, $6 - \frac{3}{4}x^{-3/2} + 240x^{-6}$

(d) $4(x-1)^{-3}$, $-12(x-1)^{-4}$

(e) $\frac{5}{9}x^{-11/6} + \frac{40}{9}x^{-8/3}$, $-\frac{55}{54}x^{-17/6} - \frac{320}{27}x^{-11/3}$

(f) $4(1-3x^2)(x^2+1)^{-3}$, $-48x(1-x^2)(x^2+1)^{-4}$

2. (a) $\dfrac{12}{(1-12y)^3}$ (b) $\dfrac{1}{3}\left(\dfrac{a^2}{x^4y}\right)^{1/3}$ (c) $-\dfrac{b^4}{a^2y^3}$

(d) $-\dfrac{144}{81(y-1)^3}$

7. (a) $-20 - 32t$, -32 (b) $8(4t + 7)$, 32

    (c) $6t^2 - 24t + 24$, $12t - 24$

    (d) $3t^2 + 12t - 21$, $6t + 12$

8. (a) $-20, 56, 24, -21$ (b) $-\frac{5}{8}, -\frac{7}{4}, 2, -2 \pm \sqrt{11}$

    (c) $-32, 32, 0, \pm 6\sqrt{11}$

    (d) $t < -\frac{5}{8}, t > -\frac{5}{8}; t > -\frac{7}{4}, t < -\frac{7}{4}; t \neq 2, \phi; t < -2 - \sqrt{11}$
         or $t > -2 + \sqrt{11}$, $-2 - \sqrt{11} < t < -2 + \sqrt{11}$

9. (a) $x - x^2$, $1 - 2x$ (b) $\frac{3}{16}, \frac{2}{9}, \frac{1}{4}, \frac{2}{9}, \frac{3}{16}$

    (d) $\frac{1}{2}, \frac{1}{3}, 0, -\frac{1}{3}, -\frac{1}{2}$

Chapter 6   Exercise 1

1. increasing on $(-\infty, -1) \cup (3, \infty)$, decreasing on $(-1, 3)$

2. increasing on $(-\infty, -2) \cup (2, \infty)$, decreasing on $(-2, 2)$

3. increasing on $(-\infty, -1) \cup (1, \infty)$, decreasing on $(-1, 1)$

4. increasing on $(-\infty, -1) \cup (\frac{1}{2}, \infty)$, decreasing on $(-1, \frac{1}{2})$

5. increasing on $(-\frac{1}{2}, 0) \cup (\frac{1}{2}, \infty)$, decreasing on $(-\infty, -\frac{1}{2}) \cup (0, \frac{1}{2})$

6. increasing on $(-\infty, -\frac{1}{2}) \cup (3, \infty)$, decreasing on $(-\frac{1}{2}, 3)$

7. increasing on $(1, 2) \cup (3, \infty)$, decreasing on $(-\infty, 1) \cup (2, 3)$

9. increasing on $(0, 2)$, decreasing on $(2, 4)$

10. decreasing

Exercise 2

1. rel. max. at x_2, x_4, x_5, x_7; rel. min. at x_1, x_3, x_5, x_6, x_8; abs. max. at x_7; abs. min. at x_3

2. no C.V.; rel. and abs. min. at 0; rel. and abs. max. at 2

3. C.V. at -2; rel. and abs. max. at -2; rel. min. at $-4, 4$; abs. min. at 4

4. C.V. at $0, 2$; rel. min. at 2; rel. max. at 0

5. C.V. at 0; rel. and abs. min. at 0

6. C.V. at $-1 \pm \sqrt{5}$; rel. max. $-1 - \sqrt{5}$, rel. min. $-1 + \sqrt{5}$

7. C.V. at $0, \pm\frac{1}{2}$; rel. max. at 0; ± 1; rel. and abs. min. at $\pm\frac{1}{2}$; abs. max. ± 1

8. C.V. at 0; rel. and abs. min. at -1; rel. and abs. max. at 1

9. C.V. at 0; rel. and abs. min. at 0

10. C.V. at $-4, -\frac{2}{5}, 2$; rel. max. at $-\frac{2}{5}$; rel. min. at 2

11. C.V. at $0, \pm 2$; rel. max. at $0, \pm 4$; rel. and abs. min. at ± 2; abs. max. at ± 4

12. C.V. at 0; rel. and abs. max. at 0

13. none

14. C.V. at $-2 \pm 2\sqrt{2}$; rel. and abs. min. at $-2 - 2\sqrt{2}$; rel. and abs. max. at $-2 + 2\sqrt{2}$

16. (a) $a = -3$, $b = -24$

(b) rel. max. at -2, rel. min. at 4

19. (a) $x < 20$ (b) $x > 20$ (c) $x = 20$

Exercise 3

1. square with side $\dfrac{P}{4}$ **2.** cube with side 4

3. $10''$ **4.** $2\sqrt{3} \times 3\sqrt{3}$ **5.** $8'' \times 8'' \times 8''$

7. $r\sqrt{2} \times \dfrac{r}{\sqrt{2}}$ **8.** $a\sqrt{2} \times b\sqrt{2}$ **9.** 3×2

10. $\dfrac{80}{\pi + 4}, \dfrac{20\pi}{\pi + 4}$ **11.** radius = height = a

12. length = width = 12′, depth = 2′

13. $300\sqrt{2}$ ft

14. length = width = height = 10′

Exercise 4

1. $N = \frac{1}{2}C$ **2.** $N = 800$ **3.** $10^{-5/2}$ **4.** $\frac{2}{3}r_0$

5. (a) $\frac{5}{3}$ (b) $500/c$ (c) $1{,}000$

6. \$5,000

7. (a) $P = 1{,}497x - 0.015x^2 - 9$

(b) $49{,}900$ (c) \$37,349,641

8. 7,000, \$11

9. marginal revenue = marginal cost

12. (a) $r(x) = 899x - 4x^2$ (b) 112   (c) \$451

Exercise 5

1. concave up everywhere

2. concave up $(-\frac{2}{3}, +\infty)$, concave down $(-\infty, -\frac{2}{3})$, pt. of inf. at $x = -\frac{2}{3}$

3. concave up $(\frac{5}{3}, +\infty)$, concave down $(-\infty, \frac{5}{3})$, pt. of inf. at $x = \frac{5}{3}$

4. concave up $(-3, +\infty)$, concave down $(-\infty, -3)$, pt. of inf. at $x = -3$

5. concave up $(-\infty, \frac{1}{6}) \cup (\frac{1}{3}, +\infty)$, concave down $(\frac{1}{6}, \frac{1}{3})$, pts. of inf. at $x = \frac{1}{6}, \frac{1}{3}$

6. concave up $(0, +\infty)$, concave down $(-\infty, 0)$
7. concave up $(-\infty, -1/\sqrt{3}) \cup (1/\sqrt{3}, +\infty)$ concave down $(-1/\sqrt{3}, 1/\sqrt{3})$, pts. of inf. at $x = -1/\sqrt{3}, 1/\sqrt{3}$
8. concave down everywhere
9. concave up $(-\sqrt{3}/3, \sqrt{3}/3)$, concave down $(-\infty, -\sqrt{3}/3) \cup (\sqrt{3}/3, +\infty)$, pts. of inf. at $x = \pm\sqrt{3}/3$
10. concave up $(6, +\infty)$, concave down $(-\infty, -3) \cup (-3, 6)$, pt. of inf. at $x = 6$
11. (a) $18x + 27y - 11 = 0$ (b) $9x + 27y - 17 = 0$
    (c) $9x + y - 27 = 0$ 12. -8
13. concave down at $x = 100$, concave up at $x = 300, 400$, pt. of inf. at $x = 200$ 14. (a) $1, -2 \pm \sqrt{3}$

Exercise 6

1. $dy = 24$, $\Delta y = 27\frac{1}{8}$, $|\Delta y - dy| = 3\frac{1}{8}$
2. $dy = 44\frac{1}{2}$, $\Delta y = 50\frac{3}{4}$, $|\Delta y - dy| = 6\frac{1}{4}$
3. $dy = 6\frac{1}{2}$, $\Delta y = 6\frac{3}{4}$, $|\Delta y - dy| = \frac{1}{4}$
4. $dy = \frac{1}{8} = 0.125$, $\Delta y = \frac{1}{2}(3\sqrt{2} - 4) \cong 0.121$, $|\Delta y - dy| = 0.004$
5. $dy = \frac{3}{49}$, $\Delta y = \frac{2}{35}$, $|\Delta y - dy| = \frac{1}{245}$
6. 2.0125
7. (a) 11.0909 (b) 0.11958 (c) 15.9687 (d) 3.2156
8. 30.4 9. 2.925
10. 2π sq in 11. $\pm 2.4\pi$ sq in; 1.7% 12. ± 0.007 ft
14. (a) $dC = (\frac{1}{50}x + 5)dx$
    (b) 7 (c) $\frac{1}{100}$ (d) $\frac{7}{800}$, $\frac{7}{8}$%

15. (a) $P(x) = \frac{2}{5}x^2 - 35x - \dfrac{x^3}{1500}$

    (b) $dP = \left(\frac{4}{5}x - 35 - \dfrac{x^2}{500}\right)dx$

    (c) 0

Chapter 7 ## Exercise 1

1. $\dfrac{x}{2} + C$ 2. $\dfrac{x^2}{4} + C$ 3. $\dfrac{3x^4}{4} + C$

4. $\dfrac{x^2}{2} + \dfrac{x^4}{4} + C$ 5. $\dfrac{x^3}{3} + x^2 + x + C$

6. $\dfrac{x^3}{3} - x^2 + C$ 7. $\dfrac{-1}{x} + C$ 8. $\dfrac{-3}{2x^2} + C$

9. $\dfrac{x^3}{3} - x^2 + x + C$ **10.** $-\dfrac{1}{2x^2} - \dfrac{4}{3x^3} + C$

11. $\frac{4}{3}x^{3/2} + C$ **12.** $2x^{1/2} + C$

13. $\frac{3}{2}x^{2/3} + C$ **14.** $\frac{9}{4}x^{4/3} + 3x^{2/3} + C$

15. $\dfrac{x^2}{2} - x + C$ **16.** $\frac{3}{2}x^{4/3} + \frac{3}{5}x^{5/3} + C$

17. $\frac{2}{7}x^{7/2} - 4x^{1/2} + C$ **18.** $\dfrac{x^3}{3} + \dfrac{3x^2}{2} + 9x + C$

19. $\frac{2}{3}x^{3/2} - 2x + C$ **20.** $\frac{3}{5}x^{5/3} - \frac{3}{4}x^{4/3} + x + C$

Exercise 2

1. $\frac{39}{16}, \frac{41}{16}$ **2.** $\frac{15}{8}, \frac{17}{8}$ **3.** $\frac{35}{2}, \frac{51}{2}$ **4.** $\frac{7}{16}, \frac{15}{16}$

5. $\frac{1}{9} + \frac{1}{10} + \frac{1}{11} + \ldots + \frac{1}{16}, \frac{1}{8} + \frac{1}{9} + \frac{1}{10} + \ldots + \frac{1}{15}$

6. $\dfrac{3}{8}\sum\limits_{k=1}^{8}\sqrt{\dfrac{45 + 3k}{8}}, \dfrac{3}{8}\sum\limits_{k=1}^{8}\sqrt{\dfrac{48 + 3k}{8}}$

11. 2 **12.** $\frac{7}{2}$ **13.** $\frac{1}{3}$ **14.** $\frac{32}{3}$

15. $\frac{15}{2}$ **16.** 12 **17.** 9 **18.** $\frac{26}{3}$

19. 3 **20.** $\frac{1}{4}$

Exercise 3

1. $\frac{8}{3}$ **2.** 12 **3.** 9

4. $-\frac{5}{3}$ **5.** 38 **6.** $\frac{4}{3}$

7. 30 **8.** $2(\sqrt{5} - \sqrt{2})$ **9.** $-\frac{32}{3}$

10. $3(\sqrt[3]{2} + \sqrt[3]{4})$ **11.** $\frac{80}{243}$ **12.** $\frac{256}{5}$

14. 10 **15.** 36 **16.** $\frac{50}{3}$

17. $\frac{4}{3}$ **18.** $\frac{9}{2}$ **19.** $\frac{1}{3}$

20. $\frac{8}{3}$ **21.** $\frac{4}{3}$ **22.** 36

23. x **24.** x^2 **25.** $(x - 1)^2$

26. $2\sqrt{x}$ **27.** $-x^2$ **28.** $\dfrac{26x^3}{3}$

29. $4(x + 1)$ **30.** 0

31. 2 **32.** $4, -\frac{3}{2}$

33. 2 **34.** $2, -\frac{3}{2}$

Exercise 4

1. $y(t) = 4 + t - \frac{3}{2}t^2$ **2.** $y(x) = x^4 + \frac{1}{3}x^3 - \frac{10}{3}$

3. $y(t) = \frac{1}{4}t^4 - t^{-1} - \frac{7}{2}$ **4.** $y(t) = \frac{2}{3}t^{3/2} + 4t - \frac{55}{3}$

5. $y(x) = -\frac{3}{2}x^{-2} - \frac{1}{3}x^{-3} + \frac{5}{6}$

6. $y(x) = 2x - \frac{2}{5}x^{5/2} + \frac{406}{5}$

7. $y(x) = \frac{3}{4}x^{4/3} + \frac{9}{5}x^{5/3} + \frac{907}{20}$

8. $y(t) = 4t^{1/2} - 1$

9. $y(x) = -\frac{2}{5}x^{5/2} + 2x^{1/2} + \frac{83}{10}$

10. $y(x) = \frac{9}{4}x^{4/3} - \frac{9}{5}x^{5/3} + \frac{103}{5}$

11. (a) $C(x) = 2x^2 - 350x + 100$

    (b) $P(x) = -2x^2 + 360x - 100$

    (c) 90 units

12. $C(x) = \frac{20}{3}x^{3/2} + 350$

13. (a) $R(x) = \frac{1}{100}x^2 + 20x$ (b) $p(x) = \frac{1}{100}x + 20$

14. (a) 9 min   (b) 48 in   (c) 18 min

15. (a) $A(t) = -0.02t^{3/2} + 350$

    (b) 674 min

16. yes, by 6 feet

17. (a) $v(t) = \frac{1}{2}t^2 + t + 2, \ s(t) = \frac{1}{6}t^3 + \frac{1}{2}t^2 + 2t + 1$

    (b) $v(t) = \frac{1}{3}t^3 - 5t - 1, \ s(t) = \frac{1}{12}t^4 - \frac{5}{2}t^2 - t + 4$

18. (a) $v(t) = -32t + 288, \ s(t) = -16t^2 + 288t$

    (b) 1296 ft

19. $3\sqrt{2}$ sec   **20.** 4 sec

Exercise 5

1. $\dfrac{(x + 4)^3}{3} + C$ **2.** $\dfrac{(x - 5)^7}{7} + C$

3. $\frac{2}{3}(3x + 4)^{3/2} + C$ **4.** $\frac{2}{3}(3x + 4)^{1/2} + C$

5. $-\frac{1}{4}(4 - x)^4 + C$ **6.** $\dfrac{(x - 3)^3}{12} + C$

7. $-\frac{1}{4}(4x - 1)^{-1} + C$ **8.** $\dfrac{-(x^2 + 3x - 1)^{-2}}{2} + C$

9. $\dfrac{(4x^5 + 6)^{101}}{2020} + C$ **10.** $\frac{1}{8}(x^2 + 2x - 5)^4 + C$

11. $\frac{1}{3}(x^2 - 4)^{3/2} + C$ **12.** $-\frac{1}{3}(4 - x^2)^{3/2} + C$

13. $\frac{2}{3}(1 - x^{-1})^{3/2} + C$ **14.** $\frac{2}{5}(2 + x^{5/3})^{3/2} + C$

15. $\frac{1}{2}$ **16.** $2\sqrt{3}$

17. $2(\sqrt{6} - \sqrt{2})$ **18.** $\frac{62}{15}$

19. $\frac{1}{4}$ **20.** 1

21. $\dfrac{2x^3}{3} - \dfrac{x^2}{2} + C$

22. $\frac{2}{3}x(1 + x)^{3/2} - \frac{4}{15}(1 + x)^{5/2} + C$

23. $\dfrac{x(x + 5)^7}{7} - \dfrac{1}{56}(x + 5)^8 + C$

24. $x(2x + 1)^{1/2} - \frac{1}{3}(2x + 1)^{3/2} + C$

25. $\frac{4}{5}3^{5/2} - \frac{4}{35}3^{7/2} + \frac{4}{35}$

26. $\frac{-x^2}{3}(1 - x^2)^{3/2} - \frac{2}{15}(1 - x^2)^{5/2} + C$

27. $\frac{x^2}{5}(1 + x^2)^{5/2} - \frac{2}{35}(1 + x^2)^{7/2} + C$

28. $-\frac{3}{4}\sqrt{3} + \frac{4}{3}$ **29.** $\frac{2644}{135}$ **30.** $\frac{1004}{15}$

Chapter 8

Exercise 1

1. 1   **2.** 2   **3.** 25   **4.** $\frac{9}{2}$ **5.** $(c/b^2)^y$

6. 1   **7.** 5   **8.** 4   **9.** $\frac{5}{6}$ **10.** $\frac{23}{20}$

11. 0   **12.** 3   **13.** 2   **14.** 3   **15.** 2

16. 1   **17.** 3   **18.** −4   **19.** 3   **20.** −5

21. $\log_b \dfrac{-1 + \sqrt{5}}{2}$ **22.** 4

23. 16   **24.** 5

25. −3   **26.** 100

Exercise 2

1. $4e^{4x}$ **2.** $32xe^{4x^2+5}$ **3.** $-4x^{-3}e^{2x^{-2}}$

4. $(6x - 4)e^{3x^2-4x+1}$ **5.** $3x^2e^{2x} + 2x^3e^{2x}$

6. $(2x^2 + 4x + 3)e^x$ **7.** $-4xe^{-2x^2}$

8. $-4e^{-4x}$ **9.** $\dfrac{2x}{x^2 + 4}$

10. $\dfrac{e^x}{e^x + 1}$ **11.** $(\ln x + 1)e^{x\ln x}$

12. $\dfrac{2e^x}{(e^x + 1)^2}$ **13.** $\dfrac{4xe^{2x^2+4}}{1 + e^{2x^2+4}}$

14. $12xe^{6x^2-1} + 40xe^{5x^2-1}$ **15.** $5^x \ln 5$

16. $-5^{-x} \ln 5$ **17.** $2^{x-1}5^x \ln 10$ **18.** $2^{x-1}5^{-x} \ln \frac{2}{5}$

19. $2e^{2x} + 2^e e x^{e-1}$ **20.** $\dfrac{2}{\sqrt{x^2 + 1}}$ **21.** $\dfrac{\log_{10} e}{x}$

22. $\dfrac{(\log_{10} e)^2}{x \log_{10} x}$ **23.** $\dfrac{(2x + e^x) \log_3 e}{x^2 + e^x}$

24. $\dfrac{(3x^2 + \frac{1}{2}x^{-1/2}) \log_2 e \log_5 e}{(x^3 + x^{1/2}) \log_5 (x^3 + x^{1/2})}$

25. $\dfrac{16}{x+1} + \dfrac{8(4x+1)}{2x^2+x} - \dfrac{x}{x^2+4}$

26. $\dfrac{14x}{x^2+5} + \dfrac{x^2}{x^3+1} - \dfrac{5x}{2(x^2+1)}$

27. $\dfrac{14xe^{2x}}{e^{2x}+6} + \dfrac{1}{2(x+4)} - \dfrac{5(e^x - e^{-x})}{e^{-x}+e^x}$

28. $\dfrac{(3x^2+1)^5(x^2+1)^{3/2}}{(x^2+4)^{10}} \left[\dfrac{30x}{3x^2+1} + \dfrac{3x}{x^2+1} - \dfrac{20x}{x^2+4} \right]$

29. $\dfrac{(x^2+x+1)^{60}(x^2-5x)^4}{\sqrt{x^2-4x+10}}$

$\times \left[\dfrac{60(2x+1)}{x^2+x+1} + \dfrac{4(2x-5)}{x^2-5x} + \dfrac{2-x}{x^2-4x+10} \right]$

30. $y \left[\dfrac{5(20x+5)}{2(10x^2+5x+1)} + \dfrac{9(2x-3)}{7(x^2-3x+1)} \right.$

$\left. - \dfrac{8(3x^2+4x)}{x^3+2x^2+5} - 3x^2 + 5 \right]$

Exercise 3

1. $\frac{1}{3}e^{3x} + C$ **2.** $\frac{1}{2}(e^5 - e)$ **3.** $\frac{1}{2}(e^4 - e)$

4. $\frac{1}{2}xe^{2x} - \frac{1}{4}e^{2x} + C$ **5.** $-2e^{1/x} + C$

6. $x^2e^x - 2xe^x + 2e^x + C$ **7.** $2(e^2 - e)$

8. $\frac{1}{2}e^{2x} + 2x - \frac{1}{2}e^{-2x} + C$ **9.** $\ln |x-2| + C$

10. $\frac{1}{4}\ln |2x^2+3| + C$ **11.** $\ln |x^2+x-6| + C$

12. $\dfrac{(\ln 2)^3}{3}$ **13.** $\frac{5}{2} - \ln 2$ **14.** $\frac{3}{2} - \ln 3$

15. $\frac{1}{3}x^3 + \frac{3}{2}x^2 + 4x - 16\ln |x+2| + C$

16. (a) $\dfrac{dp}{dt} = Kp$ (b) $p(t) = Ce^{kt}$ (c) $10^{41/12}$

17. $5 \cdot 3^{5/4} \cdot 10^4$

18. $32 \times P_0$ where P_0 is the initial population

Exercise 4

1. $\ln |(x+1)^2(x-3)^3| + C$

2. $\ln |(x+1)^5(x-3)^2| + C$

3. $\ln \left| \dfrac{x-3}{(x-2)^2(x+4)^3} \right| + C$

4. $\ln (x - 4)^2 - 5(x - 4)^{-1} + C$

5. $-5 \ln |x| + 2 \ln |x + 1| + 4 \ln |x - 1| + C$

6. $\dfrac{1}{9} \ln \left| \dfrac{x + 2}{x - 1} \right| - \dfrac{4}{3} (x - 1)^{-1} + C$

7. $\dfrac{3}{2} \ln \left| \dfrac{x + 1}{x - 1} \right| - \dfrac{1}{x - 1} + C$

8. $\ln \left| \dfrac{x^2 - 1}{x} \right| + \dfrac{1}{2} x^2 + C$

9. $\dfrac{2}{3} \ln \left| \dfrac{x - 2}{x + 1} \right| + \dfrac{1}{x + 1} + C$

10. $\dfrac{7}{27} \ln \left| \dfrac{x + 2}{x - 1} \right| - \dfrac{7}{9(x - 1)} + \dfrac{1}{6(x - 1)^2} + C$

11. $\tfrac{11}{32} \ln |x - 2| - \dfrac{13}{16(x - 2)} + \tfrac{21}{32} \ln |x + 2| + \dfrac{3}{16(x + 2)} + C$

12. $\dfrac{1}{3} \ln \left| \dfrac{(x - 2)^2 (x + 1)}{(x + 2)^2 (x - 1)} \right| + C$

13. $\tfrac{11}{8} \ln |x - 1| + \tfrac{13}{8} \ln |x + 3| - \dfrac{1}{2(x - 1)} + C$

14. $\dfrac{1}{3} \ln \left| \dfrac{(x - 2)^4}{x + 1} \right| - \dfrac{1}{(x + 1)(x - 2)} + C$

15. $q(t) = \dfrac{3^t}{1 + 3^t}$

16. (a) $x(t) = \dfrac{2(e^{kt} - 1)}{2e^{kt} - 1}$ (b) $k = \ln \tfrac{3}{2}$

17. (a) $P(t) = \dfrac{M P_0 e^{P_0 t}}{1 + M e^{P_0 t}}, \left(M = \dfrac{1}{P_0 - 1} \right)$

    (b) $t = -\dfrac{1}{P_0} \ln M$

Exercise 5

1. (a) $\approx 17\frac{1}{3}$ yr   (b) $\approx 17\frac{1}{2}$ yr

2. $\approx 6.93\%$ **3.** $\approx \$1360$ **4.** \$11,416

5. \$3297   **6.** ≈ 38 **7.** ≈ 1.34 hr

8. (a) $21.8°$ (b) 9.4 min

9. (a) ≈ 0.08664 (b) ≈ 16 min

10. $p(10) = 2.96 \times 10^8$, $p(20) = 4.2824 \times 10^8$,   double in $17\frac{2}{3}$ yr, triple in 28 yr

11. ≈ 1990 12. in the year 2022
13. (a) 1.253   (b) $p = 2229$
14. (a) 64.84%   (b) 74.91%
15. 6.738×10^9, 2.035×10^8
16. ≈ 63.4 ft   17. $(7, 2 \ln 4)$

Chapter 9

Exercise 1

1. $\{(x, y) \mid x, y \in R\}$ 2. $\left\{(x, y) \mid \dfrac{x^2}{4} + \dfrac{y^2}{9} \leq 1\right\}$

3. $\{(x, y) \mid a \leq x \leq b \text{ and } c \leq y \leq d$
4. $\{(x, y) \mid x > 0, y > 0, \text{ and } xy \neq 1\}$
5. $\{(x, y) \mid (x, y) \neq (0, 0)\}$
7. (a) yz-plane   (b) xz-plane
   (c) plane parallel to xy-plane (3 units above)
   (d) z-axis
   (e) straight line parallel to z-axis (through $(1, 0, 0)$)
   (f) plane parallel to xy-plane (2 units below)
   (g) straight line through $(0, 0, 0)$ and $(1, 1, 1)$
   (h) point $(1, -1, 2)$
8. (a) $\sqrt{a^2 + b^2 + c^2}$ (b) 3
   (c) 5   (d) 5   (e) 13   (f) $\sqrt{155}$

Exercise 3

1. (a) 3, continuous at $(2, 1)$ (b) $\frac{3}{2}$, continuous at $(1, 1)$
   (c) 3, discontinuous at $(2, 2)$
   (d) 2, discontinuous at $(1, 1)$
2. (a) $f_x = 2xy - y^2$, $f_y = x^2 - 2xy$
   (b) $f_x = 3x^2y^3$, $f_y = 3x^3y^2$ (c) $f_x = y^{-1}$, $f_y = -xy^{-2}$
   (d) $f_x = -2y(x - y)^{-2}$, $f_y = 2x(x - y)^{-2}$
3. (a) $f_x = e^y$, $f_y = xe^y$
   (b) $f_x = 2xe^{x^2+y^2}$, $f_y = 2ye^{x^2+y^2}$
   (c) $f_x = e^{x+y}[\ln(x^2 + y^2) + 2x(x^2 + y^2)^{-1}]$,
       $f_y = e^{x+y}[\ln(x^2 + y^2) + 2y(x^2 + y^2)^{-1}]$
   (d) $f_x = 3x^2 + x^{-1}y^2 + e^{xy^2+x^3y}[1 + xy^2 + 3x^3y]$,
       $f_y = 2y \ln x + e^{xy^2+x^3y}[2x^2y + x^4]$
4. (a) $f_{xx}(1, 2) = 4$, $f_{xy}(1, 2) = f_{yx}(1, 2) = -2$, $f_{yy}(1, 2) = -2$
   (b) $f_{xx}(1, 2) = 48$, $f_{xy}(1, 2) = f_{yx}(1, 2) = 36$, $f_{yy}(1, 2) = 12$
   (c) $f_{xx}(1, 2) = 0$, $f_{xy}(1, 2) = f_{yx}(1, 2) = -\frac{1}{4}$, $f_{yy}(1, 2) = \frac{1}{4}$
   (d) $f_{xx}(1, 2) = -8$, $f_{xy}(1, 2) = f_{yx}(1, 2) = 6$, $f_{yy}(1, 2) = -4$

5. (a) $f_x = 2xe^{u+v}$, $f_y = 3y^2 e^{u+v}$, $f_u = (x^2 + y^3)e^{u+v}$,
$f_v = (x^2 + y^3)e^{u+v}$
(b) $f_{xy} = 0$, $f_{xu} = 2xe^{u+v}$, $f_{uv} = (x^2 + y^3)e^{u+v}$
(c) $f_{xyu} = f_{uyx} = f_{uxy} = 0$

Exercise 4

1. $\left(\dfrac{\partial P}{\partial x}\right)_{(3,\,5)} = 6c + 5d + 5$, $\left(\dfrac{\partial P}{\partial y}\right)_{(2,\,3)} = 2d + 2$

2. $\left(\dfrac{\partial P}{\partial x}\right)_{(3,\,5)} = 18e^{23}$, $\left(\dfrac{\partial P}{\partial y}\right)_{(2,\,3)} = 9e^{11}$

3. (a) $\dfrac{\partial P}{\partial x} = \dfrac{y(x^2 + y^2 + y) - 2x}{(x^2 + y^2 + y)^2}\, e^{xy}$,

$\dfrac{\partial P}{\partial y} = \dfrac{x(x^2 + y^2 + y) - (2y + 1)}{(x^2 + y^2 + y)^2}\, e^{xy}$

(b) $\left(\dfrac{\partial P}{\partial x}\right)_{(200,\,5)} = \dfrac{199{,}750e^{1000}}{(40{,}030)^2}$,

$\left(\dfrac{\partial P}{\partial y}\right)_{(5,\,100)} = \dfrac{50424}{(10125)^2}\, e^{500}$

4. (a) $\dfrac{\partial P}{\partial x} = 4x(x^2 + e^y)$, $\dfrac{\partial P}{\partial y} = 2e^y(x^2 + e^y)$

(b) $\left(\dfrac{\partial P}{\partial x}\right)_{(200,\,5)} = 32x10^6 + 800e^5$,

$\left(\dfrac{\partial P}{\partial y}\right)_{(5,\,100)} = 50e^{100} + 2e^{200}$

5. (a) $f_x = 50y$, $f_y = 50x$　　(b) $f_x|_{y=5} = 250$
(c) $f_y|_{x=2} = 100$
6. (a) $f_x = 2x + 20y$, $f_y = 2y + 20x$
(b) $f_x|_{y=5} = 2x + 100$, $f_y|_{x=2} = 2y + 40$
7. (a) $dz = (y^2 + e^y)\,dx + x(2y + e^y)\,dy$
(b) $dz = (1 + y/x)\,dx + \ln x\,dy$
8. 2%　　9. 3%　　10. 4%　　11. 1%　　12. ≈ 27

Exercise 5

1. $(0, 0)$ saddle point　　2. $(0, 0)$ rel. min.
3. $(-1, -2)$ saddle point　　4. $(0, 0)$ saddle point
5. $(0, 0)$ rel. min.　　6. $(0, 0)$ saddle point

7. none   **8.** none   **9.** $(0, 0, 1)$

10. $4' \times 4' \times 2'$ **11.** $6'' \times 6'' \times 3''$

14. $l = w = 3.494,\ h = 5.241$

15. $l = 3.319,\ w = 3.319,\ h = 5.810$

16. $x = 5,\ y = 10$ **17.** $x = 6,\ y = 10$

Exercise 6

3. max value $\sqrt{2}$ at $\left(\dfrac{1}{\sqrt{2}}, \dfrac{1}{\sqrt{2}} \right)$,

     min value $-\sqrt{2}$ at $\left(-\dfrac{1}{\sqrt{2}}, -\dfrac{1}{\sqrt{2}} \right)$

4. max. value 1 at $(\pm 1, 0)$, min value -1 at $(0, \pm 1)$

5. min. value $\frac{1}{2}$ at $(\frac{1}{2}, \frac{1}{2})$ **11.** $\frac{134}{75}$

Chapter 10   Exercise 1

1. (a) $\dfrac{3\pi}{20}$ (b) $\dfrac{13\pi}{9}$ (c) $\dfrac{31\pi}{18}$ (d) $\dfrac{5\pi}{2}$

     (e) ≈ 0.826 (f) ≈ 1.48

2. (a) $194°48'20''$ (b) $154°41'55''$ (c) $231°25'43''$

     (d) $105°$ (e) $22°30'$ (f) $139°13'43''$

3. 0.047 rad.   **4.** $65°24'30''$

5. 22 ft/sec   **6.** $2\pi/3$

7. ≈ 4710 miles   **8.** ≈ 2094.4 miles

9. 7.088 ft, 18 sq ft   **10.** 30 sq ft

Exercise 2

1. (a) IV   (b) II   (c) I   (d) III

2. (a) I, III   (b) I, II   (c) I, IV

3. (a) $\sin \theta = 24/25$ (b) $\sin \theta = -12/\sqrt{193}$

       $\cos \theta = -7/25$ $\cos \theta = -7/\sqrt{193}$

       $\tan \theta = -24/7$ $\tan \theta = 12/7$

     (c) $\sin \theta = 4/5$

       $\cos \theta = -3/5$

       $\tan \theta = -4/3$

4. (a) $\sin \theta = \pm\sqrt{69}/13$ (b) $\sin \theta = -3/4$

       $\cot \theta = \pm 10/\sqrt{69}$ $\cot \theta = \mp\sqrt{7}/3$

       $\sec \theta = 13/10$ $\sec \theta = \pm 4/\sqrt{7}$

5. (a) $\sin 0 = \tan 0 = 0$, $\cos 0 = \sec 0 = 1$, cot 0, csc 0 undef.

(b) $\sin \pi = \tan \pi = 0$, $\cos \pi = \sec \pi = -1$, cot π, csc π undef.

(c) $\sin \dfrac{3\pi}{2} = \csc \dfrac{3\pi}{2} = -1$, $\cos \dfrac{3\pi}{2} = \cot \dfrac{3\pi}{2} = 0$,

$\tan \dfrac{3\pi}{2}$, $\sec \dfrac{3\pi}{2}$ undef.

Exercise 3

1. (a) $\dfrac{\sqrt{2}}{2}, \dfrac{-\sqrt{2}}{2}$ (b) $-\dfrac{1}{2}, \dfrac{\sqrt{3}}{2}$ (c) $-\dfrac{1}{2}, -\dfrac{\sqrt{3}}{2}$

(d) $\dfrac{\sqrt{2}}{2}, \dfrac{\sqrt{2}}{2}$ (e) $-\dfrac{1}{2}, \dfrac{\sqrt{3}}{2}$ (f) $\dfrac{\sqrt{2}}{2}, \dfrac{\sqrt{2}}{2}$

2. (a) $\dfrac{\sqrt{2}}{2}, -\dfrac{\sqrt{2}}{2}$ (b) $-\dfrac{\sqrt{3}}{2}, -\dfrac{1}{2}$ (c) $\dfrac{1}{2}, \dfrac{\sqrt{3}}{2}$

(d) $\dfrac{1}{2}, \dfrac{\sqrt{3}}{2}$

3. (a) $\cos \dfrac{\pi}{26}$ (b) $\sec \dfrac{3\pi}{14}$ (c) $\csc \dfrac{\pi}{6}$

(d) $-\cot \dfrac{\pi}{4}$ (e) $\tan \dfrac{\pi}{3}$ (f) $-\sin \dfrac{\pi}{6}$

Exercise 4

4. (a) $\dfrac{2\pi}{3}$ (b) π (c) $\dfrac{\pi}{5}$

5. $\sin \dfrac{2\pi}{5} x$ **6.** $\tan 4x$ **7.** $\dfrac{\pi}{2}$ **8.** π

Exercise 5

10. $-12/13, -12/5, 13/12, -13/5$

13. no

Exercise 6

1. (a) $\frac{1}{4}(\sqrt{6} + \sqrt{2})$ (b) $\frac{1}{4}(\sqrt{6} - \sqrt{2})$

6. (a) $\frac{1}{3}$ (b) $-2\sqrt{2}/3$

Exercise 7

3. (a) $\dfrac{\pi}{4}$ (b) $\dfrac{\pi}{3}$ (c) $\dfrac{\pi}{2}$ (d) $-\dfrac{\pi}{2}$

(e) $\dfrac{2\pi}{3}$ (f) $\dfrac{\pi}{4}$

4. (a) $-\dfrac{\pi}{2}$ (b) $\frac{3}{5}$ (c) $\sqrt{2}/2$ (d) $\frac{4}{5}$

(e) $-\dfrac{\pi}{4}$ (f) $\dfrac{\pi}{4}$ (g) $\frac{5}{13}$ (h) $\dfrac{2\pi}{7}$

Exercise 8

1. 2 **2.** 1 **3.** 0 **4.** 1 **5.** 1

Exercise 9

1. $\sin x + x \cos x$

2. $\dfrac{4 \cos x + (4x + 1) \sin x}{\cos^2 x}$

3. $\dfrac{-2}{(\sin x - \cos x)^2}$

4. $\dfrac{3 \cos (3x + 1) \cos (2x + 3) + 2 \sin (3x + 1) \sin (2x + 3)}{\cos^2 (2x + 3)}$

5. $4(14x + 3) \tan^3 (7x^2 + 3x + 9) \sec^2 (7x^2 + 3x + 9)$

6. $4 \sin (2x) \cos^4 (2x) - 6 \sin^3 (2x) \cos^2 (2x)$

7. $(\cos x + \sec^2 x)e^{\sin x + \tan x}$ 8. $\dfrac{2 \sin x + x \cos x}{x \sin x}$

9. $e^x \sin (\cos x)[\sin (\cos x) - 2 \cos (\cos x) \sin x]$

10. $\dfrac{2}{\sqrt{1 - 4x^2}}$ 11. $1 + 2x \arctan x$

12. $\dfrac{-x}{|x|\sqrt{1 - x^2}}$

13. $\left[\cos x + \dfrac{1}{1 + x^2}\right] e^{\sin x + \arctan x}$

14. $\dfrac{x\sqrt{x^2 - 1} \,[2x + \sec^2 x] - 1}{x\sqrt{x^2 - 1} \,[x^2 + \tan x + \operatorname{arccsc} x]}$

15. $3 + \dfrac{\pi}{2}$ 17. $\dfrac{2}{\pi}$ ft/min 19. 100 ft

Exercise 10

1. $\ln |1 + \tan x| + C$

2. $-\ln |1 + \cos x| + C$

3. $e^{\sin x} + C$

4. $\frac{1}{4} \tan^4 x + C$

5. $\frac{1}{4}(x^2 + 1) + \frac{1}{8} \sin 2(x^2 + 1) + C$

6. $e^{1/2 \sin 2x} + C$

7. $\dfrac{1}{\cos x} + C$

8. $\ln |\tan x + 5| + C$

9. $\frac{1}{8}x - \frac{1}{32} \sin 4x + C$

10. $\cos x + x \sin x + C$

11. $\frac{1}{2} \sec x \tan x + \frac{1}{2} \ln |\sec x + \tan x| + C$

12. $2x \sin x - (x^2 - 2) \cos x + C$

13. $x \tan x - \frac{1}{2}x^2 + \ln |\cos x| + C$

14. $\dfrac{\pi}{4}$

15. $\frac{1}{3}$

16. $-\frac{1}{4} + \frac{1}{2} \ln 2$

17. $\dfrac{\pi}{3}$

18. $\frac{1}{2}x\sqrt{a^2 - x^2} + \frac{1}{2}a^2 \arcsin\left(\dfrac{x}{a}\right) + C$

19. $\frac{1}{3} \arctan \frac{1}{3}(x + 1) + C$

20. $-\frac{1}{3}\sqrt{(9 - x^2)^3} + C$

21. $x \arctan x - \frac{1}{2} \ln (1 + x^2) + C$

22. $\frac{1}{2}(1 + x^2) \arctan x - \dfrac{x}{2} + C$

23. $\frac{1}{2}x^2 \operatorname{arcsec} x - \frac{1}{2}\sqrt{x^2 - 1} + C$

24. $\frac{1}{5} \ln |x + 1| - \frac{1}{10} \ln |x^2 + 4| + \frac{1}{10} \arctan\left(\dfrac{x}{2}\right) + C$

25. $\dfrac{1}{50} \ln \dfrac{(x + 1)^2}{x^2 + 4} - \dfrac{x}{10(x^2 + 4)} - \dfrac{2}{5(x^2 + 4)}$

$- \dfrac{1}{25} \arctan\left(\dfrac{x}{2}\right) + C$

26. $\dfrac{1}{24} \ln \dfrac{(x - 2)^2}{x^2 + 2x + 4} - \dfrac{\sqrt{3}}{12} \arctan\left(\dfrac{x + 1}{\sqrt{3}}\right) + C$

27. $\dfrac{1}{2} \arctan\left(\dfrac{2x^2 + 1}{2}\right) + C$

28. $\dfrac{1}{25} \ln \dfrac{x^2 + 4}{(x - 1)^2} - \dfrac{3}{50} \arctan\left(\dfrac{x}{2}\right) - \dfrac{1}{5(x - 1)} + C$

29. $-2 \ln |x| + \frac{3}{2} \ln |x^2 + 4| + 3 \arctan\left(\dfrac{x}{2}\right) + C$

30. $\ln |x - 2| + \frac{3}{2} \ln |x^2 + 2x + 5| + \frac{1}{2} \arctan \left(\dfrac{x + 1}{2} \right) + C$

Exercise 11

1. 5, 12   **2.** 3, 9   **3.** $\frac{3}{5}, \frac{19}{5}$ **4.** $\frac{3}{5}, \frac{24}{5}$

5. (a) $\dfrac{u^2 \sin^2 \theta}{2g}$ (b) $\dfrac{2u \sin \theta}{g}$

   (c) $\dfrac{u^2 \sin 2\theta}{g}$ (d) $\dfrac{\pi}{4}$

7. (a) 144 feet   (b) 6 sec   (c) $576\sqrt{3}$ feet
9. 40 ft/sec, $25\sqrt{3}$ ft   **13.** $4\sqrt{2}$ ft/sec, 2π sec

14. π ft/sec, $\dfrac{\sqrt{3}\,\pi}{2}$ ft/sec

Appendix 1

Exercise 1

1. $1^2 + 2^2 + 3^2 + 4^2 + 5^2 = 1 + 4 + 9 + 16 + 25$
2. $(x_3 + y_3) + (x_4 + y_4) + (x_5 + y_5) + (x_6 + y_6)$
   $+ (x_7 + y_7) + (x_8 + y_8)$
3. $7x^1 + 7x^2 + 7x^3 + 7x^4 + 7x^5$
4. $6 + 6 + 6 + 6 + 6 + 6 + 6 + 6 + 6$
5. $(x_2 + 1)^2 + (x_3 + 1)^2 + (x_4 + 1)^2$

6. $\displaystyle\sum_{i=1}^{50} i$ **7.** $\displaystyle\sum_{n=1}^{16} n(n + 1)$ **8.** $\displaystyle\sum_{i=1}^{n} 2^i$

9. $\displaystyle\sum_{i=1}^{50} x_i^2$ **10.** $\displaystyle\sum_{i=1}^{n} \frac{x_i}{i}$

Exercise 3

1. 5,040   **2.** 3,628,800   **3.** 56   **4.** 330
5. 87   **6.** $\frac{16}{15}$ **7.** 28   **8.** 2415
12. (a) $x^6 + 6x^5 + 15x^4 + 20x^3 + 15x^2 + 6x + 1$
   (b) $1 + 10x + 40x^2 + 80x^3 + 80x^4 + 32x^5$
   (c) $-32 + 80x - 80x^2 + 40x^3 - 10x^4 + x^5$

   (d) $\dfrac{1}{x^6} + \dfrac{6}{x^4} + \dfrac{15}{x^2} + 20 + 15x^2 + 6x^4 + x^6$

13. (a) $-280x^4y^3$ (b) 84
   (c) $495a^8x^4y^{16}$ (d) $372096x^{-14}$

14. (a) 4th term = 660   (b) 3rd term = 240
     (c) 3rd term = 180   (d) 7th term = 84

Appendix 2

Exercise 1

1. (a) $(3, -2)$ (b) $(-4, -2)$ (c) $(0, 4)$

     (d) $\left(\dfrac{-\sqrt{2}}{2}, \dfrac{-5}{2}\right)$ (e) $(0, 4)$ (f) $\left(\frac{2}{5}, \frac{1}{5}\right)$

     (g) $\left(\frac{23}{25}, \frac{11}{25}\right)$ (h) $\left(\frac{3}{2}, \frac{3}{2}\right)$

2. (a) $-1 - 3i$ (b) $\frac{5}{2}i$ (c) $1 + 2i$

     (d) $-8i$ (e) $1 + i$ (f) $\frac{19}{13} - \frac{4}{13}i$

3. (a) $-i$ (b) 1   (c) i (d) $-i$

9. $z_1 = 2i,\ z_2 = -2i$ **10.** $z_1 = -1 - i,\ z_2 = -1 + i$

11. $z_1 = -1 + \sqrt{3}\,i,\ z_2 = -1 - \sqrt{3}\,i$

12. $z_1 = \dfrac{1}{6} + \dfrac{\sqrt{11}}{6}\,i,\ z_2 = \dfrac{1}{6} - \dfrac{\sqrt{11}}{6}\,i$

Appendix 3

Exercise 1

1. $2x^2 - 2x + \frac{7}{2},\ -\frac{15}{2}$ **2.** $\frac{4}{3}x^2 - \frac{7}{9}x - \frac{50}{27},\ -\frac{73}{27}$

3. $2x^2 - \frac{1}{3}x + \frac{4}{9},\ \frac{32}{9}$ **4.** $9x^2 - 46x + 217,\ -1079$

5. $x^2 - 74x - 79,\ -78$ **6.** $3x^3 + 6x^2 + 6x + 15,\ 23$

7. $-x^4 + 2x^3,\ 2$ **8.** $x^4 + x^3 + x^2 + x + 1,\ 0$

9. $x^5 - x^4 + 3x^3 - 2x^2 + 2x - 2,\ 7$

10. $\frac{1}{2}x^5 - \frac{3}{4}x^4 + \frac{9}{8}x^3 - \frac{27}{16}x^2 + \frac{129}{32}x - \frac{387}{64},\ \frac{1325}{64}$

11. $p(-2) = 18$ **12.** $p(3) = 78$ **13.** $p(-1) = -3$

14. $p(1) = 9$ **15.** $p(2) = 67$ **16.** $2x + 1$

17. $x^2 - x + 3$ **18.** $\frac{1}{2}$

19. $-\frac{3}{4}$ **20.** 9

Exercise 2

1. -1 **2.** 1   **3.** 3   **4.** $\frac{1}{2}$

5. none   **6.** $1, -1, -3$

7. $2, 2, -2, -2$ **8.** $\pm\sqrt{5}, \pm i\sqrt{3}$

9. $\pm\dfrac{\sqrt{2}}{4},\ \pm i\dfrac{\sqrt{2}}{4}$ **10.** $1, 1, -2, -2$

11. 1, 1, −1, −1   **12.** 0, 0, 1, 1, −1, −1

15. (a) −3, 7   (b) $\frac{3}{2}$, $-\frac{7}{2}$

    (c) $-\frac{4}{3}$, $-\frac{2}{3}$ (d) 10, 19

16. (a) $x^2 - 5x + 9$ (b) $x^2 + 2x + 3$

    (c) $6x^2 - 9x + 8$ (d) $3x^2 - 5x + 9$

17. (a) $\frac{9}{4}$ (b) $-\frac{25}{4}$ (c) $\frac{4}{3}$ or $-\frac{4}{3}$

19. (a) $3i$, $\frac{7}{3}$ (b) $1 + 2i$, $\pm i$

Tables

**Table I
Squares and
Square Roots**

x	x^2	$\sqrt{x}$	x	x^2	$\sqrt{x}$
1	1	1.000	51	2,601	7.141
2	4	1.414	52	2,704	7.211
3	9	1.732	53	2,809	7.280
4	16	2.000	54	2,916	7.348
5	25	2.236	55	3,025	7.416
6	36	2.449	56	3,136	7.483
7	49	2.646	57	3,249	7.550
8	64	2.828	58	3,364	7.616
9	81	3.000	59	3,481	7.681
10	100	3.162	60	3,600	7.746
11	121	3.317	61	3,721	7.810
12	144	3.464	62	3,844	7.874
13	169	3.606	63	3,969	7.937
14	196	3.742	64	4,096	8.000
15	225	3.873	65	4,225	8.062
16	256	4.000	66	4,356	8.124
17	289	4.123	67	4,489	8.185
18	324	4.243	68	4,624	8.246
19	361	4.359	69	4,761	8.307
20	400	4.472	70	4,900	8.367
21	441	4.583	71	5,041	8.426
22	484	4.690	72	5,184	8.485
23	529	4.796	73	5,329	8.544
24	576	4.899	74	5,476	8.602
25	625	5.000	75	5,625	8.660
26	676	5.099	76	5,776	8.718
27	729	5.196	77	5,929	8.775
28	784	5.292	78	6,084	8.832
29	841	5.385	79	6,241	8.888
30	900	5.477	80	6,400	8.944
31	961	5.568	81	6,561	9.000
32	1,024	5.657	82	6,724	9.055
33	1,089	5.745	83	6,889	9.110
34	1,156	5.831	84	7,056	9.165
35	1,225	5.916	85	7,225	9.220
36	1,296	6.000	86	7,396	9.274
37	1,369	6.083	87	7,569	9.327
38	1,444	6.164	88	7,744	9.381
39	1,521	6.245	89	7,921	9.434
40	1,600	6.325	90	8,100	9.487
41	1,681	6.403	91	8,281	9.539
42	1,764	6.481	92	8,464	9.592
43	1,849	6.557	93	8,649	9.644
44	1,936	6.633	94	8,836	9.695
45	2,025	6.708	95	9,025	9.747
46	2,116	6.782	96	9,216	9.798
47	2,209	6.856	97	9,409	9.849
48	2,304	6.928	98	9,604	9.899
49	2,401	7.000	99	9,801	9.950
50	2,500	7.071	100	10,000	10.000

**Table II
Exponential Functions**

x	e^x	e^{-x}	x	e^x	e^{-x}
0.00	1.0000	1.0000	1.5	4.4817	0.2231
0.01	1.0101	0.9901	1.6	4.9530	0.2019
0.02	1.0202	0.9802	1.7	5.4739	0.1827
0.03	1.0305	0.9705	1.8	6.0496	0.1653
0.04	1.0408	0.9608	1.9	6.6859	0.1496
0.05	1.0513	0.9512	2.0	7.3891	0.1353
0.06	1.0618	0.9418	2.1	8.1662	0.1225
0.07	1.0725	0.9324	2.2	9.0250	0.1108
0.08	1.0833	0.9331	2.3	9.9742	0.1003
0.09	1.0942	0.9139	2.4	11.0230	0.0907
0.10	1.1052	0.9048	2.5	12.182	0.0821
0.11	1.1163	0.8958	2.6	13.464	0.0743
0.12	1.1275	0.8869	2.7	14.880	0.0672
0.13	1.1388	0.8781	2.8	16.445	0.0608
0.14	1.1503	0.8694	2.9	18.174	0.0550
0.15	1.1618	0.8607	3.0	20.086	0.0498
0.16	1.1735	0.8521	3.1	22.198	0.0450
0.17	1.1853	0.8437	3.2	24.533	0.0408
0.18	1.1972	0.8353	3.3	27.113	0.0369
0.19	1.2092	0.8270	3.4	29.964	0.0334
0.20	1.2214	0.8187	3.5	33.115	0.0302
0.21	1.2337	0.8106	3.6	36.598	0.0273
0.22	1.2461	0.8025	3.7	40.447	0.0247
0.23	1.2586	0.7945	3.8	44.701	0.0224
0.24	1.2712	0.7866	3.9	49.402	0.0202
0.25	1.2840	0.7788	4.0	54.598	0.0183
0.30	1.3499	0.7408	4.1	60.340	0.0166
0.35	1.4191	0.7047	4.2	66.686	0.0150
0.40	1.4918	0.6703	4.3	73.700	0.0136
0.45	1.5683	0.6376	4.4	81.451	0.0123
0.50	1.6487	0.6065	4.5	90.017	0.0111
0.55	1.7333	0.5769	4.6	99.484	0.0101
0.60	1.8221	0.5488	4.7	109.950	0.0091
0.65	1.9155	0.5220	4.8	121.510	0.0082
0.70	2.0138	0.4966	4.9	134.290	0.0074
0.75	2.1170	0.4724	5.0	148.41	0.0067
0.80	2.2255	0.4493	5.5	244.69	0.0041
0.85	2.3396	0.4274	6.0	403.43	0.0025
0.90	2.4596	0.4066	6.5	665.14	0.0015
0.95	2.5857	0.3867	7.0	1096.60	0.0009
1.0	2.7183	0.3679	7.5	1808.0	0.0006
1.1	3.0042	0.3329	8.0	2981.0	0.0003
1.2	3.3201	0.3012	8.5	4914.8	0.0002
1.3	3.6693	0.2725	9.0	8103.1	0.0001
1.4	4.0552	0.2466	10.0	22026.0	0.00005

Table III
Common Logarithms

x	0	1	2	3	4	5	6	7	8	9
1.0	.0000	.0043	.0086	.0128	.0170	.0212	.0253	.0294	.0334	.0374
1.1	.0414	.0453	.0492	.0531	.0569	.0607	.0645	.0682	.0719	.0755
1.2	.0792	.0828	.0864	.0899	.0934	.0969	.1004	.1038	.1072	.1106
1.3	.1139	.1173	.1206	.1239	.1271	.1303	.1335	.1367	.1399	.1430
1.4	.1461	.1492	.1523	.1553	.1584	.1614	.1644	.1673	.1703	.1732
1.5	.1761	.1790	.1818	.1847	.1875	.1903	.1931	.1959	.1987	.2014
1.6	.2041	.2068	.2095	.2122	.2148	.2175	.2201	.2227	.2253	.2279
1.7	.2304	.2330	.2355	.2380	.2405	.2430	.2455	.2480	.2504	.2529
1.8	.2553	.2577	.2601	.2625	.2648	.2672	.2695	.2718	.2742	.2765
1.9	.2788	.2810	.2833	.2856	.2878	.2900	.2923	.2945	.2967	.2989
2.0	.3010	.3032	.3054	.3075	.3096	.3118	.3139	.3160	.3181	.3201
2.1	.3222	.3243	.3263	.3284	.3304	.3324	.3345	.3365	.3385	.3404
2.2	.3424	.3444	.3464	.3483	.3502	.3522	.3541	.3560	.3579	.3598
2.3	.3617	.3636	.3655	.3674	.3692	.3711	.3729	.3747	.3766	.3784
2.4	.3802	.3820	.3838	.3856	.3874	.3892	.3909	.3927	.3945	.3962
2.5	.3979	.3997	.4014	.4031	.4048	.4065	.4082	.4099	.4116	.4133
2.6	.4150	.4166	.4183	.4200	.4216	.4232	.4249	.4265	.4281	.4298
2.7	.4314	.4330	.4346	.4362	.4378	.4393	.4409	.4425	.4440	.4456
2.8	.4472	.4487	.4502	.4518	.4533	.4548	.4564	.4579	.4594	.4609
2.9	.4624	.4639	.4654	.4669	.4683	.4698	.4713	.4728	.4742	.4757
3.0	.4771	.4786	.4800	.4814	.4829	.4843	.4857	.4871	.4886	.4900
3.1	.4914	.4928	.4942	.4955	.4969	.4983	.4997	.5011	.5024	.5038
3.2	.5051	.5065	.5079	.5092	.5105	.5119	.5132	.5145	.5159	.5172
3.3	.5185	.5198	.5211	.5224	.5237	.5250	.5263	.5276	.5289	.5302
3.4	.5315	.5328	.5340	.5353	.5366	.5378	.5391	.5403	.5416	.5428
3.5	.5441	.5453	.5465	.5478	.5490	.5502	.5514	.5527	.5539	.5551
3.6	.5563	.5575	.5587	.5599	.5611	.5623	.5635	.5647	.5658	.5670
3.7	.5682	.5694	.5705	.5717	.5729	.5740	.5752	.5763	.5775	.5786
3.8	.5798	.5809	.5821	.5832	.5843	.5855	.5866	.5877	.5888	.5899
3.9	.5911	.5922	.5933	.5944	.5955	.5966	.5977	.5988	.5999	.6010
4.0	.6021	.6031	.6042	.6053	.6064	.6075	.6085	.6096	.6107	.6117
4.1	.6128	.6138	.6149	.6160	.6170	.6180	.6191	.6201	.6212	.6222
4.2	.6232	.6243	.6253	.6263	.6274	.6284	.6294	.6304	.6314	.6325
4.3	.6335	.6345	.6355	.6365	.6375	.6385	.6395	.6405	.6415	.6425
4.4	.6435	.6444	.6454	.6464	.6474	.6484	.6493	.6503	.6513	.6522
4.5	.6532	.6542	.6551	.6561	.6571	.6580	.6590	.6599	.6609	.6618
4.6	.6628	.6637	.6646	.6656	.6665	.6675	.6684	.6693	.6702	.6712
4.7	.6721	.6730	.6739	.6749	.6758	.6767	.6776	.6785	.6794	.6803
4.8	.6812	.6821	.6830	.6839	.6848	.6857	.6866	.6875	.6884	.6893
4.9	.6902	.6911	.6920	.6928	.6937	.6946	.6955	.6964	.6972	.6981
5.0	.6990	.6998	.7007	.7016	.7024	.7033	.7042	.7050	.7059	.7067
5.1	.7076	.7084	.7093	.7101	.7110	.7118	.7126	.7135	.7143	.7152
5.2	.7160	.7168	.7177	.7185	.7193	.7202	.7210	.7218	.7226	.7235
5.3	.7243	.7251	.7259	.7267	.7275	.7284	.7292	.7300	.7308	.7316
5.4	.7324	.7332	.7340	.7348	.7356	.7364	.7372	.7380	.7388	.7396
x	0	1	2	3	4	5	6	7	8	9

Table III
(*continued*)

x	0	1	2	3	4	5	6	7	8	9
5.5	.7404	.7412	.7419	.7427	.7435	.7443	.7451	.7459	.7466	.7474
5.6	.7482	.7490	.7497	.7505	.7513	.7520	.7528	.7536	.7543	.7551
5.7	.7559	.7566	.7574	.7582	.7589	.7597	.7604	.7612	.7619	.7627
5.8	.7634	.7642	.7649	.7657	.7664	.7672	.7679	.7686	.7694	.7701
5.9	.7709	.7716	.7723	.7731	.7738	.7745	.7752	.7760	.7767	.7774
6.0	.7782	.7789	.7796	.7803	.7810	.7818	.7825	.7832	.7839	.7846
6.1	.7853	.7860	.7868	.7875	.7882	.7889	.7896	.7903	.7910	.7917
6.2	.7924	.7931	.7938	.7945	.7952	.7959	.7966	.7973	.7980	.7987
6.3	.7993	.8000	.8007	.8014	.8021	.8028	.8035	.8041	.8048	.8055
6.4	.8062	.8069	.8075	.8082	.8089	.8096	.8102	.8109	.8116	.8122
6.5	.8129	.8136	.8142	.8149	.8156	.8162	.8169	.8176	.8182	.8189
6.6	.8195	.8202	.8209	.8215	.8222	.8228	.8235	.8241	.8248	.8254
6.7	.8261	.8267	.8274	.8280	.8287	.8293	.8299	.8306	.8312	.8319
6.8	.8325	.8331	.8338	.8344	.8351	.8357	.8363	.8370	.8376	.8382
6.9	.8388	.8395	.8401	.8407	.8414	.8420	.8426	.8432	.8439	.8445
7.0	.8451	.8457	.8463	.8470	.8476	.8482	.8488	.8494	.8500	.8506
7.1	.8513	.8519	.8525	.8531	.8537	.8543	.8549	.8555	.8561	.8567
7.2	.8573	.8579	.8585	.8591	.8597	.8603	.8609	.8615	.8621	.8627
7.3	.8633	.8639	.8645	.8651	.8657	.8663	.8669	.8675	.8681	.8686
7.4	.8692	.8698	.8704	.8710	.8716	.8722	.8727	.8733	.8739	.8745
7.5	.8751	.8756	.8762	.8768	.8774	.8779	.8785	.8791	.8797	.8802
7.6	.8808	.8814	.8820	.8825	.8831	.8837	.8842	.8848	.8854	.8859
7.7	.8865	.8871	.8876	.8882	.8887	.8893	.8899	.8904	.8910	.8915
7.8	.8921	.8927	.8932	.8938	.8943	.8949	.8954	.8960	.8965	.8971
7.9	.8976	.8982	.8987	.8993	.8998	.9004	.9009	.9015	.9020	.9025
8.0	.9031	.9036	.9042	.9047	.9053	.9058	.9063	.9069	.9074	.9079
8.1	.9085	.9090	.9096	.9101	.9106	.9112	.9117	.9122	.9128	.9133
8.2	.9138	.9143	.9149	.9154	.9159	.9165	.9170	.9175	.9180	.9186
8.3	.9191	.9196	.9201	.9206	.9212	.9217	.9222	.9227	.9232	.9238
8.4	.9243	.9248	.9253	.9258	.9263	.9269	.9274	.9279	.9284	.9289
8.5	.9294	.9299	.9304	.9309	.9315	.9320	.9325	.9330	.9335	.9340
8.6	.9345	.9350	.9355	.9360	.9365	.9370	.9375	.9380	.9385	.9390
8.7	.9395	.9400	.9405	.9410	.9415	.9420	.9425	.9430	.9435	.9440
8.8	.9445	.9450	.9455	.9460	.9465	.9469	.9474	.9479	.9484	.9489
8.9	.9494	.9499	.9504	.9509	.9513	.9518	.9523	.9528	.9533	.9538
9.0	.9542	.9547	.9552	.9557	.9562	.9566	.9571	.9576	.9581	.9586
9.1	.9590	.9595	.9600	.9605	.9609	.9614	.9619	.9624	.9628	.9633
9.2	.9638	.9643	.9647	.9652	.9657	.9661	.9666	.9671	.9675	.9680
9.3	.9685	.9689	.9694	.9699	.9703	.9708	.9713	.9717	.9722	.9727
9.4	.9731	.9736	.9741	.9745	.9750	.9754	.9759	.9763	.9768	.9773
9.5	.9777	.9782	.9786	.9791	.9795	.9800	.9805	.9809	.9814	.9818
9.6	.9823	.9827	.9832	.9836	.9841	.9845	.9850	.9854	.9859	.9863
9.7	.9868	.9872	.9877	.9881	.9886	.9890	.9894	.9899	.9903	.9908
9.8	.9912	.9917	.9921	.9926	.9930	.9934	.9939	.9943	.9948	.9952
9.9	.9956	.9961	.9965	.9969	.9974	.9978	.9983	.9987	.9991	.9996
x	0	1	2	3	4	5	6	7	8	9

Table IV
Natural Logarithms of Numbers

x	$\log_e x$	x	$\log_e x$	x	$\log_e x$
	*	4.5	1.5041	9.0	2.1972
0.1	7.6974	4.6	1.5261	9.1	2.2083
0.2	8.3906	4.7	1.5476	9.2	2.2192
0.3	8.7960	4.8	1.5686	9.3	2.2300
0.4	9.0837	4.9	1.5892	9.4	2.2407
0.5	9.3069	5.0	1.6094	9.5	2.2513
0.6	9.4892	5.1	1.6292	9.6	2.2618
0.7	9.6433	5.2	1.6487	9.7	2.2721
0.8	9.7769	5.3	1.6677	9.8	2.2824
0.9	9.8946	5.4	1.6864	9.9	2.2925
1.0	0.0000	5.5	1.7047	10	2.3026
1.1	0.0953	5.6	1.7228	11	2.3979
1.2	0.1823	5.7	1.7405	12	2.4849
1.3	0.2624	5.8	1.7579	13	2.5649
1.4	0.3365	5.9	1.7750	14	2.6391
1.5	0.4055	6.0	1.7918	15	2.7081
1.6	0.4700	6.1	1.8083	16	2.7726
1.7	0.5306	6.2	1.8245	17	2.8332
1.8	0.5878	6.3	1.8405	18	2.8904
1.9	0.6419	6.4	1.8563	19	2.9444
2.0	0.6931	6.5	1.8718	20	2.9957
2.1	0.7419	6.6	1.8871	25	3.2189
2.2	0.7885	6.7	1.9021	30	3.4012
2.3	0.8329	6.8	1.9169	35	3.5553
2.4	0.8755	6.9	1.9315	40	3.6889
2.5	0.9163	7.0	1.9459	45	3.8067
2.6	0.9555	7.1	1.9601	50	3.9120
2.7	0.9933	7.2	1.9741	55	4.0073
2.8	1.0296	7.3	1.9879	60	4.0943
2.9	1.0647	7.4	2.0015	65	4.1744
3.0	1.0986	7.5	2.0149	70	4.2485
3.1	1.1314	7.6	2.0281	75	4.3175
3.2	1.1632	7.7	2.0412	80	4.3820
3.3	1.1939	7.8	2.0541	85	4.4427
3.4	1.2238	7.9	2.0669	90	4.4998
3.5	1.2528	8.0	2.0794	100	4.6052
3.6	1.2809	8.1	2.0919	110	4.7005
3.7	1.3083	8.2	2.1041	120	4.7875
3.8	1.3350	8.3	2.1163	130	4.8676
3.9	1.3610	8.4	2.1282	140	4.9416
4.0	1.3863	8.5	2.1401	150	5.0106
4.1	1.4110	8.6	2.1518	160	5.0752
4.2	1.4351	8.7	2.1633	170	5.1358
4.3	1.4586	8.8	2.1748	180	5.1930
4.4	1.4816	8.9	2.1861	190	5.2470

*Subtract 10 for $x < 1$. Thus $\log_e 0.3 = 8.7960 - 10 = -1.2040$.

**Table V
Values of
Trigonometric Functions**

Angle θ									Angle θ	
Degrees	Radians	sin θ	csc θ	tan θ	cot θ	sec θ	cos θ			
0° 00′	.0000	.0000	No value	.0000	No value	1.000	1.0000	1.5708	90° 00′	
10	029	029	343.8	029	343.8	000	000	679	50	
20	058	058	171.9	058	171.9	000	000	650	40	
30	087	087	114.6	087	114.6	000	1.0000	621	30	
40	116	116	85.95	116	85.94	000	.9999	592	20	
50	145	145	68.76	145	68.75	000	999	563	10	
1° 00′	.0175	.0175	57.30	.0175	57.29	1.000	.9998	1.5533	89° 00′	
10	204	204	49.11	204	49.10	000	998	504	50	
20	233	233	42.98	233	42.96	000	997	475	40	
30	262	262	38.20	262	38.19	000	997	446	30	
40	291	291	34.38	291	34.37	000	996	417	20	
50	320	320	31.26	320	31.24	001	995	388	10	
2° 00′	.0349	.0349	28.65	.0349	28.64	1.001	.9994	1.5359	88° 00′	
10	378	378	26.45	378	26.43	001	993	330	50	
20	407	407	24.56	407	24.54	001	992	301	40	
30	436	436	22.93	437	22.90	001	990	272	30	
40	465	465	21.49	466	21.47	001	989	243	20	
50	495	494	20.23	495	20.21	001	988	213	10	
3° 00′	.0524	.0523	19.11	.0524	19.08	1.001	.9986	1.5184	87° 00′	
10	553	552	18.10	553	18.07	002	985	155	50	
20	582	581	17.20	582	17.17	002	983	126	40	
30	611	610	16.38	612	16.35	002	981	097	30	
40	640	640	15.64	641	15.60	002	980	068	20	
50	669	669	14.96	670	14.92	002	978	039	10	
4° 00′	.0698	.0698	14.34	.0699	14.30	1.002	.9976	1.5010	86° 00′	
10	727	727	13.76	729	13.73	003	974	981	50	
20	756	765	13.23	758	13.20	003	971	952	40	
30	785	785	12.75	787	12.71	003	969	923	30	
40	814	814	12.29	816	12.25	003	967	893	20	
50	844	843	11.87	846	11.83	004	964	864	10	
5° 00′	.0873	.0872	11.47	.0875	11.43	1.004	.9962	1.4835	85° 00′	
10	902	901	11.10	904	11.06	004	959	806	50	
20	931	929	10.76	934	10.71	004	957	777	40	
30	960	958	10.43	963	10.39	005	954	748	30	
40	.0989	.0987	10.13	.0992	10.08	005	951	719	20	
50	.1018	.1016	9.839	.1022	9.788	005	948	690	10	
6° 00′	.1047	.1045	9.567	.1051	9.514	1.006	.9945	1.4661	84° 00′	
10	076	074	9.309	080	9.255	006	942	632	50	
20	105	103	9.065	110	9.010	006	939	603	40	
30	134	132	8.834	139	8.777	006	936	573	30	
40	164	161	8.614	169	8.556	007	932	544	20	
50	193	190	8.405	198	8.345	007	929	515	10	
7° 00′	.1222	.1219	8.206	.1228	8.144	1.008	.9925	1.4486	83° 00′	
10	251	248	8.016	257	7.953	008	922	457	50	
20	280	276	7.834	287	7.770	008	918	428	40	
30	309	305	7.661	317	7.596	009	914	399	30	
40	338	334	7.496	346	7.429	009	911	370	20	
50	367	363	7.337	376	7.269	009	907	341	10	
8° 00′	.1396	.1392	7.185	.1405	7.115	1.010	.9903	1.4312	82° 00′	
		cos θ	sec θ	cot θ	tan θ	csc θ	sin θ	Radians	Degrees	
									Angle θ	

Table V
(*continued*)

Angle θ

Degrees	Radians	$\sin \theta$	$\csc \theta$	$\tan \theta$	$\cot \theta$	$\sec \theta$	$\cos \theta$		
8° 00′	.1396	.1392	7.185	.1405	7.115	1.010	.9930	1.4312	82° 00′
10	425	421	7.040	435	6.968	010	899	283	50
20	454	449	6.900	465	827	011	894	254	40
30	484	478	765	495	691	011	890	224	30
40	513	507	636	524	561	012	886	195	20
50	542	536	512	554	435	012	881	166	10
9° 00′	.1571	.1564	6.392	.1584	6.314	1.012	.9877	1.4137	81° 00′
10	600	593	277	614	197	013	872	108	50
20	629	622	166	644	6.084	013	868	079	40
30	658	650	6.059	673	5.976	014	863	050	30
40	687	679	5.955	703	871	014	858	1.4021	20
50	716	708	855	733	769	015	853	1.3992	10
10° 00′	.1745	.1736	5.759	.1763	5.671	1.015	.9848	1.3963	80° 00′
10	774	765	665	793	576	016	843	934	50
20	804	794	575	823	485	016	838	904	40
30	833	822	487	853	396	017	833	875	30
40	862	851	403	883	309	018	827	846	20
50	891	880	320	914	226	018	822	817	10
11° 00′	.1920	.1908	5.241	.1944	5.145	1.019	.9816	1.3788	79° 00′
10	949	937	164	.1974	5.066	019	811	759	50
20	.1978	965	089	.2004	4.989	020	805	730	40
30	.2007	.1994	5.016	035	915	020	799	701	30
40	036	.2022	4.945	065	843	021	793	672	20
50	065	051	876	095	773	022	787	643	10
12° 00′	.2094	.2079	4.810	.2126	4.705	1.022	.9781	1.3614	78° 00′
10	123	108	745	156	638	023	775	584	50
20	153	136	682	186	574	024	769	555	40
30	182	164	620	217	511	024	763	526	30
40	211	193	560	247	449	025	757	497	20
50	240	221	502	278	390	026	750	468	10
13° 00′	.2269	.2250	4.445	.2309	4.331	1.026	.9744	1.3439	77° 00′
10	298	278	390	339	275	027	737	410	50
20	327	306	336	370	219	028	730	381	40
30	356	334	284	401	165	028	724	352	30
40	385	363	232	432	113	029	717	323	20
50	414	391	182	462	061	030	710	294	10
14° 00′	.2443	.2419	4.134	.2493	4.011	1.031	.9703	1.3265	76° 00′
10	473	447	086	524	3.962	031	696	235	50
20	502	476	4.039	555	914	032	689	206	40
30	531	504	3.994	586	867	033	681	177	30
40	560	532	950	617	821	034	674	148	20
50	589	560	906	648	776	034	667	119	10
15° 00′	.2618	.2588	3.864	.2679	3.732	1.035	.9659	1.3090	75° 00′
10	647	616	822	711	689	036	652	061	50
20	676	644	782	742	647	037	644	032	40
30	705	672	742	773	606	038	636	1.3003	30
40	734	700	703	805	566	039	628	1.2974	20
50	763	728	665	836	526	039	621	945	10
16° 00′	.2793	.2756	3.628	.2867	3.487	1.040	.9613	1.2915	74° 00′

| | | $\cos \theta$ | $\sec \theta$ | $\cot \theta$ | $\tan \theta$ | $\csc \theta$ | $\sin \theta$ | Radians | Degrees |

Angle θ

Table V
(*continued*)

Angle θ Degrees	Radians	sin θ	csc θ	tan θ	cot θ	sec θ	cos θ		
16° 00'	.2793	.2756	3.628	.2867	3.487	1.040	.9613	1.2915	74° 00'
10	822	784	592	899	450	041	605	886	50
20	851	812	556	931	412	042	596	857	40
30	880	840	521	962	376	043	588	828	30
40	909	868	487	.2944	340	044	580	799	20
50	938	896	453	.3026	305	045	572	770	10
17° 00'	.2967	.2924	3.420	.3057	3.271	1.046	.9563	1.2741	73° 00'
10	.2996	952	388	089	237	047	555	712	50
20	.3025	.2979	357	121	204	048	546	683	40
30	054	.3007	326	153	172	048	537	654	30
40	083	035	295	185	140	049	528	625	20
50	113	062	265	217	108	050	520	595	10
18° 00'	.3142	.3090	3.236	.3249	3.078	1.051	.9511	1.2566	72° 00'
10	171	118	207	281	047	052	502	537	50
20	200	145	179	314	3.018	053	492	508	40
30	229	173	152	346	2.989	054	483	479	30
40	258	201	124	378	960	056	474	450	20
50	287	228	098	411	932	057	465	421	10
19° 00'	.3316	.3256	3.072	.3443	2.904	1.058	.9455	1.2392	71° 00'
10	345	283	046	476	877	059	446	363	50
20	374	311	3.021	508	850	060	436	334	40
30	403	338	2.996	541	824	061	426	305	30
40	432	365	971	574	798	062	417	275	20
50	462	393	947	607	773	063	407	246	10
20° 00'	.3491	.3420	2.924	.3640	2.747	1.064	.9397	1.2217	70° 00'
10	520	448	901	673	723	065	387	188	50
20	549	475	878	706	699	066	377	159	40
30	578	502	855	739	675	068	367	130	30
40	607	529	833	772	651	069	356	101	20
50	636	557	812	805	628	070	346	072	10
21° 00'	.3665	.3584	2.790	.3839	2.605	1.071	.9336	1.2043	69° 00'
10	694	611	769	872	583	072	325	1.2014	50
20	723	638	749	906	560	074	315	1.1985	40
30	752	665	729	939	539	075	304	956	30
40	782	692	709	.3973	517	076	293	926	20
50	811	719	689	.4006	496	077	283	897	10
22° 00'	.3840	.3746	2.669	.4040	2.475	1.079	.9272	1.1868	68° 00'
10	869	773	650	074	455	080	261	839	50
20	898	800	632	108	434	081	250	810	40
30	927	827	613	142	414	082	239	781	30
40	956	854	595	176	394	084	228	752	20
50	985	881	577	210	375	085	216	723	10
23° 00'	.4014	.3907	2.559	.4245	2.356	1.086	.9205	1.1694	67° 00'
10	043	934	542	279	337	088	194	665	50
20	072	961	525	314	318	089	182	636	40
30	102	.3987	508	348	300	090	171	606	30
40	131	.4014	491	383	282	092	159	577	20
50	160	041	475	417	264	093	147	548	10
24° 00'	.4189	.4067	2.459	.4452	2.246	1.095	.9135	1.1519	66° 00'
		cos θ	sec θ	cot θ	tan θ	csc θ	sin θ	Radians	Degrees

Angle θ

Table V
(*continued*)

Angle θ									
Degrees	Radians	sin θ	csc θ	tan θ	cot θ	sec θ	cos θ		
24° 00′	.4189	.4067	2.459	.4452	2.246	1.095	.9135	1.1519	66° 00′
10	218	094	443	487	229	096	124	490	50
20	247	120	427	522	211	097	112	461	40
30	276	147	411	557	194	099	100	432	30
40	305	173	396	592	177	100	088	403	20
50	334	200	381	628	161	102	075	374	10
25° 00′	.4363	.4226	2.366	.4663	2.145	1.103	.9063	1.1345	65° 00′
10	392	253	352	699	128	105	051	316	50
20	422	279	337	734	112	106	038	286	40
30	451	305	323	770	097	108	026	257	30
40	480	331	309	806	081	109	013	228	20
50	509	358	295	841	066	111	.9001	199	10
26° 00′	.4538	.4384	2.281	.4877	2.050	1.113	.8988	1.1170	64° 00′
10	567	410	268	913	035	114	975	141	50
20	596	436	254	950	020	116	962	112	40
30	625	462	241	.4986	2.006	117	949	083	30
40	654	488	228	.5022	1.991	119	936	054	20
50	683	514	215	059	977	121	923	1.1025	10
27° 00′	.4712	.4540	2.203	.5095	1.963	1.122	.8910	1.0996	63° 00′
10	741	566	190	132	949	124	897	966	50
20	771	592	178	169	935	126	884	937	40
30	800	617	166	206	921	127	870	908	30
40	829	643	154	243	907	129	857	879	20
50	858	669	142	280	894	131	843	850	10
28° 00′	.4887	.4695	2.130	.5317	1.881	1.133	.8829	1.0821	62° 00′
10	916	720	118	354	868	134	816	792	50
20	945	746	107	392	855	136	802	763	40
30	.4974	772	096	430	842	138	788	734	30
40	.5003	797	085	467	829	140	774	705	20
50	032	823	074	505	816	142	760	676	10
29° 00′	.5061	.4848	2.063	.5543	1.804	1.143	.8746	1.0647	61° 00′
10	091	874	052	581	792	145	732	617	50
20	120	899	041	619	780	147	718	588	40
30	149	924	031	658	767	149	704	559	30
40	178	950	020	696	756	151	689	530	20
50	207	.4975	010	735	744	153	675	501	10
30° 00′	.5236	.5000	2.000	.5774	1.732	1.155	.8660	1.0472	60° 00′
10	265	025	1.990	812	720	157	646	443	50
20	294	050	980	851	709	159	631	414	40
30	323	075	970	890	698	161	616	385	30
40	352	100	961	930	686	163	601	356	20
50	381	125	951	.5969	675	165	587	327	10
31° 00′	.5411	.5150	1.942	.6009	1.664	1.167	.8572	1.0297	59° 00′
10	440	175	932	048	653	169	557	268	50
20	469	200	923	088	643	171	542	239	40
30	498	225	914	128	632	173	526	210	30
40	527	250	905	168	621	175	511	181	20
50	556	275	896	208	611	177	496	152	10
32° 00′	.5585	.5299	1.887	.6249	1.600	1.179	.8480	1.0123	58° 00′
		cos θ	sec θ	cot θ	tan θ	csc θ	sin θ	Radians	Degrees
								Angle θ	

Table V
(*continued*)

Angle θ

Degrees	Radians	sin θ	csc θ	tan θ	cot θ	sec θ	cos θ	Radians	Degrees
32° 00′	.5585	.5299	1.887	.6249	1.600	1.179	.8480	1.0123	58° 00′
10	614	324	878	289	590	181	465	094	50
20	643	348	870	330	580	184	450	065	40
30	672	373	861	371	570	186	434	036	30
40	701	398	853	412	560	188	418	1.0007	20
50	730	422	844	452	550	190	403	.9977	10
33° 00′	.5760	.5446	1.836	.6494	1.540	1.192	.8387	.9948	57° 00′
10	789	471	828	536	530	195	371	919	50
20	818	495	820	577	520	197	355	890	40
30	847	519	812	619	511	199	339	861	30
40	876	544	804	661	501	202	323	832	20
50	905	568	796	703	492	204	307	803	10
34° 00′	.5934	.5592	1.788	.6745	1.483	1.206	.8290	.9774	56° 00′
10	963	616	781	787	473	209	274	745	50
20	.5992	640	773	830	464	211	258	716	40
30	.6021	664	766	873	455	213	241	687	30
40	050	688	758	916	446	216	225	657	20
50	080	712	751	.6959	437	218	208	628	10
35° 00′	.6109	.5736	1.743	.7002	1.428	1.221	.8192	.9599	55° 00′
10	138	760	736	046	419	223	175	570	50
20	167	783	729	089	411	226	158	541	40
30	196	807	722	133	402	228	141	512	30
40	225	831	715	177	393	231	124	483	20
50	254	854	708	221	385	233	107	454	10
36° 00′	.6283	.5878	1.701	.7265	1.376	1.236	.8090	.9425	54° 00′
10	312	901	695	310	368	239	073	396	50
20	341	925	688	355	360	241	056	367	40
30	370	948	681	400	351	244	039	338	30
40	400	972	675	445	343	247	021	308	20
50	429	.5995	668	490	335	249	.8004	279	10
37° 00′	.6458	.6018	1.662	.7536	1.327	1.252	.7986	.9250	53° 00′
10	487	041	655	581	319	255	696	221	50
20	516	065	649	627	311	258	951	192	40
30	545	088	643	673	303	260	934	163	30
40	574	111	636	720	295	263	916	134	20
50	603	134	630	766	288	266	898	105	10
38° 00′	.6632	.6157	1.624	.7813	1.280	1.269	.7880	.9076	52° 00′
10	661	180	618	860	272	272	862	047	50
20	690	202	612	907	265	275	844	.9018	40
30	720	225	606	.7954	257	278	826	.8988	30
40	749	248	601	.8002	250	281	808	959	20
50	778	271	595	050	242	284	790	930	10
39° 00′	.6807	.6293	1.589	.8098	1.235	1.287	.7771	.8901	51° 00′
10	836	316	583	146	228	290	753	872	50
20	865	338	578	195	220	293	735	843	40
30	894	361	572	243	213	296	716	814	30
40	923	383	567	292	206	299	698	785	20
50	952	406	561	342	199	302	679	756	10
40° 00′	.6981	.6428	1.556	.8391	1.192	1.305	.7660	.8727	50° 00′
		cos θ	sec θ	cot θ	tan θ	csc θ	sin θ	Radians	Degrees

Angle θ

Table V
(*continued*)

Angle θ									
Degrees	Radians	sin θ	csc θ	tan θ	cot θ	sec θ	cos θ		
40° 00′	.6981	.6428	1.556	.8391	1.192	1.305	.7660	.8727	50° 00′
10	.7010	450	550	441	185	309	642	698	50.
20	039	472	545	491	178	312	623	668	40
30	069	494	540	541	171	315	604	639	30
40	098	517	535	591	164	318	585	610	20
50	127	539	529	642	157	322	566	581	10
41° 00′	.7156	.6561	1.524	.8693	1.150	1.325	.7547	.8552	49° 00′
10	185	583	519	744	144	328	528	523	50
20	214	604	514	796	137	332	509	494	40
30	243	626	509	847	130	335	490	465	30
40	272	648	504	899	124	339	470	436	20
50	301	670	499	.8952	117	342	451	407	10
42° 00′	.7330	.6691	1.494	.9004	1.111	1.346	.7431	.8378	48° 00′
10	359	713	490	057	104	349	412	348	50
20	389	734	485	110	098	353	392	319	40
30	418	756	480	163	091	356	373	290	30
40	447	777	476	217	085	360	353	261	20
50	476	799	471	271	079	364	333	232	10
43° 00′	.7505	.6820	1.466	.9325	1.072	1.367	.7314	.8203	47° 00′
10	534	841	462	380	066	371	294	174	50
20	563	862	457	435	060	375	274	145	40
30	592	884	453	490	054	379	254	116	30
40	621	905	448	545	048	382	234	087	20
50	650	926	444	601	042	386	214	058	10
44° 00′	.7679	.6947	1.440	.9657	1.036	1.390	.7193	.8029	46° 00′
10	709	967	435	713	030	394	173	.7999	50
20	738	.6988	431	770	024	398	153	970	40
30	767	.7009	427	827	018	402	133	941	30
40	796	030	423	884	012	406	112	912	20
50	825	050	418	.9942	006	410	092	883	10
45° 00′	.7854	.7071	1.414	1.000	1.000	1.414	.7071	.7854	45° 00′
		cos θ	sec θ	cot θ	tan θ	csc θ	sin θ	Radians	Degrees

Angle θ

Photo Credits

Index